Synthesis Lectures on Ocean Systems Engineering

Series Editor

Nikolas Xiros, University of New Orleans, New Orleans, USA

The series publishes short books on state-of-the-art research and applications in related and interdependent areas of design, construction, maintenance and operation of marine vessels and structures as well as ocean and oceanic engineering.

Shengzhe Wang

Applied Ocean Mechanics for Structural Engineering and Climate Adaptation

Springer

Shengzhe Wang
Department of Civil Engineering
and Construction
University of Colorado Denver
Denver, CO, USA

ISSN 2692-4420 ISSN 2692-4471 (electronic)
Synthesis Lectures on Ocean Systems Engineering
ISBN 978-3-032-18198-5 ISBN 978-3-032-18199-2 (eBook)
https://doi.org/10.1007/978-3-032-18199-2

This Springer imprint is published by the registered company Springer Nature Switzerland AG
The registered company address is: Gewerbestrasse 11, 6330 Cham, Switzerland

Preface

The 21st century and beyond will likely bear witness to unprecedented changes in Earth's climate, the consequences of which may prove catastrophic to our existing way of life. In particular, sea level rise exacerbated by a warming planet will see coastal communities, from remote settlements to global megacities, becoming more vulnerable to ocean hazards such as extreme waves, storm surges, and tsunamis. In adaptation to this uncertain future, the structural engineering profession must look beyond its traditionally well-defined roles to foster transformative infrastructure solutions that reimagine the complex and evolving relationship between water and the built environment.

The current state of practice within the structural engineering discipline places little emphasis on the effects of water and waves interacting with large-scale structures. The analysis and design of structures not supported by geotechnical foundations, such as floating buildings, are also seldom featured within a traditional engineering education. To foster innovations that lead to more adaptable and resilient future cities, our profession must promote the active integration of coastal and oceanic hydrodynamics into the structural engineering curriculum. The objective of this book is to introduce these topics to individuals with little or no background knowledge of hydrodynamics, water wave mechanics, or related subjects, in an accessible and intuitive manner. Through this process, I hope that readers will be better positioned to leverage such knowledge in order to develop more resilient infrastructure solutions that enable vulnerable communities to adapt to an increasingly unpredictable climate.

The chapters of this book can be broadly categorized into four distinct themes that successively build upon one another. Chapters 1 and 2 introduce the mathematical language of fluid dynamics that underscore many of the derivations presented throughout the text. Chapters 3–9 apply this knowledge to elucidate the fundamental mechanics of water waves that are necessary to describe their behavior in nature. By introducing the theoretical foundations of wave propagation and their statistics, the reader will be equipped with an understanding of the conditions under which engineering design must be performed. Armed with these tools, Chaps. 10–16 present methods to estimate the forces produced by surface waves, both wind-driven and tsunamis, interacting with fixed structures. Examples include artificial "gray" infrastructure like piles, breakwaters, and seawalls, in addition to natural "green" systems such as coastal vegetation. Finally, Chaps. 17–21 introduce the static and dynamic analysis of idealized floating structures while discussing their potential role in future cities. The goal is to shed light on some promising but unconventional pathways to climate resilience that have yet to gain widespread recognition amongst designers or policymakers.

Many of these chapters were adapted from a collection of lecture notes which accompanied a new course that I developed with Prof. Maria Garlock at Princeton University, where it was first offered by the Department of Civil and Environmental Engineering in spring 2022. The target audience of this book includes senior undergraduate and graduate students, university instructors and researchers, as well as architectural or engineering practitioners interested in the theory and applications of ocean mechanics within the context of structural design. While a basic understanding of vector calculus, partial differential equations, and structural analysis is encouraged, no prior knowledge of fluid dynamics or wave theory will be assumed.

Please contact me with any suggestions that you feel might improve the quality/clarity of this text or to offer corrections to errors that may have been overlooked. Your time and interest are greatly appreciated.

Denver, USA Shengzhe Wang

Competing Interests The author has no competing interests to declare that are relevant to the content of this manuscript.

Contents

Part I Mathematical Fundamentals

1 Introduction to Fluid Dynamics 3
1.1 Principles of Fluid Dynamics 3
1.2 Review of Vector Algebra 5
1.2.1 Vector Addition 5
1.2.2 Scalar Multiplication 6
1.2.3 Dot Product 6
1.2.4 Vector Magnitude 7
1.2.5 Cross Product 7
1.2.6 Matrix Multiplication 8
1.3 Review of Partial Differential Equations 9
1.4 The Gradient Operator 12
1.5 The Divergence Operator 12
1.6 The Laplace Operator 15
1.7 The Curl Operator 15
Reference 17

2 The Navier–Stokes and Bernoulli Equations 19
2.1 The Continuity Equation 20
2.2 The Momentum Equations 22
2.3 The Unsteady Bernoulli Equation 26
Reference 28

Part II Mechanics of Surface Waves

3 Linear Wave Theory and Surface Gravity Waves 31
3.1 Properties of Linear Waves 31
3.2 Assumptions and Boundary Conditions 33
3.2.1 Dynamic Condition 34

3.2.2 Kinematic Condition 34
3.2.3 Seabed Condition 34
3.3 The Velocity Potential 35
3.4 Dynamic Pressure and the Dispersion Relation 37
3.5 Kinematics of Fluid Particles 40
3.6 Velocity of Surface Waves 43
3.7 Classification of Surface Waves 45
3.8 Nonlinear Waves and Wave Breaking 48
References 49

4 Wave Energetics 51
4.1 Energy Components 51
4.2 Principles of Energy Conservation 53
4.3 Derivation of Kinetic and Potential Components 55
4.3.1 Kinetic Energy 56
4.3.2 Potential Energy 57
4.4 Velocity of Energy Propagation 58
References 60

5 The Shallow-Water Equations 61
5.1 Derivation of the Shallow-Water Equations 62
5.2 Energetics of Shallow-Water Waves 65
5.3 Transformation of Shallow-Water Waves 66
5.4 Example: Tsunami Propagation onto a Continental Shelf 70
References 74

6 Irregular Waves and the Wave Spectrum 75
6.1 The Wave Record 75
6.2 Irregular Wave Statistics 77
6.3 The Wave Spectrum 82
6.4 Parametric Wave Spectrum for Deep Water 86
6.5 Parametric Wave Spectrum for Shallow Water 92
6.6 Example: Synthesis of Artificial Wave Record 94
References 98

7 Experimental Generation of Surface Waves 99
7.1 Generation of Regular Waves 99
7.2 Generation of Irregular Waves 102
7.3 Generation of Focused Waves 104
7.4 Generation of Solitary Waves 104
7.5 Scaling Relationships 106
References 107

8 Wind and Waves in Tropical Cyclones 109
8.1 Terminology 109
8.2 Hurricane Intensity 111
8.3 Hurricane Wind Field 112
8.4 Example: Parametric Wind Field of Hurricane Katrina (2005) 114
8.5 Hurricane Wind Fetch 118
8.6 Hurricane Wave Spectrum 119
8.7 Summary: Hurricane Wave Spectrum at Radius to Maximum Wind 121
References 122

9 Wave Shoaling and Breaking 125
9.1 Wave Shoaling 125
9.2 Wave Breaking 127
9.3 Example: Wave Shoaling and Breaking on a Sloping Beach 131
9.3.1 Wave Shoaling 132
9.3.2 Wave Breaking 133
References 134

Part III Wave Interactions with Fixed Structures

10 Hydrostatic Forces 139
10.1 Derivation of Hydrostatic Principles 139
10.2 Hydrostatics of Submerged Bodies 141
10.2.1 Buoyancy Force on Rectangular Bodies 141
10.2.2 Buoyancy Force on Arbitrary Bodies 142
10.3 Hydrostatics of Surface-Piercing Bodies 143
10.4 Example: Hydrostatic Forces on a Rotating Gate 146
Reference 148

11 Wave Forces on Walls 149
11.1 Breaking, Nonbreaking, and Broken Waves 149
11.2 Historical Wave Pressure Formulations 153
11.2.1 Hiroi's Formula 153
11.2.2 Sainflou's Formula 153
11.2.3 Minikin's Formula 154
11.3 The Goda Equation 155
11.3.1 Wave Pressure Distribution 155
11.3.2 Design Wave Parameters 158
11.3.3 Resultant Forces and Moments 158
11.3.4 Bearing Stress 161
11.4 The Goda Equation for Inclined Surfaces 163
11.5 The Goda Equation for Elevated Structures 164

11.6 Modified Goda Equation for Impulsive Loading 167
11.7 Structural Modification Factors for the Goda Equation 169
11.8 Example: Wave Forces on an Offshore Breakwater 170
11.8.1 Design Wave Parameters 171
11.8.2 Wave Pressures 172
11.8.3 Resultant Forces and Overturning Moments 175
11.8.4 Factors of Safety 176
11.8.5 Bearing Stress 177
11.9 Summary: Goda Wave Forces on Vertical Walls 178
References 180

12 Extra Topics on Breaking Wave Forces on Walls 183
12.1 Breaking Wave Forces from PROVERBS 183
12.1.1 Horizontal Impulsive Force 185
12.1.2 Uplift Impulsive Force 188
12.2 Breaking Wave Forces from VOWS 190
12.3 Broken Wave Forces 191
12.4 Example: Breaking Wave Forces on a Vertical Wall 193
12.4.1 Horizontal Wave Force and Moment 193
12.4.2 Uplift Force and Moment 195
12.4.3 Distribution of Impact Forces on Structure 196
12.5 Summary: Breaking Wave Forces from PROVERBS Procedure 197
12.6 ASCE 7 Guidelines for Nonbreaking and Breaking Waves 197
12.6.1 Nonbreaking Waves 198
12.6.2 Breaking Waves 200
References 202

13 Wave Forces on Piles 203
13.1 Forcing Components 203
13.2 Linear Wave Forces on Piles 205
13.2.1 Drag and Inertia Forces 205
13.2.2 Total Pile Force and Base Moment 206
13.3 Nonlinear Wave Forces on Piles 208
13.4 Breaking Wave Forces on Piles 210
13.4.1 Dynamic Approach 210
13.4.2 Equivalent Static Approach 214
13.4.3 Comparison Between Dynamic and Static Forces 214
13.5 Example: Wave Forces on an Isolated Pile 216
13.6 Summary: Wave Forces on Vertical Piles 221
13.7 Wave Forces on Pile Groups 222
References 229

14 Tsunami Forces on Overland Structures ... 231
14.1 Tsunami Runup ... 231
14.2 Tsunami Inundation Depth and Flow Velocity ... 234
14.3 Tsunami Pressure ... 238
14.3.1 Flow Conditions ... 239
14.3.2 Water Density ... 239
14.3.3 Tsunami Importance Factor ... 239
14.3.4 Drag Coefficient ... 240
14.3.5 Amplification Factor for Tsunami Bores ... 241
14.4 Simplified Procedure for Tsunami Pressure ... 241
14.5 Hydrostatic Pressure ... 242
14.6 Tsunami Forces ... 243
14.7 Example: Tsunami Forces on a Frame Building ... 244
14.7.1 Maximum Inundation and Velocity ... 245
14.7.2 Hydrodynamic Pressure and Forces ... 247
14.8 Summary: Tsunami Forces on Overland Structures ... 250
14.9 Impact Forces Due to Floating Debris ... 251
14.9.1 Dynamic Force from Debris Impact ... 255
14.9.2 Equivalent Static Force from Debris Impact ... 257
14.9.3 Example of Debris Impact ... 258
14.10 Impact Forces Based on Debris Type Per ASCE 7 Guidelines ... 260
14.10.1 Wooden Logs and Poles ... 260
14.10.2 Boulders and Concrete Debris ... 260
14.10.3 Passenger Vehicles ... 261
14.10.4 Shipping Containers ... 261
References ... 261

15 Wave Attenuation Over Vegetated Terrain ... 263
15.1 Overview of Wave Attenuation ... 264
15.2 Navier–Stokes Equations and Boundary Conditions ... 267
15.2.1 Navier–Stokes Equations ... 267
15.2.2 Boundary Conditions ... 268
15.3 Fluid Dynamics Above Vegetated Zone ... 268
15.4 Fluid Dynamics Within Vegetated Zone ... 271
15.5 Dispersion Relation and Decay Coefficient ... 273
15.6 Damping Coefficient ... 274
15.7 Drag Coefficient ... 276
15.8 Example: Attenuation of Regular Waves Over Submerged Vegetation ... 277
15.9 Summary: Wave Attenuation Over Rigid Vegetation for Regular Waves ... 279
References ... 281

16 Extra Topics on Wave Attenuation Over Vegetated Terrain 283
16.1 Attenuation of Shallow-Water Waves on Sloping Bathymetry 283
16.2 Attenuation of Irregular Waves 289
16.3 Example: Attenuation of Irregular Waves Over Submerged Vegetation 293
16.4 Summary: Wave Attenuation Over Rigid Vegetation for Irregular Waves 296
16.5 Wave Attenuation Over Highly Flexible Vegetation 297
References 298

Part IV Wave Interactions with Floating Structures

17 Static Analysis and Stability of Floating Structures 301
17.1 Floating Structures for Urban Development 301
17.2 Heel and Trim 303
17.3 Principles of Buoyancy 305
17.4 Stability of Submerged Bodies 307
17.5 Stability of Floating Bodies 307
17.5.1 Derivation of Stability Requirements 309
17.5.2 Angles of Heel and Trim 312
17.5.3 Hydrostatic Stiffness 314
17.6 Stability Requirements for Small-Scale Floating Dwellings 314
17.7 Example: Hydrostatic Stability of a Floating House 316
17.7.1 Metacentric Height 316
17.7.2 Angles of Heel and Trim 317
17.7.3 Freeboard After Rotation 318
17.8 Summary: Analysis Procedure for Hydrostatic Stability 319
17.9 Structural Analysis of Floating Pontoons 319
17.9.1 Pontoon Loads 321
17.9.2 Local Response 322
17.9.3 Global Response 323
References 325

18 Rigid-Body Dynamics of Floating Structures 327
18.1 Classification of Floating Body Dynamics 327
18.2 Degrees of Freedom 328
18.3 Principles of Dynamic Behavior 329
18.3.1 Equation of Motion 331
18.3.2 Dynamic Properties 331
18.3.3 General Solution for "Stiffness Dominated" Degrees of Freedom 332

18.3.4 Solution for "Mass Dominated" Degrees of Freedom 337
18.3.5 Response Amplitude Operator 337
18.4 Waterplane Stiffness 338
18.5 Forcing Amplitude 339
18.6 Added Mass 341
18.7 Center of Rotation 346
18.8 Example: Dynamic Properties of a Floating House 346
18.8.1 Waterplane Stiffness 348
18.8.2 Added Mass 348
18.8.3 Total Mass 351
18.8.4 Natural Frequency 352
References 352

19 Dynamics of Floating Structures Under Regular Waves 355
19.1 The Response Amplitude Operator 355
19.2 Geometric Definitions of a Floating Structure 358
19.3 Dynamic Response in Surge 360
19.3.1 Froude-Krylov Force 360
19.3.2 Added Mass Force 362
19.3.3 Response Amplitude Operator 363
19.4 Dynamic Response in Sway 363
19.4.1 Froude-Krylov Force 363
19.4.2 Added Mass Force 364
19.4.3 Response Amplitude Operator 365
19.5 Dynamic Response in Heave 365
19.5.1 Froude-Krylov Force 365
19.5.2 Added Mass Force 366
19.5.3 Response Amplitude Operator 367
19.6 Dynamic Response in Roll 368
19.6.1 Froude-Krylov Force 368
19.6.2 Added Mass Force 370
19.6.3 Response Amplitude Operator 370
19.7 Dynamic Response in Pitch 371
19.7.1 Froude-Krylov Force 371
19.7.2 Added Mass Force 372
19.7.3 Response Amplitude Operator 374
19.8 Dynamic Response in Yaw 374
19.8.1 Froude-Krylov Force 374
19.8.2 Added Mass Force 375
19.8.3 Response Amplitude Operator 377

19.9 Summary: Rigid-Body Dynamics Under Regular Waves 378
19.10 Internal Pontoon Forces Induced by Regular Waves 379
References 385

20 Dynamics of Floating Structures Under Irregular Waves 387
20.1 Dynamic Response Time History 387
20.2 Most Probable Maximum Response 390
20.3 Structural Accelerations Under Irregular Waves 393
20.4 Acceleration Thresholds for Occupant Comfort 396
20.4.1 ISO 6897 396
20.4.2 ISO 10137 397
20.4.3 Netherlands Organization for Applied Scientific Research 397
20.4.4 NordForsk 398
20.4.5 Summary of Acceleration Limits 398
20.5 Summary: Comfort Assessment of Floating Structures for Human Habitation 399
References 401

21 Floating Breakwaters for Wave Attenuation 403
21.1 Overview of Floating Breakwaters 403
21.2 Transmission Coefficient Under Regular Waves 406
21.3 Transmission Coefficient Under Irregular Waves 408
21.4 Wave Forces on Floating Pontoon 411
21.5 Mooring Forces 412
21.6 Example: Simplified Analysis of a Box-Type Floating Breakwater 413
21.6.1 Transmitted Wave Height 414
21.6.2 Maximum Wave Force on Breakwater 414
21.6.3 Maximum Force in Mooring Lines 416
21.6.4 Required Weight of Anchor Blocks 417
References 417

Correction to: Applied Ocean Mechanics for Structural Engineering and Climate Adaptation C1

Index 419

Part I

Mathematical Fundamentals

Introduction to Fluid Dynamics 1

The aim of this book is to explore how fluids in motion (e.g., waves produced by a hurricane) interact with and exert loads upon fixed and floating structures to ultimately inform their structural design. As such, we must first introduce the study of fluid dynamics to become familiar with the underlying physics that govern natural phenomena such as the propagation of surface gravity waves. This chapter introduces the fundamentals of fluid dynamics at its most conceptual level, before summarizing some important mathematical concepts that are required to engage with many of the derivations presented throughout this text. Interested readers are encouraged to consult Batchelor [1] for a more detailed introduction to the topic.

1.1 Principles of Fluid Dynamics

We begin by recognizing that fluid dynamics is intrinsically a field theory. That is, for a given subset in space filled with fluid (e.g., Fig. 1.1), we can assign vector or scalar quantities to every point within that space as a function of time. Figure 1.1 provides an example of a vector and scalar field in 2D generated by waves propagating over a body of water. Here, we take a vertical "slice" along the direction of wave propagation where x and z denote the horizontal and vertical axes, respectively. Note that the wave is shown via the red solid line while the dashed line denotes the still water level. The left figure shows the velocity vector (both magnitude and direction) associated with the motion of fluid particles at a fixed snapshot in time while the right plot displays the dynamic pressure (a scalar with only magnitude) at every point within the domain at the same instant in time.

S. Wang, *Applied Ocean Mechanics for Structural Engineering and Climate Adaptation*,
Synthesis Lectures on Ocean Systems Engineering,
https://doi.org/10.1007/978-3-032-18199-2_1

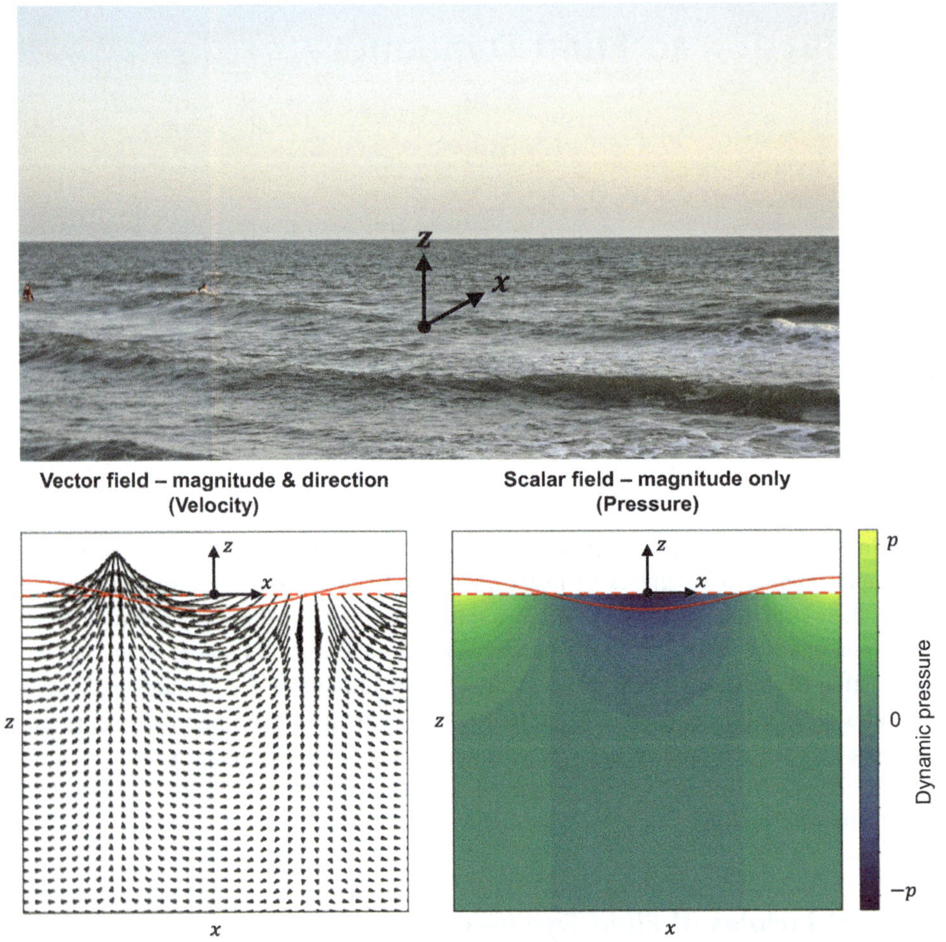

Fig. 1.1 Top: waves near a beach in Ostia, Italy. Bottom: vector (velocity) and scalar (pressure) fields for surface waves in 2D space

In classical solid mechanics, we can often assign properties to discrete objects that describe their motion through space. For example, we can easily determine the trajectory of a simple pendulum based on the length of the rod from which a mass is suspended. However, when characterizing the motion of fluids, we assign properties such as density, velocity, pressure to individual "fluid particles" that collectively constitute the entire fluid domain (i.e., "field") we are interested in. A fluid particle essentially represents an arbitrary volume of fluid that is small enough to be considered a discrete point in space, yet large enough that it can represent well defined properties. Technically speaking, this means that the size of a fluid particle cannot be smaller than a water molecule, if the fluid in question is water. In this sense, the term "fluid property" is simply a quantity that

Table 1.1 Examples of vector and scalar properties

Vector properties	Scalar properties
• Displacement • Velocity • Acceleration • Force • Momentum	• Temperature • Pressure • Density • Volume • Time

describes the behavior of a collection of molecules that make up a fluid particle. Fluid properties can either be vector or scalar quantities to which a few common examples are presented in Table 1.1. Finally, we note that the concept of absolute mass (with units of kg) has no meaning when working with fluids. Instead, the mass per unit volume (i.e., density with units of kg/m^3) is used when assigning mass to a fluid.

In order to derive equations that enable us to compute vector or scalar properties arising from fluid motion (in the form of a wave on the sea surface for example), we must first review a few mathematical principles that will be important when carrying out these derivations in later chapters.

1.2 Review of Vector Algebra

In this book, all vectors will be denoted in lower case **bold**. We will also adopt curly brackets {∗} to represent vectors and square brackets [∗] for matrices. The most common vector quantity we will encounter is the velocity vector $\boldsymbol{u}$ which represents the velocity of a fluid particle along the Cartesian x, y, and z-axis:

$$\boldsymbol{u} = \begin{Bmatrix} u \\ v \\ w \end{Bmatrix} \tag{1.1}$$

where u, v, and w denote the magnitude of the velocity along x, y, and z, respectively, as shown in Fig. 1.2. Vectors are also known as "first-order tensors".

1.2.1 Vector Addition

Only vectors with the same dimension may be added together. By adding two vectors:

$$\boldsymbol{a} = \begin{Bmatrix} a_1 \\ a_2 \\ a_3 \end{Bmatrix}, \quad \boldsymbol{b} = \begin{Bmatrix} b_1 \\ b_2 \\ b_3 \end{Bmatrix} \tag{1.2}$$

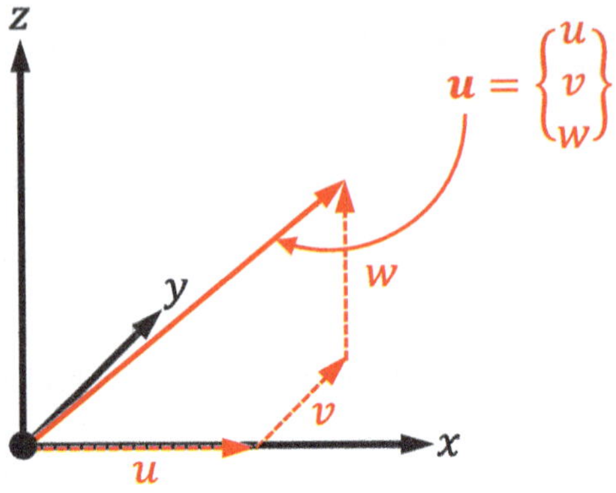

Fig. 1.2 Illustration of a vector $\boldsymbol{u}$

both with dimension = 3, we obtain:

$$\boldsymbol{a} + \boldsymbol{b} = \begin{Bmatrix} a_1 \\ a_2 \\ a_3 \end{Bmatrix} + \begin{Bmatrix} b_1 \\ b_2 \\ b_3 \end{Bmatrix} = \begin{Bmatrix} a_1 + b_1 \\ a_2 + b_2 \\ a_3 + b_3 \end{Bmatrix} \tag{1.3}$$

1.2.2 Scalar Multiplication

By multiplying a scalar a with vector $\boldsymbol{b}$, we obtain:

$$a\boldsymbol{b} = a \begin{Bmatrix} b_1 \\ b_2 \\ b_3 \end{Bmatrix} = \begin{Bmatrix} ab_1 \\ ab_2 \\ ab_3 \end{Bmatrix} \tag{1.4}$$

1.2.3 Dot Product

The dot product (also called "scalar product" or "inner product") between two vectors returns a scalar value and represents the projection of one vector onto another. Using $\boldsymbol{a} \cdot \boldsymbol{b}$ as an example:

$$\boldsymbol{a} \cdot \boldsymbol{b} = \begin{Bmatrix} a_1 \\ a_2 \\ a_3 \end{Bmatrix} \cdot \begin{Bmatrix} b_1 \\ b_2 \\ b_3 \end{Bmatrix} = a_1 b_1 + a_2 b_2 + a_3 b_3 \tag{1.5}$$

The dot product can also be used to compute the angle between two vectors as shown in Fig. 1.3 via:

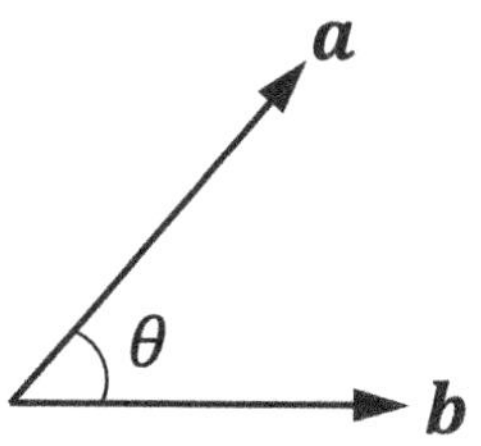

Fig. 1.3 Angle θ between two vectors $\boldsymbol{a}$ and $\boldsymbol{b}$

$$\theta = \arccos\left(\frac{\boldsymbol{a} \cdot \boldsymbol{b}}{|\boldsymbol{a}||\boldsymbol{b}|}\right) \tag{1.6}$$

where $|\boldsymbol{a}|$ and $|\boldsymbol{b}|$ denote the magnitudes of vectors $\boldsymbol{a}$ and $\boldsymbol{b}$. Note that if the two vectors are orthogonal ($\theta = 90° = \pi/2$), then $\boldsymbol{a} \cdot \boldsymbol{b} = 0$.

1.2.4 Vector Magnitude

The magnitude (absolute value) of a vector computes its length. The magnitude of vector $\boldsymbol{a}$ is:

$$|\boldsymbol{a}| = \sqrt{\boldsymbol{a} \cdot \boldsymbol{a}} = \sqrt{a_1^2 + a_2^2 + a_3^2} \tag{1.7}$$

Therefore, the square of the vector magnitude is simply the vector dotted with itself:

$$|\boldsymbol{a}|^2 = \boldsymbol{a} \cdot \boldsymbol{a} = \begin{Bmatrix} a_1 \\ a_2 \\ a_3 \end{Bmatrix} \cdot \begin{Bmatrix} a_1 \\ a_2 \\ a_3 \end{Bmatrix} = a_1^2 + a_2^2 + a_3^2 \tag{1.8}$$

1.2.5 Cross Product

The cross product (also called "vector product" or "outer product") of two vectors $\boldsymbol{a}$ and $\boldsymbol{b}$ yields a third vector that is orthogonal to the plane formed by $\boldsymbol{a}$ and $\boldsymbol{b}$:

$$\boldsymbol{a} \times \boldsymbol{b} = \begin{Bmatrix} a_1 \\ a_2 \\ a_3 \end{Bmatrix} \times \begin{Bmatrix} b_1 \\ b_2 \\ b_3 \end{Bmatrix} = \begin{Bmatrix} a_2 b_3 - a_3 b_2 \\ a_3 b_1 - a_1 b_3 \\ a_1 b_2 - a_2 b_1 \end{Bmatrix} \tag{1.9}$$

The direction that the cross-product vector is oriented follows the right-hand rule, and for this reason, $\boldsymbol{a} \times \boldsymbol{b} = -\boldsymbol{b} \times \boldsymbol{a}$. Its magnitude is equal to the area of the parallelogram formed by $\boldsymbol{a}$ and $\boldsymbol{b}$ as shown in Fig. 1.4.

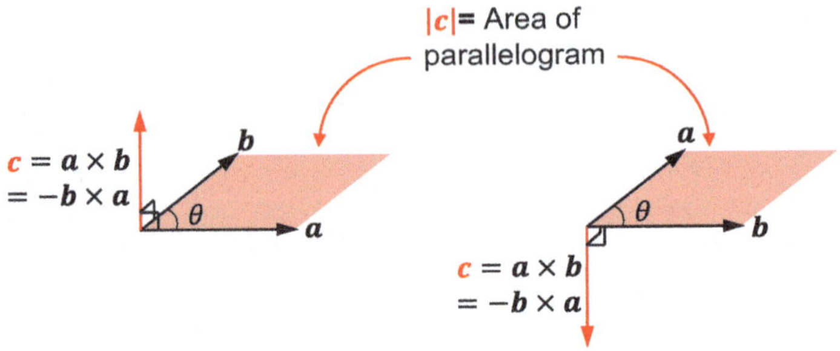

Fig. 1.4 Illustration of the cross product $\boldsymbol{a} \times \boldsymbol{b}$

1.2.6 Matrix Multiplication

A matrix (also called "second-order tensor") $\boldsymbol{A}$ given as:

$$\boldsymbol{A} = \begin{bmatrix} A_{11} & A_{12} & A_{13} \\ A_{21} & A_{22} & A_{23} \\ A_{31} & A_{32} & A_{33} \end{bmatrix} \tag{1.10}$$

multiplied with vector $\boldsymbol{b}$ gives:

$$\boldsymbol{A}\boldsymbol{b} = \begin{bmatrix} A_{11} & A_{12} & A_{13} \\ A_{21} & A_{22} & A_{23} \\ A_{31} & A_{32} & A_{33} \end{bmatrix} \begin{Bmatrix} b_1 \\ b_2 \\ b_3 \end{Bmatrix} = \begin{Bmatrix} A_{11}b_1 + A_{12}b_2 + A_{13}b_3 \\ A_{21}b_1 + A_{22}b_2 + A_{23}b_3 \\ A_{31}b_1 + A_{32}b_2 + A_{33}b_3 \end{Bmatrix} \tag{1.11}$$

This operation can only be performed if the number of columns in $\boldsymbol{A}$ matches the number of rows in $\boldsymbol{b}$. On the other hand, a row vector $\boldsymbol{b}^T$ can be multiplied with a matrix following:

$$\boldsymbol{b}^T\boldsymbol{A} = \begin{Bmatrix} b_1 & b_2 & b_3 \end{Bmatrix} \begin{bmatrix} A_{11} & A_{12} & A_{13} \\ A_{21} & A_{22} & A_{23} \\ A_{31} & A_{32} & A_{33} \end{bmatrix} = \begin{Bmatrix} A_{11}b_1 + A_{21}b_2 + A_{31}b_3 \\ A_{12}b_1 + A_{22}b_2 + A_{32}b_3 \\ A_{13}b_1 + A_{23}b_2 + A_{33}b_3 \end{Bmatrix}^T \tag{1.12}$$

where:

$$\boldsymbol{b}^T = \begin{Bmatrix} b_1 \\ b_2 \\ b_3 \end{Bmatrix}^T = \begin{Bmatrix} b_1 & b_2 & b_3 \end{Bmatrix} \tag{1.13}$$

denotes the transpose of $\boldsymbol{b}$, which converts it from a column into row vector. The following relationship also applies for the multiplication between a matrix $\boldsymbol{A}$ and a column ($\boldsymbol{b}$) or row ($\boldsymbol{b}^T$) vector:

$$Ab = \left(b^T A^T\right)^T \tag{1.14}$$

where:

$$A^T = \begin{bmatrix} A_{11} & A_{12} & A_{13} \\ A_{21} & A_{22} & A_{23} \\ A_{31} & A_{32} & A_{33} \end{bmatrix}^T = \begin{bmatrix} A_{11} & A_{21} & A_{31} \\ A_{12} & A_{22} & A_{32} \\ A_{13} & A_{23} & A_{33} \end{bmatrix} \tag{1.15}$$

is the transpose of matrix $\boldsymbol{A}$. The transpose of a matrix effectively swaps the row with its respective column, i.e., row 1 becomes column 1, row 2 becomes column 2, and row 3 becomes column 3.

1.3 Review of Partial Differential Equations

A partial differential equation (PDE) expresses the relationship between a dependent variable (e.g., temperature) and two or more independent variables (e.g., position and time). Since many phenomena found in nature are described using PDEs, including that of water waves, it is helpful to provide a conceptual overview and example that illustrates how they can be used to represent field properties that change across multiple dimensions. PDEs are often classified based on their linearity and order:

- A PDE is linear if the dependent variable and its derivatives are not multiplied with each other, raised to higher powers, or feature within nonlinear functions such as $\cos(*)$ or $\exp(*)$. Otherwise, it is nonlinear.
- The order of a PDE is given by the order of the highest partial derivative featured within the equation.

Table 1.2 gives a few examples of PDEs sorted by their order and linearity, where T is the dependent variable. Note that when working with nonlinear PDEs, it is often desirable to linearize them by eliminating all nonlinear terms. This is because it is much easier to find analytical solutions to linear PDEs than their nonlinear counterparts, as will be demonstrated throughout this text.

For this example, we will consider a square 2D plate with each side having a length of 1. Let's say we wish to determine the temperature T at every location on the plate from $x = 0$ to 1 and $y = 0$ to 1. Our dependent variable is therefore the temperature T and the independent variables are the x and y coordinates. The steady-state temperature distribution in 2D must satisfy the following second-order linear PDE:

$$\frac{\partial^2 T}{\partial x^2} + \frac{\partial^2 T}{\partial y^2} = 0 \tag{1.16}$$

Table 1.2 Examples of PDEs classified by order and linearity

Order	Linear	Nonlinear[a]
1st order	$\frac{\partial T}{\partial x} + T = 0$	$\frac{\partial T}{\partial x} + T^2 = 0$
	$x\frac{\partial T}{\partial x} + \frac{\partial T}{\partial y} = 0$	$x\frac{\partial T}{\partial x} + T\frac{\partial T}{\partial y} = 0$
2nd order	$\frac{\partial^2 T}{\partial x^2} + T = 0$	$\frac{\partial^2 T}{\partial x^2} + T\frac{\partial T}{\partial y} = 0$
	$x\frac{\partial T}{\partial x} + \frac{\partial^2 T}{\partial x \partial y} = 0$	$\left(\frac{\partial T}{\partial x}\right)^2 + \frac{\partial^2 T}{\partial x \partial y} = 0$

[a] The nonlinear terms here are: T^2, $T(\partial T/\partial y)$, and $(\partial T/\partial x)^2$

which is also known as the 2D Laplace equation. To solve this PDE, we must find an equation for T as a function of x and y based on its boundary conditions. These boundary conditions could be the temperature distribution along the four edges of the plate, such as that shown in Fig. 1.5. To keep things simple, suppose that the solution to this PDE is given by:

$$T = x^3 - 3xy^2 \tag{1.17}$$

We can confirm that Eq. (1.17) indeed satisfies (1.16) by computing the partial double derivative of T with respect to x and y. Note that when taking the partial derivative of a multivariable function with respect to one variable, all other independent variables are treated as constants. This gives:

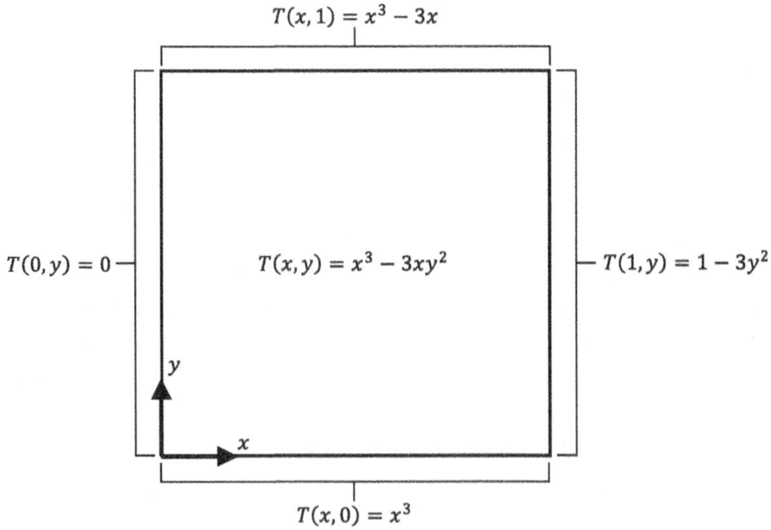

Fig. 1.5 Spatial distribution and boundary conditions of T

$$\frac{\partial^2 T}{\partial x^2} = 6x, \quad \frac{\partial^2 T}{\partial y^2} = -6x \tag{1.18}$$

which when substituted back into Eq. (1.16) returns 0 as expected. When working with PDEs, we use $\partial(*)/\partial x$ to denote the derivative of some function $*$ with respect to x, where ∂ ("del") signifies that the dependent variable $*$ is a function of two or more independent variables. This is in contrast to the $d(*)/dx$ used for ordinary differential equations (ODEs) where $*$ is a function of only a single variable (i.e., x).

By plotting T for all values of x and y between its boundaries, we can visualize how the temperature changes across the entire plate (see Fig. 1.6). Suppose now we wish to understand the rate at which T is changing along the x and y directions at any given location on the plate. To do so, we simply take the first partial derivative of T with respect to x and y, respectively:

$$\frac{\partial T}{\partial x} = 3x^2 - 3y^2, \quad \frac{\partial T}{\partial y} = -6xy \tag{1.19}$$

Hence, by substituting different values of x and y into Eq. (1.19), we can find the rate at which T is changing along either the positive x or y direction at that particular location. For example, the rate of change at the center of the plate ($x = y = 0.5$) is:

$$\frac{\partial T}{\partial x} = 3 \times (0.5)^2 - 3 \times (0.5)^2 = 0, \quad \frac{\partial T}{\partial y} = -6 \times 0.5 \times 0.5 = -1.5 \tag{1.20}$$

which signifies that the temperature is not locally changing along the x direction while decreasing along the positive y direction at a rate of 1.5 per unit length. Equation (1.20) therefore represents the slope of the tangent line to T at $x = y = 0.5$ as shown in Fig. 1.6.

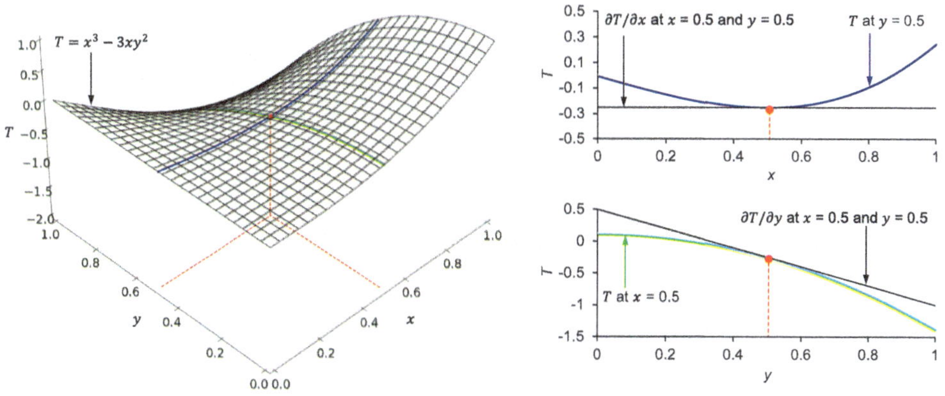

Fig. 1.6 Demonstration of a differential function T with two independent variables (x and y) along with its spatial rate of change at $x = y = 0.5$

1.4 The Gradient Operator

The gradient (∇) measures the spatial rate of change of a scalar field at a given point in space along the x, y, and z directions. The gradient of a scalar returns a vector. If we use the pressure p as an example, the gradient of p is:

$$\nabla p = \begin{Bmatrix} \frac{\partial}{\partial x} \\ \frac{\partial}{\partial y} \\ \frac{\partial}{\partial z} \end{Bmatrix} p = \begin{Bmatrix} \frac{\partial p}{\partial x} \\ \frac{\partial p}{\partial y} \\ \frac{\partial p}{\partial z} \end{Bmatrix} \tag{1.21}$$

where:

$$\nabla = \begin{Bmatrix} \frac{\partial}{\partial x} \\ \frac{\partial}{\partial y} \\ \frac{\partial}{\partial z} \end{Bmatrix} \tag{1.22}$$

is termed the "vector differential operator" (the symbol ∇ is the Greek letter "nabla"). The gradient of p simply produces a vector with each component being the derivative of p with respect to the x, y, and z coordinates and is equal to the scalar multiplication between the vector differential operator and p.

If we consider the horizontal 2D case with only x and y coordinates, Fig. 1.7 presents a physical interpretation of the pressure gradient at the center of four smoothly-varying pressure fields superimposed onto a 3×3 grid in space. In this example, the pressure measured at each point simply takes values of 1, 2, or 3. Here, the pressure gradient at p_1 is zero along both the x and y directions since the pressure is uniform at all locations. Likewise, the pressure gradient at p_2 is zero because it represents a local maxima such that the slopes along the x and y directions at that particular location are flat (i.e., the slopes have a gradient of zero). On the other hand, the pressure at p_3 is decreasing along the positive x direction while increasing along the positive y direction, resulting in their gradients being negative and positive, respectively. At p_4, the pressure is increasing along both x and y directions, hence their gradients are both positive. This example is similar to that illustrated in Fig. 1.6.

1.5 The Divergence Operator

The divergence ($\nabla \cdot$) measures the extent to which vectors are pointing outwards from a given point in space. The divergence of a vector returns a scalar. If using the velocity vector field $\boldsymbol{u}$ as an example, the divergence of $\boldsymbol{u}$ is:

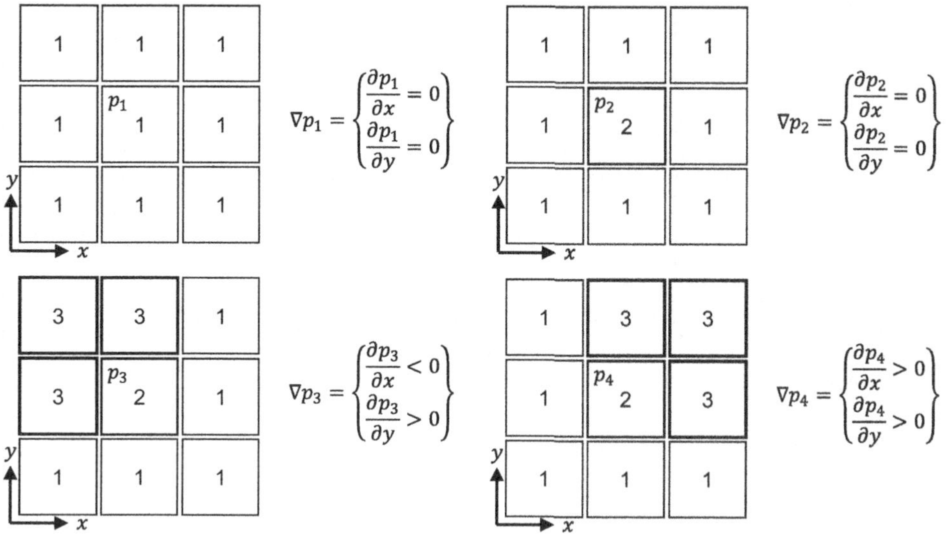

Fig. 1.7 Gradient of 2D scalar fields p

$$\nabla \cdot \boldsymbol{u} = \begin{Bmatrix} \frac{\partial}{\partial x} \\ \frac{\partial}{\partial y} \\ \frac{\partial}{\partial z} \end{Bmatrix} \cdot \begin{Bmatrix} u \\ v \\ w \end{Bmatrix} = \frac{\partial u}{\partial x} + \frac{\partial v}{\partial y} + \frac{\partial w}{\partial z} \tag{1.23}$$

The divergence of $\boldsymbol{u}$ is simply the sum of the derivatives of each component of the velocity vector and equal to the dot product between the vector differential operator and $\boldsymbol{u}$. A positive divergence implies a net outflow from a point (source) whereas a negative divergence implies a net inflow (sink).

If we consider the horizontal 2D case where $\boldsymbol{u} = \{u, v\}$, Fig. 1.8 presents a physical interpretation of $\nabla \cdot \boldsymbol{u}$ at a given point in space across four velocity fields where the length of each arrow represents the velocity magnitude. In the first example, the flow velocity is increasing along x and y which means that more fluid is leaving than entering the point. This location therefore serves as a source with $\nabla \cdot \boldsymbol{u} > 0$. The opposite occurs in the second example with the point acting like a sink resulting in $\nabla \cdot \boldsymbol{u} < 0$. In the third example, the flow velocity is increasing along x but decreasing along y. If we assume that the gradients of u and v are equal, then the point will experience no net inflow or outflow and $\nabla \cdot \boldsymbol{u} = 0$. The same principle also applies to the fourth example.

Since the divergence is a linear operator, the following property applies:

$$\nabla \cdot (a\boldsymbol{b} + c\boldsymbol{d}) = a\nabla \cdot \boldsymbol{b} + c\nabla \cdot \boldsymbol{d} \tag{1.24}$$

Similarly, the product rule pertaining to the divergence of a scalar multiplied with a vector is:

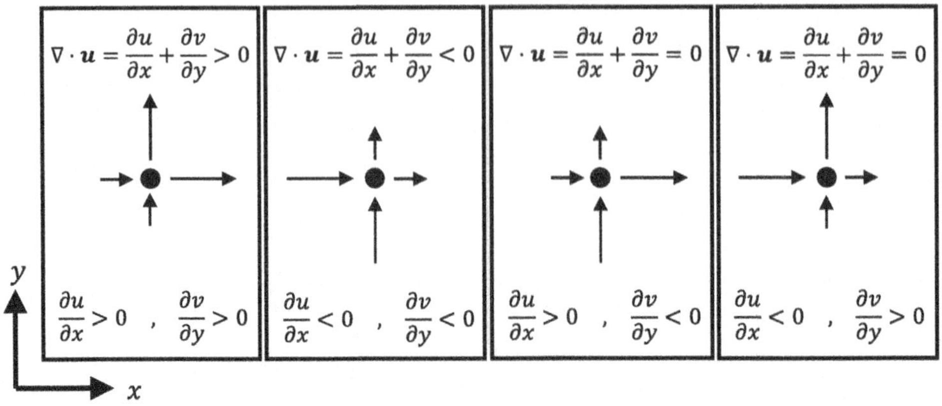

Fig. 1.8 Divergence of velocity vector $\boldsymbol{u}$

$$\nabla \cdot (a\boldsymbol{b}) = a\nabla \cdot \boldsymbol{b} + \boldsymbol{b} \cdot \nabla a \tag{1.25}$$

where ∇a denotes the gradient of the scalar a. We can easily prove Eq. (1.25) by recognizing that the left-hand side yields:

$$\nabla \cdot (a\boldsymbol{b}) = \begin{Bmatrix} \frac{\partial}{\partial x} \\ \frac{\partial}{\partial y} \\ \frac{\partial}{\partial z} \end{Bmatrix} \cdot \begin{Bmatrix} ab_1 \\ ab_2 \\ ab_3 \end{Bmatrix} = \frac{\partial (ab_1)}{\partial x} + \frac{\partial (ab_2)}{\partial y} + \frac{\partial (ab_3)}{\partial z} \tag{1.26}$$

when taking $\boldsymbol{b} = \{b_1, b_2, b_3\}$. Equation (1.26) can be further manipulated by applying the product rule of differentiation which states that:

$$\frac{\partial (ab)}{\partial x} = a\frac{\partial b}{\partial x} + b\frac{\partial a}{\partial x} \tag{1.27}$$

Applying the product rule then leads to:

$$\frac{\partial (ab_1)}{\partial x} + \frac{\partial (ab_2)}{\partial y} + \frac{\partial (ab_3)}{\partial z} = a\frac{\partial b_1}{\partial x} + b_1\frac{\partial a}{\partial x} + a\frac{\partial b_2}{\partial y} + b_2\frac{\partial a}{\partial y} + a\frac{\partial b_3}{\partial z} + b_3\frac{\partial a}{\partial z} \tag{1.28}$$

which can be rearranged into:

$$a\left(\frac{\partial b_1}{\partial x} + \frac{\partial b_2}{\partial y} + \frac{\partial b_3}{\partial z}\right) + b_1\frac{\partial a}{\partial x} + b_2\frac{\partial a}{\partial y} + b_3\frac{\partial a}{\partial z} = a\nabla \cdot \boldsymbol{b} + \boldsymbol{b} \cdot \nabla a \tag{1.29}$$

as consistent with Eq. (1.25).

In addition, the divergence of a second-order tensor (i.e., a matrix) $\boldsymbol{A}$ is:

$$\nabla \cdot \boldsymbol{A} = \begin{Bmatrix} \frac{\partial}{\partial x} \\ \frac{\partial}{\partial y} \\ \frac{\partial}{\partial z} \end{Bmatrix} \cdot \begin{bmatrix} A_{11} & A_{12} & A_{13} \\ A_{21} & A_{22} & A_{23} \\ A_{31} & A_{32} & A_{33} \end{bmatrix} = \begin{Bmatrix} \frac{\partial A_{11}}{\partial x} + \frac{\partial A_{21}}{\partial y} + \frac{\partial A_{31}}{\partial z} \\ \frac{\partial A_{12}}{\partial x} + \frac{\partial A_{22}}{\partial y} + \frac{\partial A_{32}}{\partial z} \\ \frac{\partial A_{13}}{\partial x} + \frac{\partial A_{23}}{\partial y} + \frac{\partial A_{33}}{\partial z} \end{Bmatrix} \tag{1.30}$$

Hence, we see that the divergence of a second-order tensor returns a vector (i.e., first-order tensor).

1.6 The Laplace Operator

When the divergence operator is applied to the gradient of a scalar field, this operator is called the Laplacian (∇^2) and is expressed as follows:

$$\nabla^2 p = \nabla \cdot \nabla p = \left(\frac{\partial^2}{\partial x^2} + \frac{\partial^2}{\partial y^2} + \frac{\partial^2}{\partial z^2} \right) p = \frac{\partial^2 p}{\partial x^2} + \frac{\partial^2 p}{\partial y^2} + \frac{\partial^2 p}{\partial z^2} \tag{1.31}$$

where:

$$\nabla^2 = \nabla \cdot \nabla = \begin{Bmatrix} \frac{\partial}{\partial x} \\ \frac{\partial}{\partial y} \\ \frac{\partial}{\partial z} \end{Bmatrix} \cdot \begin{Bmatrix} \frac{\partial}{\partial x} \\ \frac{\partial}{\partial y} \\ \frac{\partial}{\partial z} \end{Bmatrix} = \frac{\partial^2}{\partial x^2} + \frac{\partial^2}{\partial y^2} + \frac{\partial^2}{\partial z^2} \tag{1.32}$$

signifies the vector differential operator dotted with itself. The Laplace operator is sometimes written as Δ. As can be seen, taking the Laplacian of a scalar returns a scalar.

A physical interpretation of the Laplacian is that it measures how much the value of a given field (e.g., p) at a particular location differs from its average value when taken over the surrounding points. A positive Laplacian indicates that the field at a given location is lower than its surrounding average, while a negative Laplacian infers a larger value than its neighbors. As an example, consider the 1D scalar field $p = \sin x$ shown in Fig. 1.9, with the corresponding Laplacian $\nabla^2 p = -\sin x$. As can be seen, the Laplacian reaches a maximum positive value at the "valley" ($x = 3\pi/2$) and is most negative at the "peak" ($x = \pi/2$). At the inflection point ($x = \pi$), the Laplacian is zero. Another way to interpret the Laplacian is that it essentially measures the "curvature" of the function at any given location.

1.7 The Curl Operator

The curl ($\nabla \times$) measures the magnitude and direction of rotation at a given point within a vector field. The curl of a vector returns a vector. Using the velocity vector $\boldsymbol{u} = \{u, v, w\}$ as an example, the curl of $\boldsymbol{u}$ is:

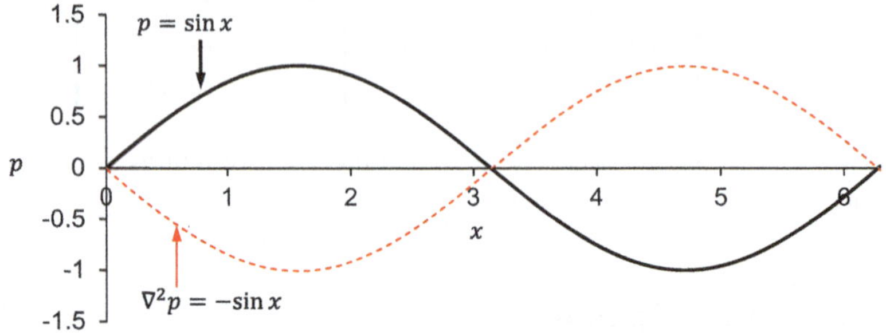

Fig. 1.9 Laplacian of a 1D scalar field p

$$\nabla \times \boldsymbol{u} = \begin{Bmatrix} \frac{\partial}{\partial x} \\ \frac{\partial}{\partial y} \\ \frac{\partial}{\partial z} \end{Bmatrix} \times \begin{Bmatrix} u \\ v \\ w \end{Bmatrix} = \begin{Bmatrix} \frac{\partial w}{\partial y} - \frac{\partial v}{\partial z} \\ \frac{\partial u}{\partial z} - \frac{\partial w}{\partial x} \\ \frac{\partial v}{\partial x} - \frac{\partial u}{\partial y} \end{Bmatrix} \tag{1.33}$$

The curl of $\boldsymbol{u}$ is therefore equal to the cross product between the vector differential operator and $\boldsymbol{u}$. A physical interpretation of $\nabla \times \boldsymbol{u}$ is illustrated via Fig. 1.10. The axis of rotation of a fluid particle is given by the curl of its velocity vector while the speed of its rotation is given by the magnitude of the curl $|\nabla \times \boldsymbol{u}|$. The direction of rotation is obtained via the right-hand rule.

In addition, special identities exist for the curl, with the divergence of the curl of a vector field always being zero:

$$\nabla \cdot (\nabla \times \boldsymbol{u}) = 0 \tag{1.34}$$

This identity can be demonstrated by recognizing that:

$$\nabla \cdot (\nabla \times \boldsymbol{u}) = \begin{Bmatrix} \frac{\partial}{\partial x} \\ \frac{\partial}{\partial y} \\ \frac{\partial}{\partial z} \end{Bmatrix} \cdot \begin{Bmatrix} \frac{\partial w}{\partial y} - \frac{\partial v}{\partial z} \\ \frac{\partial u}{\partial z} - \frac{\partial w}{\partial x} \\ \frac{\partial v}{\partial x} - \frac{\partial u}{\partial y} \end{Bmatrix}$$

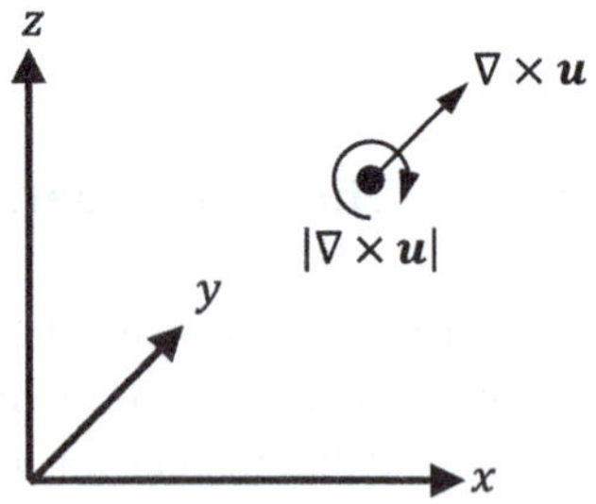

Fig. 1.10 Curl of vector $\boldsymbol{u}$

$$= \frac{\partial}{\partial x}\left(\frac{\partial w}{\partial y} - \frac{\partial v}{\partial z}\right) + \frac{\partial}{\partial y}\left(\frac{\partial u}{\partial z} - \frac{\partial w}{\partial x}\right) + \frac{\partial}{\partial z}\left(\frac{\partial v}{\partial x} - \frac{\partial u}{\partial y}\right) \tag{1.35}$$

which can be expanded to yield:

$$\frac{\partial^2 w}{\partial x \partial y} - \frac{\partial^2 v}{\partial x \partial z} + \frac{\partial^2 u}{\partial y \partial z} - \frac{\partial^2 w}{\partial y \partial x} + \frac{\partial^2 v}{\partial z \partial x} - \frac{\partial^2 u}{\partial z \partial y} = 0 \tag{1.36}$$

Furthermore, another identity is that the curl of the gradient of a scalar is always zero:

$$\nabla \times (\nabla \phi) = 0 \tag{1.37}$$

where ϕ denotes a scalar field. This can be demonstrated via:

$$\nabla \times (\nabla \phi) = \begin{Bmatrix} \frac{\partial}{\partial x} \\ \frac{\partial}{\partial y} \\ \frac{\partial}{\partial z} \end{Bmatrix} \times \begin{Bmatrix} \frac{\partial \phi}{\partial x} \\ \frac{\partial \phi}{\partial y} \\ \frac{\partial \phi}{\partial z} \end{Bmatrix} = \begin{Bmatrix} \frac{\partial^2 \phi}{\partial y \partial z} - \frac{\partial^2 \phi}{\partial z \partial y} \\ \frac{\partial^2 \phi}{\partial z \partial x} - \frac{\partial^2 \phi}{\partial x \partial z} \\ \frac{\partial^2 \phi}{\partial x \partial y} - \frac{\partial^2 \phi}{\partial y \partial x} \end{Bmatrix} = 0 \tag{1.38}$$

In this text, we will assume that fluids are irrotational such that the curl of the velocity is zero everywhere, i.e., $\nabla \times \boldsymbol{u} = 0$. This assumption is crucial as it enables us to derive analytical solutions that describe wave propagation.

Reference

1. Batchelor, G.K., 2000. An introduction to fluid dynamics. Cambridge University Press.

2 The Navier–Stokes and Bernoulli Equations

The Navier–Stokes equations are a set of partial differential equations that describe the behavior of Newtonian fluids (e.g., water and air). The equations were formulated by French engineer Claude-Louis Navier[1] and British-Irish physicist George Gabriel Stokes (Fig. 2.1) over a period spanning almost three decades. When used in conjunction with the Bernoulli equation (named for Swiss physicist and mathematician Daniel Bernoulli—Fig. 2.1), they represent the pillars of fluid mechanics and are some of the most well-known equations within science and engineering.

The aforementioned equations are so important since they describe the three conservation laws (mass, momentum, and energy) that fluids must abide by. The Navier–Stokes equations comprise a set of four separate equations that impose the following physical laws:

- **Conservation of mass** (i.e., continuity equation), given by one equation.
- **Conservation of momentum** (i.e., momentum equations), given by three equations—one for motion in each spatial dimension (x, y, and z) in 3D. Note that the momentum equations are commonly condensed into a single equation expressed in vector form.

The Bernoulli equation can be derived using the Navier-Stokes equations to address the remaining law (under incompressible, inviscid, and irrotational assumptions):

- **Conservation of energy**, given by one equation.

[1] Navier was also responsible for formulating the theory of elasticity and for this reason is sometimes referred to as the "father of structural analysis". His name is inscribed on the Eiffel Tower.

S. Wang, *Applied Ocean Mechanics for Structural Engineering and Climate Adaptation*, Synthesis Lectures on Ocean Systems Engineering,
https://doi.org/10.1007/978-3-032-18199-2_2

Fig. 2.1 Left: Claude-Louis Navier. Center: George Gabriel Stokes. Right: Daniel Bernoulli

This chapter will introduce these equations from both a conceptual and mathematical perspective.

2.1 The Continuity Equation

The continuity equation stipulates that the rate at which mass enters a fixed volume of fluid is equal to the rate at which mass leaves the volume plus the accumulation of mass within the volume (mass is conserved). Essentially, this is stating that for a fixed volume of fluid (Ω) with density ρ, the rate at which its mass increases is equal to the net inflow rate of fluid entering the volume. Mathematically, this relationship is expressed as:

$$\iiint_{\Omega} \frac{\partial \rho}{\partial t} dV = -\iiint_{\Omega} \nabla \cdot (\rho \boldsymbol{u}) dV \tag{2.1}$$

Note that the triple integral simply implies that we are summing the mass within a 3D body of fluid of fixed volume Ω, referred to as the control volume. $\boldsymbol{u} = \{u, v, w\}$ is the velocity of the fluid with respects to the three Cartesian coordinates x, y, and z, and $\nabla \cdot (\rho \boldsymbol{u})$ is the divergence of the rate of mass flow. Equation (2.1) can be visualized in Fig. 2.2.

Hence, in physical terms, $\nabla \cdot (\rho \boldsymbol{u})$ simply measures the extent to which a fluid is flowing outwards from some fixed volume. Equation (2.1) can be rearranged as follows:

$$\iiint_{\Omega} \left\{ \frac{\partial \rho}{\partial t} + \nabla \cdot (\rho \boldsymbol{u}) \right\} dV = 0 \tag{2.2}$$

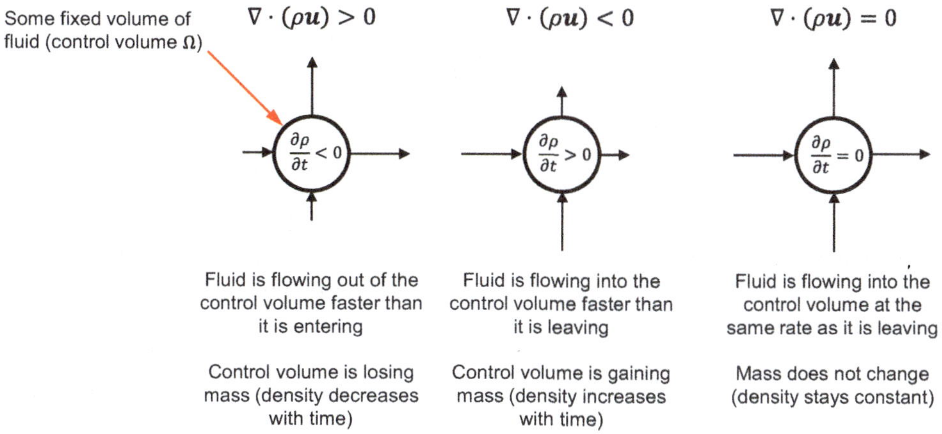

Fig. 2.2 Divergence of mass flow rate

which implies that:

$$\frac{\partial \rho}{\partial t} + \nabla \cdot (\rho \boldsymbol{u}) = 0 \tag{2.3}$$

When expanding the divergence operator ($\nabla \cdot$) via the product rule, we obtain:

$$\frac{\partial \rho}{\partial t} + \boldsymbol{u} \cdot \nabla \rho + \rho \nabla \cdot \boldsymbol{u} = 0 \tag{2.4}$$

Note that the first two terms in (2.4), $\partial \rho/\partial t + \boldsymbol{u} \cdot \nabla \rho$, are often expressed as $D\rho/Dt$ and is called the material derivative of ρ. This concept will be explained later in the chapter.

In this book, we assume that the density of water (ρ) remains constant (i.e., incompressible), which is generally an acceptable assumption in the absence of Mach shocks. The term $\partial \rho/\partial t$ describes the change in fluid density with time. Since we take the density to be constant in time, the following condition must hold:

$$\frac{\partial \rho}{\partial t} = 0 \tag{2.5}$$

In addition, the term $\nabla \rho$ (referred to as the density gradient) describes the change in fluid density across space. Mathematically, the gradient operator (∇) acting on ρ is expressed as follows:

$$\nabla \rho = \begin{Bmatrix} \frac{\partial \rho}{\partial x} \leftarrow \text{Change in density along } x \\ \frac{\partial \rho}{\partial y} \leftarrow \text{Change in density along } y \\ \frac{\partial \rho}{\partial z} \leftarrow \text{Change in density along } z \end{Bmatrix} \tag{2.6}$$

Since we take the density to be constant everywhere in space:

$$\nabla\rho = \begin{Bmatrix} 0 \\ 0 \\ 0 \end{Bmatrix} \tag{2.7}$$

Substituting (2.7) and (2.5) into (2.4) therefore yields the following:

$$\rho\nabla \cdot \boldsymbol{u} = 0 \tag{2.8}$$

We know that the fluid density ρ cannot be zero (this is the trivial solution). The divergence of the velocity ($\nabla\cdot\boldsymbol{u}$) must therefore equal zero to satisfy (2.8). In mathematical terms, the divergence of the velocity is:

$$\nabla \cdot \boldsymbol{u} = \frac{\partial u}{\partial x} + \frac{\partial v}{\partial y} + \frac{\partial w}{\partial z} \tag{2.9}$$

The continuity equation therefore reduces to:

$$\nabla \cdot \boldsymbol{u} = \frac{\partial u}{\partial x} + \frac{\partial v}{\partial y} + \frac{\partial w}{\partial z} = 0 \tag{2.10}$$

Equation (2.10) is referred to as the incompressible continuity equation.

2.2 The Momentum Equations

The Navier–Stokes momentum equations depict Newton's second law assuming constant mass (force $=$ mass $\times$ acceleration) applied to a fluid traveling along the x, y, and z directions with velocity u, v, and w, respectively. Mathematically, the three equations (condensed into vector form) are expressed as:

$$\rho\frac{D\boldsymbol{u}}{Dt} = -\nabla p + \rho\nu_k\nabla^2\boldsymbol{u} + \rho\boldsymbol{g} \tag{2.11}$$

where the left-hand side is the mass multiplied by the acceleration of a fluid particle, and the right-hand side is the sum of internal and external forces acting on the particle. Equation (2.11) can be broken down into the following components shown in Table 2.1.

Note that ν_k denotes the kinematic viscosity of the fluid ($\approx 10^{-6}$ m^2 s^{-1} for water) and $g = 9.81$ ms^{-2} is the gravitational acceleration. ∇p is the gradient of the pressure, which measures the rate and direction of the change in pressure and is expressed mathematically as follows:

$$\nabla p = \begin{Bmatrix} \frac{\partial p}{\partial x} \leftarrow \text{Change in pressure along } x \\ \frac{\partial p}{\partial y} \leftarrow \text{Change in pressure along } y \\ \frac{\partial p}{\partial z} \leftarrow \text{Change in pressure along } z \end{Bmatrix} \tag{2.12}$$

Table 2.1 Components of the Navier–Stokes momentum equations

Left-hand side		Right-hand side		
Mass (per unit volume)	ρ	**Internal force** (per unit volume)	$-\nabla p$	Force resulting from differences in fluid pressure in space (i.e., pressure gradients)
			$\rho\nu_k\nabla^2\boldsymbol{u}$	Force resulting from fluid viscosity, where $\boldsymbol{u} = \begin{Bmatrix} u \\ v \\ w \end{Bmatrix}$
Acceleration (in x, y, and z directions)	$\frac{Du}{Dt}$	**External force** (per unit volume)	$\rho\boldsymbol{g}$	Force due to gravity acting on the fluid, where $\boldsymbol{g} = \begin{Bmatrix} 0 \\ 0 \\ -g \end{Bmatrix}$

To understand how the pressure gradient $-\nabla p$ can act as a force, consider the experiment shown in Fig. 2.3. A gas is separated into two different chambers via a movable gate with the left chamber sustaining a larger pressure than the right (as it has more molecules occupying the same area). Since the pressure is decreasing towards the right (in the direction of positive x), a negative pressure gradient $-\nabla p$ exists. If the gate is rapidly removed, molecules in the left chamber will flow towards the right as if acted upon by a force towards the positive x direction. This experiment thus provides a physical interpretation of the pressure gradient force.

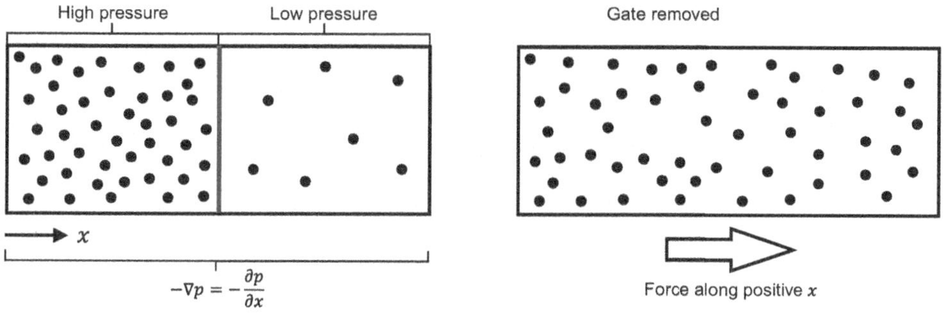

Fig. 2.3 Visual representation of force induced by a pressure gradient

Equation (2.11) can be expanded into its three constituents in x, y, and z, respectively:

$$\begin{aligned}\rho\frac{Du}{Dt} &= -\frac{\partial p}{\partial x} + \rho\nu_k\nabla^2 u\\ \rho\frac{Dv}{Dt} &= -\frac{\partial p}{\partial y} + \rho\nu_k\nabla^2 v\\ \rho\frac{Dw}{Dt} &= -\frac{\partial p}{\partial z} + \rho\nu_k\nabla^2 w - \rho g\end{aligned} \tag{2.13}$$

Notice that the gravitational acceleration g features only in z since the gravity vector is $\boldsymbol{g} = \{0, 0, -g\}$. The operation $D(*)/Dt$ is referred to as the material derivative and depicts the change in some quantity $*$ when following its motion through time and space. Mathematically, the material derivative is expressed as:

$$\frac{D(*)}{Dt} = \frac{\partial(*)}{\partial t} + \boldsymbol{u}\cdot\nabla(*) \tag{2.14}$$

where the $\partial(*)/\partial t$ term describes the change in $*$ with time and $\boldsymbol{u}\cdot\nabla(*)$ depicts its change in space. Du/Dt, Dv/Dt, and Dw/Dt is therefore the change in velocity (i.e., the acceleration) when following a fluid particle along the x, y, and z coordinates and is expressed as follows:

$$\begin{aligned}\frac{Du}{Dt} &= \frac{\partial u}{\partial t} + \boldsymbol{u}\cdot\nabla u = \frac{\partial u}{\partial t} + \begin{Bmatrix} u\\ v\\ w\end{Bmatrix}\cdot\begin{Bmatrix}\frac{\partial u}{\partial x}\\ \frac{\partial u}{\partial y}\\ \frac{\partial u}{\partial z}\end{Bmatrix} = \frac{\partial u}{\partial t} + u\frac{\partial u}{\partial x} + v\frac{\partial u}{\partial y} + w\frac{\partial u}{\partial z}\\ \frac{Dv}{Dt} &= \frac{\partial v}{\partial t} + \boldsymbol{u}\cdot\nabla v = \frac{\partial v}{\partial t} + \begin{Bmatrix} u\\ v\\ w\end{Bmatrix}\cdot\begin{Bmatrix}\frac{\partial v}{\partial x}\\ \frac{\partial v}{\partial y}\\ \frac{\partial v}{\partial z}\end{Bmatrix} = \frac{\partial v}{\partial t} + u\frac{\partial v}{\partial x} + v\frac{\partial v}{\partial y} + w\frac{\partial v}{\partial z}\\ \frac{Dw}{Dt} &= \frac{\partial w}{\partial t} + \boldsymbol{u}\cdot\nabla w = \frac{\partial w}{\partial t} + \begin{Bmatrix} u\\ v\\ w\end{Bmatrix}\cdot\begin{Bmatrix}\frac{\partial w}{\partial x}\\ \frac{\partial w}{\partial y}\\ \frac{\partial w}{\partial z}\end{Bmatrix} = \frac{\partial w}{\partial t} + u\frac{\partial w}{\partial x} + v\frac{\partial w}{\partial y} + w\frac{\partial w}{\partial z}\end{aligned} \tag{2.15}$$

For the treatment of water waves in the ocean, the role of fluid viscosity (ν_k) is often negligible (i.e., inviscid). Hence:

$$\rho\nu_k\nabla^2 u = \rho\nu_k\nabla^2 v = \rho\nu_k\nabla^2 w = 0 \tag{2.16}$$

Substituting (2.15) and (2.16) into (2.13) therefore yields the incompressible and inviscid Navier–Stokes momentum equations:

$$\rho\left(\frac{\partial u}{\partial t} + u\frac{\partial u}{\partial x} + v\frac{\partial u}{\partial y} + w\frac{\partial u}{\partial z}\right) = -\frac{\partial p}{\partial x} \tag{2.17}$$

$$\rho\left(\frac{\partial v}{\partial t}+u\frac{\partial v}{\partial x}+v\frac{\partial v}{\partial y}+w\frac{\partial v}{\partial z}\right)=-\frac{\partial p}{\partial y} \tag{2.18}$$

$$\rho\left(\frac{\partial w}{\partial t}+u\frac{\partial w}{\partial x}+v\frac{\partial w}{\partial y}+w\frac{\partial w}{\partial z}\right)=-\frac{\partial p}{\partial z}-\rho g \tag{2.19}$$

Equations (2.10), (2.17), (2.18), and (2.19) comprise the complete Navier–Stokes equations governing the behavior of incompressible and inviscid fluids (standard assumptions when dealing with surface gravity waves). They are also known as the Euler equations (Table 2.2).

We will assume that the velocity of fluid particles are sufficiently small such that all nonlinear terms in the momentum equations (2.17)–(2.19) can be ignored (this operation is called linearization since we are only retaining the linear terms). In addition, we will only consider 2D waves propagating within the xz-plane which allows us to eliminate Eq. (2.18) that deals with motion along the y-axis. Finally, it is seen that Eq. (2.19) contains a constant hydrostatic component ρg which exists irrespective of the presence of dynamic pressure gradients induced by wave propagation. We can therefore remove this term from (2.19) for the treatment of dynamic behavior. After performing these operations, the Navier–Stokes equations can be further simplified to that shown in Table 2.3.

Table 2.2 Navier–Stokes equations assuming incompressible and inviscid behavior (Euler equations)

Conservation of mass (continuity equation)	$\nabla\cdot\boldsymbol{u}=0\rightarrow\frac{\partial u}{\partial x}+\frac{\partial v}{\partial y}+\frac{\partial w}{\partial z}=0$
Conservation of momentum (momentum equations)	$\rho\frac{D\boldsymbol{u}}{Dt}=-\nabla p+\rho\boldsymbol{g}\rightarrow \begin{cases}\rho\left(\frac{\partial u}{\partial t}+u\frac{\partial u}{\partial x}+v\frac{\partial u}{\partial y}+w\frac{\partial u}{\partial z}\right)=-\frac{\partial p}{\partial x}\\ \rho\left(\frac{\partial v}{\partial t}+u\frac{\partial v}{\partial x}+v\frac{\partial v}{\partial y}+w\frac{\partial v}{\partial z}\right)=-\frac{\partial p}{\partial y}\\ \rho\left(\frac{\partial w}{\partial t}+u\frac{\partial w}{\partial x}+v\frac{\partial w}{\partial y}+w\frac{\partial w}{\partial z}\right)=-\frac{\partial p}{\partial z}-\rho g\end{cases}$

Table 2.3 Linearized Euler equations for 2D wave propagation

Conservation of mass (continuity equation)	$\nabla\cdot\boldsymbol{u}=0\rightarrow\frac{\partial u}{\partial x}+\frac{\partial w}{\partial z}=0$
Conservation of momentum (momentum equations)	$\rho\frac{\partial\boldsymbol{u}}{\partial t}=-\nabla p\rightarrow\begin{cases}\rho\frac{\partial u}{\partial t}=-\frac{\partial p}{\partial x}\\ \rho\frac{\partial w}{\partial t}=-\frac{\partial p}{\partial z}\end{cases}$

2.3 The Unsteady Bernoulli Equation

We have used u, v, and w to describe the fluid velocity along the Cartesian x, y, and z coordinates. However, if we wish to derive explicit expressions that enable us to actually calculate the velocity as a function of space and time within some domain (e.g., along a flooded beach during a storm surge), we must introduce an additional function ϕ. This function is called the "velocity potential" and serves as a mathematical point of reference that can be used to determine u, v, and w (among other useful properties) depending on how it is manipulated [1]. While the function ϕ has no physical interpretation, it represents a mathematically elegant way to describe the velocity at any point within a field via:

$$\frac{\partial \phi}{\partial x} = u, \quad \frac{\partial \phi}{\partial y} = v, \quad \frac{\partial \phi}{\partial z} = w \tag{2.20}$$

The 3D velocity vector $\boldsymbol{u} = \{u, v, w\}$ can thus be expressed as the gradient of the potential:

$$\boldsymbol{u} = \nabla \phi = \begin{Bmatrix} \frac{\partial \phi}{\partial x} = u \\ \frac{\partial \phi}{\partial y} = v \\ \frac{\partial \phi}{\partial z} = w \end{Bmatrix} \tag{2.21}$$

Substituting (2.21) into the continuity equation (2.10) therefore yields:

$$\nabla \cdot \boldsymbol{u} = \nabla \cdot (\nabla \phi) = \nabla^2 \phi = 0 \tag{2.22}$$

Likewise, the momentum equations (2.17)–(2.19) can be expressed as:

$$\rho \left[\frac{\partial}{\partial t} \frac{\partial \phi}{\partial x} + \nabla \phi \cdot \nabla \left(\frac{\partial \phi}{\partial x} \right) \right] + \frac{\partial p}{\partial x} = 0 \tag{2.23}$$

$$\rho \left[\frac{\partial}{\partial t} \frac{\partial \phi}{\partial y} + \nabla \phi \cdot \nabla \left(\frac{\partial \phi}{\partial y} \right) \right] + \frac{\partial p}{\partial y} = 0 \tag{2.24}$$

$$\rho \left[\frac{\partial}{\partial t} \frac{\partial \phi}{\partial z} + \nabla \phi \cdot \nabla \left(\frac{\partial \phi}{\partial z} \right) \right] + \frac{\partial p}{\partial z} + \rho g = 0 \tag{2.25}$$

Using (2.23) as an example, the gradient of the potential derivative is:

$$\nabla \left(\frac{\partial \phi}{\partial x} \right) = \begin{Bmatrix} \frac{\partial}{\partial x} \frac{\partial \phi}{\partial x} \\ \frac{\partial}{\partial y} \frac{\partial \phi}{\partial x} \\ \frac{\partial}{\partial z} \frac{\partial \phi}{\partial x} \end{Bmatrix} = \frac{\partial}{\partial x} \begin{Bmatrix} \frac{\partial \phi}{\partial x} \\ \frac{\partial \phi}{\partial y} \\ \frac{\partial \phi}{\partial z} \end{Bmatrix} = \frac{\partial}{\partial x} \nabla \phi \tag{2.26}$$

which when taking the dot product with $\nabla \phi$ gives:

$$\nabla\phi \cdot \frac{\partial}{\partial x}\nabla\phi = \frac{\partial}{\partial x}\left(\frac{1}{2}\nabla\phi \cdot \nabla\phi\right) \tag{2.27}$$

Incorporating this definition into (2.23)–(2.25) produces:

$$\rho\frac{\partial}{\partial t}\frac{\partial\phi}{\partial x} + \frac{\partial}{\partial x}\left(\frac{1}{2}\rho\nabla\phi \cdot \nabla\phi\right) + \frac{\partial p}{\partial x} = \frac{\partial}{\partial x}\left(\rho\frac{\partial\phi}{\partial t} + \frac{1}{2}\rho\nabla\phi \cdot \nabla\phi + p\right) = 0 \tag{2.28}$$

$$\rho\frac{\partial}{\partial t}\frac{\partial\phi}{\partial y} + \frac{\partial}{\partial y}\left(\frac{1}{2}\rho\nabla\phi \cdot \nabla\phi\right) + \frac{\partial p}{\partial y} = \frac{\partial}{\partial y}\left(\rho\frac{\partial\phi}{\partial t} + \frac{1}{2}\rho\nabla\phi \cdot \nabla\phi + p\right) = 0 \tag{2.29}$$

$$\rho\frac{\partial}{\partial t}\frac{\partial\phi}{\partial z} + \frac{\partial}{\partial z}\left(\frac{1}{2}\rho\nabla\phi \cdot \nabla\phi\right) + \frac{\partial p}{\partial z} + \rho g = \frac{\partial}{\partial z}\left(\rho\frac{\partial\phi}{\partial t} + \frac{1}{2}\rho\nabla\phi \cdot \nabla\phi + p + \rho g z\right) = 0 \tag{2.30}$$

Equations (2.28) to (2.30) can therefore be expressed as:

$$\nabla\left(\rho\frac{\partial\phi}{\partial t} + \frac{1}{2}\rho\nabla\phi \cdot \nabla\phi + p + \rho g z\right) = 0 \tag{2.31}$$

which infers that the terms inside the parenthesis must sum to a constant (since the derivative is zero):

$$\rho\frac{\partial\phi}{\partial t} + \frac{1}{2}\rho\nabla\phi \cdot \nabla\phi + p + \rho g z = C_0(t) \tag{2.32}$$

Here, C_0 is a constant that is dependent only on time t and must be selected based on the application. Equation (2.32) is called the "unsteady Bernoulli equation", where the term "unsteady" is used to denote the time-dependent nature of the velocity potential on account of the $\partial\phi/\partial t$ term. This equation essentially expresses the sum of the kinetic, potential, and pressure energy pertaining to fluid flows. Just like the Navier–Stokes momentum equations, (2.32) can be linearized by assuming $\nabla\phi \cdot \nabla\phi \approx 0$ to yield the following:

$$\rho\frac{\partial\phi}{\partial t} + p + \rho g z = C_0(t) \tag{2.33}$$

We now wish to eliminate the unknown constant $C_0(t)$ from the linearized equation. To do so, we can restate our existing velocity potential ϕ as being composed of two parts such that $\phi = \phi_1 + \phi_2$ with the condition that the time derivative of ϕ_2 is exactly equal to $C_0(t)/\rho$:

$$\rho\frac{\partial(\phi_1 + \phi_2)}{\partial t} + p + \rho g z = C_0(t), \quad \frac{\partial\phi_2}{\partial t} = \frac{C_0(t)}{\rho} \tag{2.34}$$

Taking advantage of the linearity of differentiation, (2.34) becomes:

$$\rho \frac{\partial \phi_1}{\partial t} + C_0(t) + p + \rho g z = C_0(t) \tag{2.35}$$

This allows us to eliminate $C_0(t)$ to yield:

$$\rho \frac{\partial \phi}{\partial t} + p + \rho g z = 0 \tag{2.36}$$

where we have now simply used the variable ϕ to represent ϕ_1. The unsteady Bernoulli equation in conjunction with the Navier–Stokes equations are essential to the derivation of expressions that describe the propagation of surface gravity waves in the next chapter.

Reference

1. Dean, R.G., Dalrymple, R.A., 1984. Water wave mechanics for engineers and scientists. World Scientific Publishing Co., Singapore.

Part II
Mechanics of Surface Waves

Linear Wave Theory and Surface Gravity Waves

3

Fundamentally, a wave is simply a signal propagating through a medium (e.g., sound waves travel through the air via the vibration of air molecules). Surface gravity waves are waves propagating at the ocean–atmosphere interface, with gravity acting as the restoring force on perturbations of the free surface. Wind-driven seas and tsunamis are examples of surface gravity waves (Fig. 3.1). The simplest description of such waves is through linear wave theory, which is also referred to as "Airy wave theory" after its formulation by the mathematician George Biddell Airy in the nineteenth century [1]. This chapter will discuss the properties of linear waves, in addition to the assumptions and boundary conditions under which their mechanical behavior can be described.

3.1 Properties of Linear Waves

Linear waves transport energy and momentum but do not carry mass (i.e., fluid particles), and can be characterized according to the following parameters (see Fig. 3.2):

- Wave height (H): the difference in elevation between a crest and neighboring trough.
- Wave amplitude (A): half of the wave height, $H/2$.
- Wavelength (L): distance between two neighboring crests or troughs.
- Wave period (T): the time taken for one wave cycle.
- Wave frequency (f) or angular frequency (ω): inverse of the wave period, $f = 1/T$ or $\omega = 2\pi f = 2\pi/T$, representing the number of wave cycles per second.
- Wavenumber (k): spatial frequency of a wave, $k = 2\pi/L$.

S. Wang, *Applied Ocean Mechanics for Structural Engineering and Climate Adaptation*, Synthesis Lectures on Ocean Systems Engineering,
https://doi.org/10.1007/978-3-032-18199-2_3

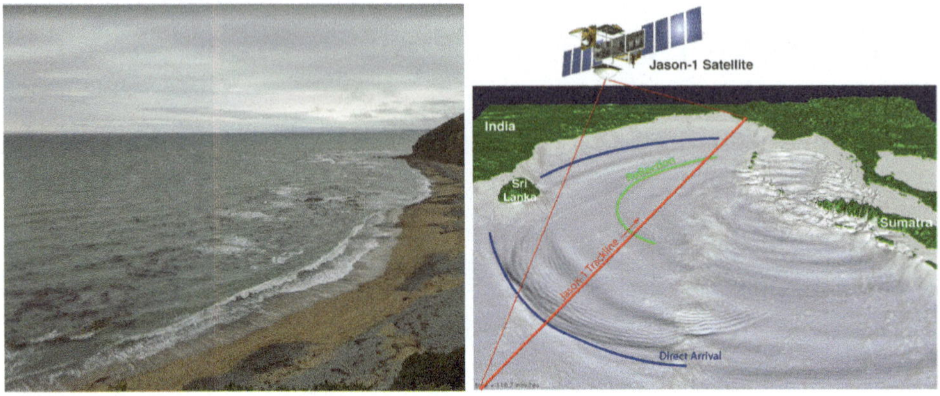

Fig. 3.1 Left: Wind waves approaching a beach near Oamaru, New Zealand. Right: 2004 Indian Ocean tsunami measured by the NASA Jason-1 satellite (courtesy of USGS)

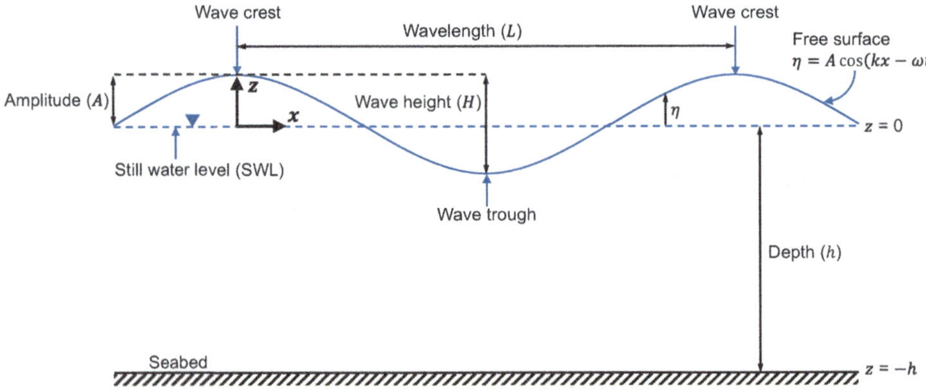

Fig. 3.2 Components of a surface wave

In 2D, the x and z axes are used to denote the horizontal and vertical profile of a wave as depicted in Fig. 3.2. The origin of this coordinate system lies at the still water level (SWL) below a wave crest. The SWL is therefore located at $z = 0$ while the seabed is at $z = -h$, where h is the water depth. The vertical deviation of the wave profile from the SWL is denoted using η which follows the form:

$$\eta(x, t) = A\cos(kx - \omega t) \tag{3.1}$$

Hence, if we know the amplitude (A), wavenumber (k), and frequency (ω) of wave propagation, we can fully describe the profile of a linear wave across space (x) and time

(t). The wavenumber and frequency are further related to each other via the dispersion relation which will be discussed later in this chapter.

3.2 Assumptions and Boundary Conditions

Linear wave theory operates on the assumption that the wave height (H) is small relative to the water depth (h) and wavelength (L). Furthermore, it is assumed that the water medium itself is *inviscid*, *incompressible*, and *irrotational* which are defined as follows:

- **Inviscid**: the influence of viscosity on fluid flow is neglected.
- **Incompressible**: the fluid density (ρ) does not change. This implies that the divergence of velocity field is zero ($\nabla \cdot \boldsymbol{u} = 0$).
- **Irrotational**: individual fluid particles do not exhibit any net rotation with respect to a given coordinate system. A physical example is the carriages on a Ferris wheel. While each carriage follows a circular path, they do not rotate relative to the earth and passengers inside continue to face the same direction throughout their journey. This implies that the curl of the velocity field is zero ($\nabla \times \boldsymbol{u} = 0$).

We wish to derive equations that describe the motion of fluid particles in response to a wave propagating on the free surface. To do so, we require the assignment of three special boundary conditions referred to as *dynamic*, *kinematic*, and *seabed* which are defined as follows:

- **Dynamic**: the dynamic pressure (p) at the free surface (η) equals zero.
- **Kinematic**: water particles on the free surface remain on the free surface, i.e., the velocity of a particle on the free surface follows the velocity of the surface itself.
- **Seabed**: a particle on the seabed exhibits zero velocity in the z direction, i.e., particles are prohibited from penetrating the seabed.

The motion of fluid particles is governed by the velocity potential (ϕ). Recall that the velocity potential describes the fluid velocity (in 2D) according to:

$$\frac{\partial \phi}{\partial x} = u, \quad \frac{\partial \phi}{\partial z} = w \tag{3.2}$$

where u and w depict the velocity of a fluid particle in the x and z directions, respectively. The three boundary conditions mentioned above are necessary to determine an equation for ϕ.

3.2.1 Dynamic Condition

Recall the linearized Bernoulli equation for unsteady flow (see Chap. 2):

$$\rho \frac{\partial \phi}{\partial t} + \rho g z + p = 0 \tag{3.3}$$

At the free surface ($z = \eta$), the dynamic boundary condition stipulates that the dynamic pressure p induced by wave propagation must be zero:

$$\rho \frac{\partial \phi}{\partial t} + \rho g \eta = 0 \tag{3.4}$$

By assuming that η is close to the SWL ($z = 0$), the following relation is produced

$$\eta = -\frac{1}{g} \frac{\partial \phi}{\partial t} \quad \text{at} \quad z = 0 \tag{3.5}$$

3.2.2 Kinematic Condition

The kinematic boundary condition stipulates that the vertical velocity w of a fluid particle on the free surface must follow the velocity of the free surface itself ($D\eta/Dt$). This condition is mathematically formalized as:

$$w = \frac{D\eta}{Dt} = \frac{\partial \eta}{\partial t} + \boldsymbol{u} \cdot \nabla \eta \quad \text{at} \quad z = \eta \tag{3.6}$$

We can further simplify (3.6) through linearization to eliminate the nonlinear $\boldsymbol{u} \cdot \nabla \eta$ term, which then allows us to evoke the approximation that the free surface η is very close to $z = 0$ as done for (3.5):

$$w = \frac{\partial \phi}{\partial z} = \frac{\partial \eta}{\partial t} \quad \text{at} \quad z = 0 \tag{3.7}$$

3.2.3 Seabed Condition

The seabed boundary condition effectively prevents fluid penetration of the seabed by imposing zero vertical velocity at $z = -h$. This is expressed via:

$$w = \frac{\partial \phi}{\partial z} = 0 \quad \text{at} \quad z = -h \tag{3.8}$$

We can also combine the dynamic and kinematic conditions into a single equation by differentiating (3.5) with respect to time t then substituting into (3.7):

$$\frac{\partial^2 \phi}{\partial t^2} + g\frac{\partial \phi}{\partial z} = 0 \quad \text{at} \quad z = 0 \tag{3.9}$$

3.3 The Velocity Potential

We must find a function for the velocity potential ϕ that satisfies Eq. (3.8), which represent the seabed condition, and (3.9), which represent the dynamic and kinematic conditions, thus adhering to all three aforementioned boundary conditions. We begin with the solution describing the free surface in space x and time t as shown in Fig. 3.2:

$$\eta = A\cos(kx - \omega t) \tag{3.10}$$

where $\omega = 2\pi f$ is the angular frequency. From Eq. (3.5), we see that the time derivative of ϕ must yield $-g\eta$ (given the condition $z = 0$) with η given by (3.10). This allows us to postulate that the velocity potential must follow a similar form being:

$$\phi = R\sin(kx - \omega t) \tag{3.11}$$

with R being a function of the depth z. To obtain an expression for ϕ, we must solve for R. From the continuity equation (2.22):

$$\nabla^2\phi = \frac{\partial^2 \phi}{\partial x^2} + \frac{\partial^2 \phi}{\partial z^2} = \left(\frac{\partial^2 R}{\partial z^2} - k^2 R\right)\sin(kx - \omega t) = 0 \tag{3.12}$$

we can infer that:

$$\frac{\partial^2 R}{\partial z^2} - k^2 R = 0 \tag{3.13}$$

This is a second-order ordinary differential equation (ODE) with roots k and $-k$. A solution to R therefore takes the form:

$$R = Be^{kz} + Ce^{-kz} \tag{3.14}$$

where we must find expressions for B and C. After applying the seabed boundary condition (3.8) we obtain:

$$\frac{\partial \phi}{\partial z} = \left(Bke^{-kh} + Cke^{kh}\right)\sin(kx - \omega t) = 0 \tag{3.15}$$

which implies that $Bke^{-kh} + Cke^{kh} = 0$. Rearranging for C:

$$C = Be^{-2kh} \tag{3.16}$$

and applying the kinematic boundary condition (3.7):

$$\frac{\partial A\cos(kx-\omega t)}{\partial t} = \left.\frac{\partial\left(Be^{kz}+Ce^{-kz}\right)\sin(kx-\omega t)}{\partial z}\right|_{z=0} \tag{3.17}$$

we obtain:

$$A\omega\sin(kx-\omega t) = k(B-C)\sin(kx-\omega t) \tag{3.18}$$

leading to:

$$\frac{A\omega}{k} = B - C \tag{3.19}$$

By substituting (3.16) into (3.19), we can now obtain an expression for B:

$$B = \frac{A\omega}{k\left(1-e^{-2kh}\right)} \tag{3.20}$$

which when substituted back into (3.16) gives:

$$C = \frac{A\omega e^{-2kh}}{k\left(1-e^{-2kh}\right)} \tag{3.21}$$

Equation (3.14) can now be expanded to yield:

$$R = \frac{A\omega e^{kz}}{k\left(1-e^{-2kh}\right)} + \frac{A\omega e^{-2kh}e^{-kz}}{k\left(1-e^{-2kh}\right)} = \frac{A\omega}{k}\left(\frac{e^{kz}+e^{-2kh}e^{-kz}}{1-e^{-2kh}}\right) \tag{3.22}$$

utilizing the hyperbolic trigonometric definition:

$$\sinh x = \frac{1-e^{-2x}}{2e^{-x}} \tag{3.23}$$

we can rewrite (3.22) as:

$$R = \frac{A\omega}{k\sinh(kh)}\left(\frac{e^{kz}+e^{-2kh}e^{-kz}}{2e^{-kh}}\right) = \frac{A\omega}{k\sinh(kh)}\left[\frac{e^{kz+kh}+e^{-(kz+kh)}}{2}\right] \tag{3.24}$$

Further applying the definition:

$$\cosh x = \frac{e^{x}+e^{-x}}{2} \tag{3.25}$$

Equation (3.24) is now:

$$R = \frac{A\omega \cosh[k(h+z)]}{k \sinh(kh)} \tag{3.26}$$

which when substituted back into (3.11) yields:

$$\phi = \frac{A\omega}{k}\left(\frac{\cosh[k(h+z)]}{\sinh(kh)}\right)\sin(kx - \omega t) \tag{3.27}$$

as the velocity potential function. In addition, if we apply the dynamic boundary condition to (3.27), we obtain:

$$A\cos(kx - \omega t) = \frac{A\omega^2}{gk}\left(\frac{\cosh[k(h+z)]}{\sinh(kh)}\right)\cos(kx - \omega t)\bigg|_{z=0} \tag{3.28}$$

giving:

$$\omega^2 = gk \tanh(kh) \tag{3.29}$$

Equation (3.29) is known as the dispersion relation which expresses the relationship between the wave frequency ω and wavenumber k, and by extension the wavelength $L = 2\pi/k$. By substituting (3.29) back into (3.27), we can obtain another form of the velocity potential ϕ:

$$\phi = \frac{gA}{\omega}\left(\frac{\cosh[k(h+z)]}{\cosh(kh)}\right)\sin(kx - \omega t) \tag{3.30}$$

Note that while Eqs. (3.27) and (3.30) are equivalent to each other, this text will adopt the form given by (3.30) as the velocity potential.

3.4 Dynamic Pressure and the Dispersion Relation

To derive an expression for the dynamic pressure p, we must first express p in terms of ϕ. At any point below the free surface, the pressure induced by the motion of fluid particles is independent of the hydrostatic component $\rho g z$ appearing within the linearized unsteady Bernoulli equation (2.36). We can therefore eliminate this hydrostatic term to yield:

$$p = -\rho\frac{\partial \phi}{\partial t} \tag{3.31}$$

Substituting (3.30) into (3.31) therefore gives:

$$p = \rho g A\left(\frac{\cosh[k(h+z)]}{\cosh(kh)}\right)\cos(kx - \omega t) \tag{3.32}$$

which shows that the dynamic pressure is at a maximum anywhere under a wave crest and minimum under a wave trough. Substituting $z = 0$ into (3.32) also yields the boundary condition at the free surface pertaining to the pressure:

$$p = \rho g \eta \quad \text{at} \quad z = 0 \tag{3.33}$$

Note that the dynamic pressure p is sometimes referred to as the "perturbation pressure". As an extra exercise, we can also use our solution for p to reproduce the dispersion relation given by Eq. (3.29). To do so, we first rearrange (3.33) in terms of η and differentiate with respect to time, yielding:

$$\frac{\partial \eta}{\partial t} = \frac{1}{\rho g}\frac{\partial p}{\partial t} \quad \text{at} \quad z = 0 \tag{3.34}$$

Substituting (3.34) into (3.7) and again differentiating with respect to time t gives:

$$\frac{\partial w}{\partial t} = \frac{1}{\rho g}\frac{\partial^2 p}{\partial t^2} \quad \text{at} \quad z = 0 \tag{3.35}$$

Next, we adopt the linearized momentum equation for w (see chapter on Navier–Stokes):

$$\frac{\partial w}{\partial t} = -\frac{1}{\rho}\frac{\partial p}{\partial z} \tag{3.36}$$

Substituting (3.35) into (3.36) produces:

$$-\frac{\partial p}{\partial z} = \frac{1}{g}\frac{\partial^2 p}{\partial t^2} \quad \text{at} \quad z = 0 \tag{3.37}$$

We can now substitute (3.32) into (3.37) and set $z = 0$:

$$-\rho g k A\left[\frac{\sinh(kh)}{\cosh(kh)}\right]\cos(kx - \omega t) = -\rho \omega^2 A\left[\frac{\cosh(kh)}{\cosh(kh)}\right]\cos(kx - \omega t) \tag{3.38}$$

which simplifies to:

$$gk\frac{\sinh(kh)}{\cosh(kh)} = \omega^2 \tag{3.39}$$

which returns the dispersion relation for the propagation of surface gravity waves:

$$\omega = \sqrt{gk\tanh(kh)} \tag{3.40}$$

Equation (3.40) is often used to determine the wavenumber k for a given wave frequency ω and water depth h, which could then be used to determine the wavelength via $L = 2\pi/k$. However, (3.40) must be solved numerically since k appears both inside and outside the hyperbolic tangent term. Alternatively, L and k may be directly estimated from ω via the following approximate solution by Eckart [3] which is generally accurate to within 10%:

$$L = \frac{2\pi}{k} \approx \frac{gT^2}{2\pi}\sqrt{\tanh\left(\frac{4\pi^2 h}{T^2 g}\right)} \tag{3.41}$$

where $T = 2\pi/\omega$ is the wave period. Furthermore, the wavelength L can also be approximated using the method given by Nielsen [7]:

$$\frac{2\pi h}{L} \approx \sqrt{\frac{2\pi h}{L_0}}\left[1 + \frac{2\pi h}{6L_0} + \frac{11}{360}\left(\frac{2\pi h}{L_0}\right)^2\right] \tag{3.42}$$

with:

$$L_0 = \frac{gT^2}{2\pi} \tag{3.43}$$

denoting the deep-water wavelength. Figure 3.3 presents the relationship between the wave period T and wavenumber k predicted using Eqs. (3.40), (3.41), and (3.42) when assuming a water depth of $h = 10$ m. We see that the solution given by Nielsen [7] works best only for periods greater than approximately 4 s while Eckart [3] provides a reliable estimate over the entire range.

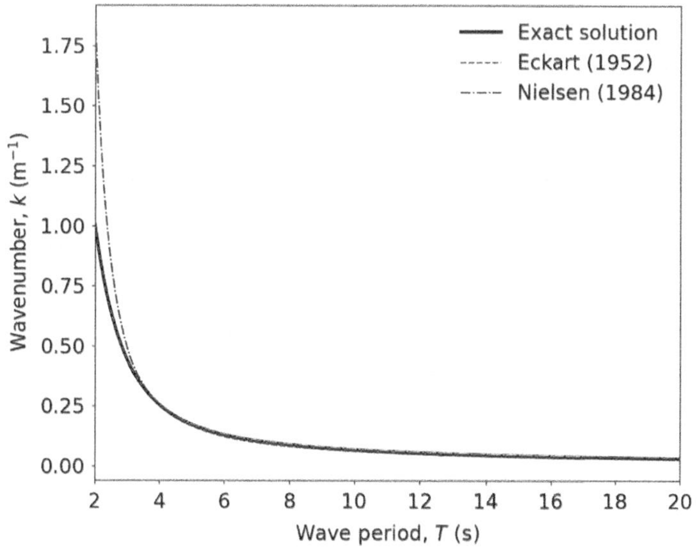

Fig. 3.3 Wavenumber versus wave period predicted using different methods

3.5 Kinematics of Fluid Particles

We can now obtain expressions describing the motion of fluid particles that drive the propagation of Airy (linear) waves on the free surface. Differentiating Eq. (3.30) with respect to x and z gives the horizontal and vertical particle velocities:

$$u = \frac{\partial \phi}{\partial x} = \frac{gkA}{\omega}\left(\frac{\cosh[k(h+z)]}{\cosh(kh)}\right)\cos(kx - \omega t) \tag{3.44}$$

$$w = \frac{\partial \phi}{\partial z} = \frac{gkA}{\omega}\left(\frac{\sinh[k(h+z)]}{\cosh(kh)}\right)\sin(kx - \omega t) \tag{3.45}$$

Equations (3.44) and (3.45) may be simplified by substituting in (3.40):

$$u = \frac{\partial \phi}{\partial x} = A\omega\left(\frac{\cosh[k(h+z)]}{\sinh(kh)}\right)\cos(kx - \omega t) \tag{3.46}$$

$$w = \frac{\partial \phi}{\partial z} = A\omega\left(\frac{\sinh[k(h+z)]}{\sinh(kh)}\right)\sin(kx - \omega t) \tag{3.47}$$

Further differentiating u and w with respect to time t yields the fluid acceleration:

$$a_x = \frac{\partial u}{\partial t} = A\omega^2\left(\frac{\cosh[k(h+z)]}{\sinh(kh)}\right)\sin(kx - \omega t) \tag{3.48}$$

$$a_z = \frac{\partial w}{\partial t} = -A\omega^2\left(\frac{\sinh[k(h+z)]}{\sinh(kh)}\right)\cos(kx - \omega t) \tag{3.49}$$

The horizontal and vertical displacements are produced by integrating u and w, respectively, giving:

$$\delta_x = \int u dt = -A\left(\frac{\cosh[k(h+z)]}{\sinh(kh)}\right)\sin(kx - \omega t) \tag{3.50}$$

$$\delta_z = \int w dt = A\left(\frac{\sinh[k(h+z)]}{\sinh(kh)}\right)\cos(kx - \omega t) \tag{3.51}$$

where the constants of integration are set to zero since $\delta_x = \delta_z = 0$ when $A = 0$. It is seen that for Eqs. (3.46)–(3.51), the first term describes the magnitude, the second describes the variation with depth z, and the third describe cyclical variations along x and with time t.

We can use the Airy wave solutions to help us better understand the kinematics of fluid particles below the free surface. As an example, consider a wave with amplitude $A = 0.1$ m propagating in water with a depth of $h = 10$ m. Here, we consider three different wave periods of $T = 3, 8$, and 25 s. These periods correspond to three different wave regimes (shallow, intermediate, and deep), which will be explained later in Sect. 3.7.

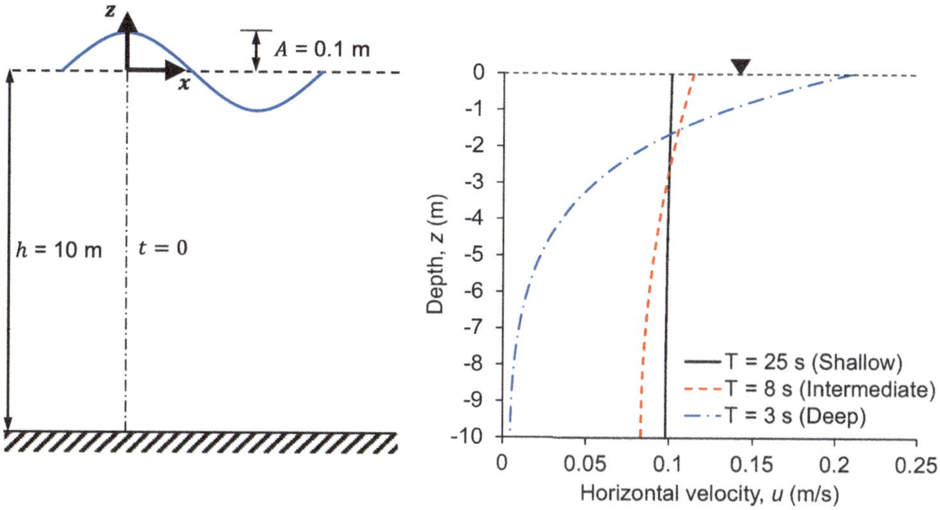

Fig. 3.4 Distribution of horizontal fluid velocity with depth under a wave crest ($x = t = 0$)

By plotting Eq. (3.46) below a wave crest, where $x = t = 0$, Fig. 3.4 reveals how the horizontal velocity of the fluid changes as a function of depth z. Since $\cos(kx - \omega t) = 1$ when $x = t = 0$, the horizontal velocity u will always be at a maximum at all depths below a wave crest. We see that as the wave period increases; the horizontal velocity profile becomes more uniform with depth.

Note that the vertical velocity w directly under a wave crest will be zero at all depths since the sine term in Eq. (3.47) equals zero when $x = t = 0$. Similar to u, the vertical velocity reaches its maximum value when $\sin(kx - \omega t) = \pm 1$. This condition occurs when the wave is halfway between a crest and trough (i.e. at the inflection point) when $t = \pm T/4$. Figure 3.5 plots the distribution of the vertical velocity w with depth at $t = -T/4$ for the three different wave periods. Here, the horizontal velocity u is zero at all depths since the cosine term in (3.46) is always zero when $x = 0$ and $t = \pm T/4$. We notice a reduction in w with increasing T at the free surface, while w always remains at zero at the seabed. This observation is to be expected since the seabed boundary condition given in Eq. (3.8) must be satisfied.

We can also comment on the orbital trajectories of an individual fluid particle using Eqs. (3.50) and (3.51). From its neutral position, the maximum distance a fluid particle can travel along the x or z direction at any given depth z is when the sine and cosine terms in (3.50) and (3.51), respectively, equal 1 or -1. Since we are only interested in the magnitude of this motion, this maximum horizontal and vertical distance is given as:

$$\delta'_x = \left| A\left(\frac{\cosh[k(h+z)]}{\sinh(kh)}\right) \right| \tag{3.52}$$

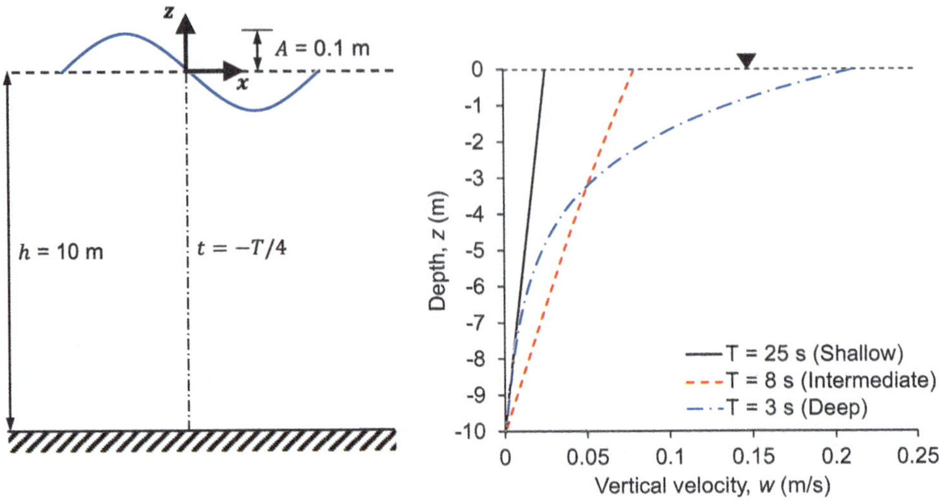

Fig. 3.5 Distribution of vertical fluid velocity with depth at the inflection point ($x = 0, t = -T/4$)

$$\delta'_z = \left| A\left(\frac{\sinh[k(h+z)]}{\sinh(kh)}\right) \right| \tag{3.53}$$

At the free surface ($z = 0$), δ'_x and δ'_z then becomes:

$$\delta'_x = A\coth(kh), \quad \delta'_z = A \tag{3.54}$$

which are shown in Fig. 3.6 for the three wave periods considered. The trajectory of a fluid particle therefore takes the form of an ellipse with δ'_x and δ'_z representing the semi-major and semi-minor axes, respectively. We also notice that at short periods, particles follow circular orbitals, which become increasingly elongated as the wave period increases. The shape of particle trajectories in 2D can be simply described using the equation for an ellipse given in Cartesian coordinates as:

$$\frac{x'^2}{\delta'^2_x} + \frac{z'^2}{\delta'^2_z} = 1 \tag{3.55}$$

where x' and z' represent local coordinates taken relative to the center of each ellipse orbital.

At the seabed ($z = -h$), Eqs. (3.52) and (3.53) become:

$$\delta'_x = \frac{A}{\sinh(kh)}, \quad \delta'_z = 0 \tag{3.56}$$

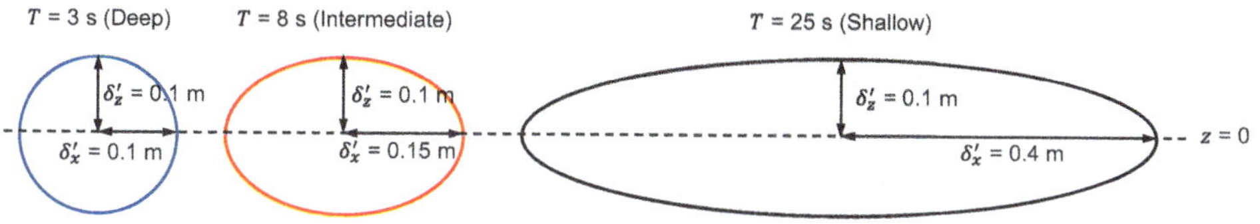

Fig. 3.6 Orbital trajectories at the free surface ($z = 0$) for different wave regimes

which indicate that fluid particles simply oscillate along the horizontal plane. The amplitudes of this oscillation are 0.0023, 0.106, and 0.390 m for $T = 3, 8$, and 25 s, respectively. When compared with the δ'_x values shown in Fig. 3.6 at $z = 0$, we notice that waves with longer periods are more capable of maintaining their horizontal trajectory with depth. On the other hand, the trajectories of fluid particles for short period waves are significantly suppressed with increasing depth.

3.6 Velocity of Surface Waves

At this stage, it is important to understand that the equations presented in (3.46)–(3.51) describe the behavior of an individual fluid particle. On the other hand, the velocity of the surface wave itself is:

$$c_p = \frac{L}{T} = \frac{\omega}{k} = \sqrt{\frac{g}{k}\tanh(kh)} \tag{3.57}$$

and is referred to as the phase velocity, c_p. However, surface waves in reality usually arise from the superposition of multiple sinusoidal wave trains such that they add together to form a wave envelope that may travel at a different speed than its individual components. To illustrate this notion, consider two waves η_1 and η_2 propagating in the same direction both with amplitude A but with different wavelengths. From (3.10), they can be described as:

$$\eta_1 = A\cos(k_1 x - \omega_1 t) \tag{3.58}$$

$$\eta_2 = A\cos(k_2 x - \omega_2 t) \tag{3.59}$$

Their summation ($\eta_g = \eta_1 + \eta_2$) then becomes:

$$\eta_g = A\cos(k_1 x - \omega_1 t) + A\cos(k_2 x - \omega_2 t) \tag{3.60}$$

From trigonometric principles, we know that:

$$\cos(a) + \cos(b) = 2\cos\left(\frac{a}{2} - \frac{b}{2}\right)\cos\left(\frac{a}{2} + \frac{b}{2}\right) \tag{3.61}$$

Applying this to (3.60) yields:

$$\eta_g = 2A\cos\left(\frac{k_2 - k_1}{2}x - \frac{\omega_2 - \omega_1}{2}t\right)\cos\left(\frac{k_2 + k_1}{2}x - \frac{\omega_2 + \omega_1}{2}t\right) \tag{3.62}$$

If we adopt the following definitions:

$$\begin{aligned} \Delta k &= \frac{k_2 - k_1}{2}, \quad \Delta\omega = \frac{\omega_2 - \omega_1}{2} \\ \overline{k} &= \frac{k_2 + k_1}{2}, \quad \overline{\omega} = \frac{\omega_2 + \omega_1}{2} \end{aligned} \tag{3.63}$$

then (3.62) becomes:

$$\eta_g = 2A\cos(\Delta kx - \Delta\omega t)\cos\left(\overline{k}x - \overline{\omega}t\right) \tag{3.64}$$

Figure 3.7 illustrates how η_1 and η_2 combine to form η_g as defined in (3.64). If k_1 and k_2 are very similar, (3.64) essentially describes a wave with a slowly varying amplitude (A_g) that forms an envelope over η_g. To explicitly account for this envelope, Eq. (3.64) can be written as:

$$\eta_g = A_g\cos\left(\overline{k}x - \overline{\omega}t\right) \tag{3.65}$$

where the amplitude A_g is:

$$A_g = 2A\cos(\Delta kx - \Delta\omega t) \tag{3.66}$$

As seen from Fig. 3.7, the change in A_g (red dashed line) between successive wave crests (or troughs) is small, hence the term "slowly varying" is used. From (3.65), we see that the wave crests and troughs represented via η_g propagates at the phase velocity c_p:

$$c_p = \frac{\overline{\omega}}{\overline{k}} \tag{3.67}$$

However, (3.66) tells us that the wave envelope A_g actually propagates at a different velocity dependent upon $\Delta\omega$ and Δk. This is referred to as the group velocity c_g:

$$c_g = \frac{\Delta\omega}{\Delta k} \tag{3.68}$$

Since we impose the condition that the wavenumbers and frequencies are very close such that $k_1 \approx k_2 \approx k$ and $\omega_1 \approx \omega_2 \approx \omega$, Eqs. (3.67) and (3.68) become the following:

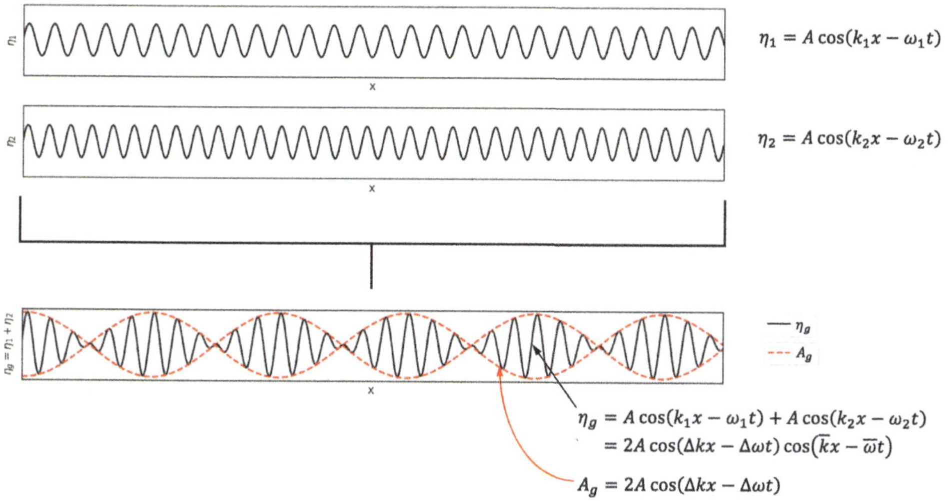

Fig. 3.7 Slowly varying wave amplitude from the superpositioning of two waves

$$c_p = \frac{\overline{\omega}}{\overline{k}} = \frac{\omega}{k}, \quad c_g = \frac{\Delta\omega}{\Delta k} = \frac{\partial\omega}{\partial k} \tag{3.69}$$

with c_p given by Eq. (3.57). When using the dispersion relation (3.40), the group velocity is therefore:

$$c_g = \frac{\partial\omega}{\partial k} = \frac{g}{2\omega}\left[\tanh(kh) + \frac{kh}{\cosh^2(kh)}\right] \tag{3.70}$$

where ω is per Eq. (3.40).

3.7 Classification of Surface Waves

The propagation of surface waves can be broadly classified into three regimes based on the water depth to wavelength ratio (h/L):

- **Deep water**: $h/L \geq 0.5$ or $kh \gg 1$
- **Intermediate water**: $0.05 < h/L < 0.5$
- **Shallow water**: $h/L \leq 0.05$ or $kh \ll 1$.

From Table 3.1, we see that $\tanh(kh) \approx kh$ and $\tanh(kh) \approx 1$ for shallow and deep-water conditions, respectively. The dispersion relation (3.40) can therefore be further simplified to:

Table 3.1 Properties of hyperbolic functions for shallow and deep-water waves

Hyperbolic function	Shallow water ($kh \ll 1$)	Deep water ($kh \gg 1$)
$\sinh(kh)$	$\approx kh$	$\approx \exp(kh)/2$
$\cosh(kh)$	≈ 1	$\approx \exp(kh)/2$
$\tanh(kh)$	$\approx kh$	≈ 1

$$\omega = \begin{cases} \sqrt{gk}, & \text{Deep water} \\ k\sqrt{gh}, & \text{Shallow water} \end{cases} \tag{3.71}$$

with phase velocity:

$$c_p = \begin{cases} \sqrt{\frac{g}{k}}, & \text{Deep water} \\ \sqrt{gh}, & \text{Shallow water} \end{cases} \tag{3.72}$$

and group velocity:

$$c_g = \begin{cases} \frac{1}{2}\sqrt{\frac{g}{k}} = \frac{1}{2}c_p, & \text{Deep water} \\ \sqrt{gh} = c_p, & \text{Shallow water} \end{cases} \tag{3.73}$$

We notice that in deep water, waves with longer wavelengths (smaller k) travel faster. Thus, deep-water waves are referred to as "dispersive waves" where longer, faster waves will separate (disperse) from waves with shorter wavelengths that travel slower. In shallow water, c_p is independent of wavelength such that waves across all frequencies and wavelengths travel at the same speed. Therefore, shallow-water waves are referred to as "nondispersive waves". Plotting Eqs. (3.50) and (3.51) also reveals the behavior of fluid particles under deep, intermediate, and shallow regimes (Fig. 3.8). It is immediately evident that all particles follow an elliptical path, becoming more elongated as waves transition from the deep to shallow regime. This demonstrates that while linear waves carry information and energy since they have velocity, they do not transport fluid particles (i.e., mass) due to their closed orbitals. In reality however, waves in nature are not perfectly linear, and wave orbitals do not completely close. The result is that fluid particles exhibit a slight forward velocity, with a magnitude much smaller than the wave velocity, in the direction of their propagation. This nonlinear phenomenon is termed "Stokes drift" after being formulated by George Gabriel Stokes in the nineteenth century [9].

A summary of important formulae pertaining to the properties of surface gravity waves is presented in Table 3.2. Note that the definitions of "deep" and "shallow" water are not absolute and depend only on the ratio between the water depth h and wavelength L. As an example, wind waves with a wavelength of 100 m propagating over the Pacific Ocean with an average depth of 4000 m results in $h/L = 40$, which is clearly much larger than the 0.5 required for them to be classified as deep-water waves. However, the 2011

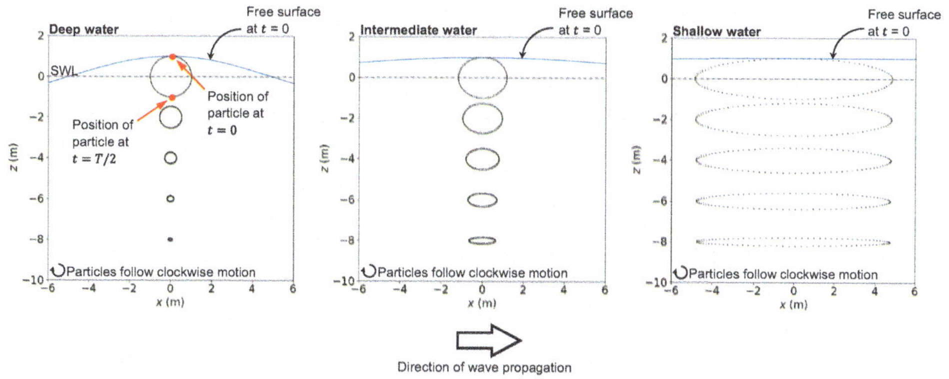

Fig. 3.8 Fluid particle orbitals in deep, intermediate, and shallow water

Tohoku Earthquake generated a wave (tsunami) with an estimated wavelength of 360 km (360,000 m). This results in $h/L = 0.011$ which is much smaller than the 0.05 ratio required for its classification as a shallow-water wave. Hence, tsunamis in the Pacific typically exist as shallow-water waves despite their propagation over the world's deepest ocean.

Table 3.2 Properties of surface gravity waves

Wave property	Shallow water $h \leq 0.05L$ $kh \ll 1$	Deep water $h \geq 0.5L$ $kh \gg 1$	All depths
Frequency (ω)	$\omega = k\sqrt{gh}$	$\omega = \sqrt{gk}$	$\omega = \sqrt{gk\tanh(kh)}$
Phase velocity (c_p)	$c_p = \sqrt{gh}$	$c_p = \sqrt{\frac{g}{k}}$	$c_p = \sqrt{\frac{g}{k}\tanh(kh)}$
Group velocity (c_g)	$c_g = \sqrt{gh}$	$c_g = \frac{1}{2}\sqrt{\frac{g}{k}}$	$c_g = \frac{g}{2\omega}\left[\tanh(kh) + \frac{kh}{\cosh^2(kh)}\right]$
Wavelength (L)	$L = T\sqrt{gh}$	$L = \frac{gT^2}{2\pi}$	$L = Tc_p = T\sqrt{\frac{g}{k}\tanh(kh)}$ Exact solution (solve numerically): $L = \frac{gT^2}{2\pi}\tanh\left(\frac{2\pi h}{L}\right)$ Approximate solution: $L \approx \frac{gT^2}{2\pi}\sqrt{\tanh\left(\frac{4\pi^2 h}{T^2 g}\right)}$

3.8 Nonlinear Waves and Wave Breaking

We started this chapter by introducing the primary assumption of linear waves—that its wave height (H) must be small relative to both the water depth (h) and its wavelength (L). However, we did not expand upon this definition of "small". For waves to be classified as linear, the relationship between its wave height, wave period (which is correlated to the wavelength—see Table 3.2), and the local water depth must fall within the zone demarcated in Fig. 3.9 (zone A in the lower right). Figure 3.9 is known as "Le Méhauté's abacus" [5] and illustrates what wave theories should be used for waves propagating with a given height H, period T, and in water of depth h. Note that the water depth and wave height represented by the horizontal and vertical axes, respectively, are both normalized by the wave period. It is seen that for waves with larger periods relative to the water depth, i.e., smaller h/gT^2, cnoidal wave theory is appropriate. This is a nonlinear theory used to describe waves with long wavelengths traveling over shallow water depths. On the contrary, waves with larger wave heights relative to the period, i.e., steeper waves with larger H/gT^2, are more aptly described via the nonlinear Stokes theories.

As an example, Fig. 3.9 compares the shape of Stokes 2nd order (Wave 2) and cnoidal (Wave 3) waves with linear wave theory (Wave 1) when assuming a constant water depth of $h = 10$ m. While many cnoidal theories exist, the wave shown in Fig. 3.9 was described using the nonlinear shallow-water Level 1 Green–Naghdi equations as described by Hayatdavoodi et al. [4]. Unlike Airy waves, the shape of nonlinear waves cannot be modeled

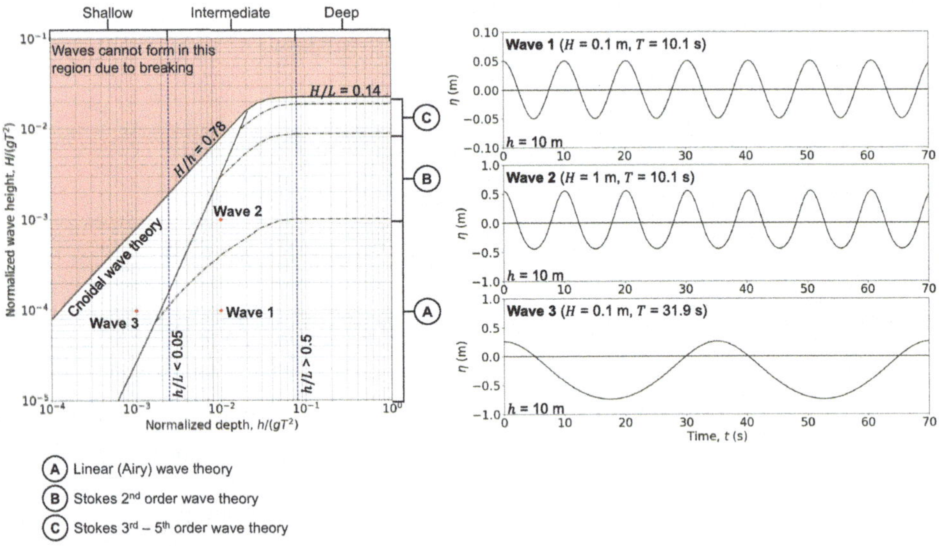

Fig. 3.9 Left: Le Méhauté's abacus showing the applicability of various wave theories. Right: Three representative waves in the linear (Wave 1), Stokes 2nd order (Wave 2), and cnoidal (Wave 3) regime

Fig. 3.10 Left: "The Great Wave off Kanagawa" by Katsushika Hokusai. Right: A breaking wave off the coast of Half Moon Bay, California (courtesy of Steve Jurvetson CC BY 2.0)

by a strictly sinusoidal profile given by Eq. (3.10). An in-depth description of nonlinear waves is beyond the scope of this text. Interested readers are encouraged to consult Dean and Dalrymple [2] or McCormick [6] for a more detailed discussion into nonlinear wave theories.

Figure 3.9 also reveals a shaded region beyond the boundaries representing $H/h = 0.78$ and $H/L = 0.14$. Waves cannot exist in this zone due to wave breaking, which is defined as waves whose height have reached a critical level such that their crest overturns and collapses back into the water. Wave breaking is a highly nonlinear and complex phenomenon, the physics of which is still not fully understood [8]. In practice, however, wave breaking is generally believed to occur when either of the following conditions are met:

- **Depth-induced wave breaking** when the wave height becomes approximately 78% of the local water depth ($H = 0.78h$) when assuming the bathymetry is sufficiently flat. This condition generally occurs in shallow or intermediate water.
- **Steepness-induced wave breaking** (also called "white-capping" or "top-breaking") when the wave height becomes approximately 14% of the wavelength ($H = 0.14L$). This condition generally occurs in deep water.

These two conditions represent the boundary shown in Fig. 3.9. Beyond their engineering implications, which will be discussed in later chapters, waves in the process of breaking have served as inspirations for artists and surfers alike (Fig. 3.10).

References

1. Airy, G.B., 1845. On tides and waves. Encyclopaedia Metropolitania 5, 241–396.

2. Dean, R.G., Dalrymple, R.A., 1984. Water wave mechanics for engineers and scientists. World Scientific Publishing Co., Singapore.
3. Eckart, C., 1952. The propagation of gravity waves from deep to shallow water. National Bureau of Standards, Washington D.C.
4. Hayatdavoodi, M., Seiffert, B., Ertekin, R.C., 2015. Experiments and calculations of cnoidal wave loads on a flat plate in shallow-water. Journal of Ocean Engineering and Marine Energy 1, 77–99.
5. Le Mehaute, B., 1976. An introduction to hydrodynamics and water waves. Springer, New York.
6. McCormick, M., 2010. Ocean engineering mechanics with applications. Cambridge University Press.
7. Nielsen, P., 1984. Explicit solutions to practical wave problems, in: Proceedings of the 19th Conference on Coastal Engineering. Houston, Texas.
8. Perlin, M., Choi, W., Tian, Z., 2013. Breaking waves in deep and intermediate waters. Annual Review of Fluid Mechanics. 45, 115–145.
9. Stokes, G.G., 1847. On the theory of oscillatory waves. Transactions of the Cambridge Philosophical Society 8, 441–455.

Wave Energetics

4

In our introduction to linear wave theory, we briefly mentioned that while Airy (i.e., linear) waves do not transport matter due to the closed orbitals of fluid particles, they do in fact carry energy. This chapter will mathematically demonstrate this principle via the continuity and momentum equations to elucidate the speed at which energy is transported via linear waves. Our solution will also reveal why waves become higher, which eventually leads to breaking, as they approach the shoreline (e.g., Fig. 4.1). In addition, understanding the energy content of linear waves is crucial when describing the highly irregular and chaotic sea states frequently observed in nature as will be demonstrated in later chapters.

4.1 Energy Components

Energy in the context of surface gravity waves involves the combination of kinetic (E_k) and potential (E_p) components arising from the motion of fluid particles. From classical mechanics, we know that the kinetic energy of an object is equal to half the mass times its velocity squared. For waves, the kinetic energy per unit volume is formulated as:

$$E_k = \frac{1}{2}\rho|\boldsymbol{u}|^2 \tag{4.1}$$

where $|\boldsymbol{u}|^2 = \boldsymbol{u}\cdot\boldsymbol{u} = u^2 + v^2 + w^2$ in 3D or $|\boldsymbol{u}|^2 = \boldsymbol{u}\cdot\boldsymbol{u} = u^2 + w^2$ in 2D. Since we wish to know the kinetic energy per unit area of the water surface, Eq. (4.1) is integrated over the entire depth h:

S. Wang, *Applied Ocean Mechanics for Structural Engineering and Climate Adaptation*, Synthesis Lectures on Ocean Systems Engineering,
https://doi.org/10.1007/978-3-032-18199-2_4

Fig. 4.1 Left: Waves breaking on a beach in Napier, New Zealand. Right: Energetic waves in Mallorca, Spain (courtesy of Chenxi Zhao. Reproduced with permission)

$$E_k = \int_{-h}^{0} \frac{1}{2}\rho|\boldsymbol{u}|^2 dz \tag{4.2}$$

The potential energy from classical mechanics is equal to the mass of a body multiplied by the gravitational acceleration and height above some datum. For waves, the total potential energy per unit horizontal area is:

$$E_p = \int_{-h}^{\eta} \rho g z \, dz = \frac{\rho g \eta^2}{2} - \frac{\rho g h^2}{2} \tag{4.3}$$

Note that (4.3) gives the total potential energy which includes the hydrostatic component ($\rho g h^2/2$) that exists even when waves are not present. Since we are only interested in the potential energy arising from the propagation of surface gravity waves, the hydrostatic component is ignored and (4.3) becomes:

$$E_p = \frac{\rho g \eta^2}{2} \tag{4.4}$$

Also note that unlike (4.2), we must integrate (4.3) up to the free surface $z = \eta$ instead of the still water level $z = 0$ in order to obtain a nontrivial solution for E_p that accounts for the potential energy produced by perturbations of the water surface.

4.2 Principles of Energy Conservation

To derive an expression that describes the conservation of kinetic and potential energy pertaining to wave propagation, we leverage the linearized Navier–Stokes continuity and momentum equations in 2D:

$$\nabla \cdot \boldsymbol{u} = 0 \tag{4.5}$$

$$\rho \frac{\partial \boldsymbol{u}}{\partial t} = -\nabla p \tag{4.6}$$

where $\boldsymbol{u} = \{u, w\}$ contains only velocities in the x and z directions. Next, we take the dot product with $\boldsymbol{u}$ for both sides of (4.6):

$$\rho\left(\boldsymbol{u} \cdot \frac{\partial \boldsymbol{u}}{\partial t}\right) = -\boldsymbol{u} \cdot \nabla p \tag{4.7}$$

Using the product rule, we know that:

$$\nabla \cdot (p\boldsymbol{u}) = \boldsymbol{u} \cdot \nabla p + p\nabla \cdot \boldsymbol{u} = \boldsymbol{u} \cdot \nabla p \tag{4.8}$$

since $\nabla \cdot \boldsymbol{u} = 0$ as stated by the continuity equation (4.5). The product rule can also be applied to the left-hand side of (4.7):

$$\frac{\partial(\boldsymbol{u} \cdot \boldsymbol{u})}{\partial t} = \boldsymbol{u} \cdot \frac{\partial \boldsymbol{u}}{\partial t} + \boldsymbol{u} \cdot \frac{\partial \boldsymbol{u}}{\partial t} = 2\boldsymbol{u} \cdot \frac{\partial \boldsymbol{u}}{\partial t} \tag{4.9}$$

which can be rearranged as follows:

$$\frac{1}{2}\frac{\partial |\boldsymbol{u}|^2}{\partial t} = \boldsymbol{u} \cdot \frac{\partial \boldsymbol{u}}{\partial t} \tag{4.10}$$

Substituting (4.8) and (4.10) into (4.7) and placing all terms on the left-hand side yields:

$$\frac{\rho}{2}\frac{\partial |\boldsymbol{u}|^2}{\partial t} + \nabla \cdot (p\boldsymbol{u}) = 0 \tag{4.11}$$

Integrating (4.11) over the entire depth h and splitting $\nabla \cdot (p\boldsymbol{u})$ into its constituents gives:

$$\int_{-h}^{0} \frac{\partial}{\partial t}\left(\frac{1}{2}\rho|\boldsymbol{u}|^2\right)dz + \int_{-h}^{0} \frac{\partial(up)}{\partial x}dz + \int_{-h}^{0} \frac{\partial(wp)}{\partial z}dz = 0 \tag{4.12}$$

The first term on the left-hand side of (4.12) clearly shows the time rate of change of the kinetic energy whereas the last term on the LHS can be evaluated to ultimately yield the rate of change of the potential energy:

$$\int_{-h}^{0} \frac{\partial (wp)}{\partial z} dz = [wp]_{z=-h}^{z=0} \tag{4.13}$$

At $z = 0$ and $z = -h$, the boundary conditions are (recall from Chap. 3 on linear wave theory):

$$p = \rho g \eta \quad \text{at} \quad z = 0$$

$$w = \frac{\partial \eta}{\partial t} \quad \text{at} \quad z = 0 \tag{4.14}$$

$$w = 0 \quad \text{at} \quad z = -h$$

Substituting (4.14) into the RHS of (4.13) subsequently enables the evaluation of the limits of integration and produces:

$$\int_{-h}^{0} \frac{\partial (wp)}{\partial z} dz = [wp]_{z=-h}^{z=0} = \rho g \frac{\partial \eta}{\partial t} \eta = \frac{\rho g}{2} \frac{\partial \eta^2}{\partial t} \tag{4.15}$$

which is substituted back into (4.12) to yield:

$$\int_{-h}^{0} \frac{\partial}{\partial t}\left(\frac{1}{2}\rho|\mathbf{u}|^2\right) dz + \int_{-h}^{0} \frac{\partial (up)}{\partial x} dz + \frac{\rho g}{2} \frac{\partial \eta^2}{\partial t} = 0 \tag{4.16}$$

giving:

$$\frac{\partial}{\partial t}\left[\int_{-h}^{0} \frac{1}{2}\rho|\mathbf{u}|^2 dz + \frac{\rho g \eta^2}{2}\right] + \int_{-h}^{0} \frac{\partial (up)}{\partial x} dz = 0 \tag{4.17}$$

which can be expressed as:

$$\frac{\partial}{\partial t}(E_k + E_p) + \frac{\partial \mathcal{F}}{\partial x} = 0 \tag{4.18}$$

where E_k and E_p are given by Eqs. (4.2) and (4.4), respectively, and:

$$\frac{\partial \mathcal{F}}{\partial x} = \int_{-h}^{0} \frac{\partial (up)}{\partial x} dz \tag{4.19}$$

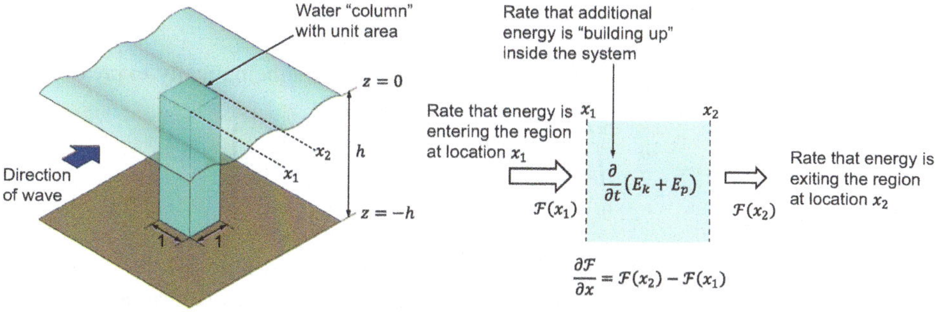

Fig. 4.2 Concept of energy flux through a water "column" with unit area

is the divergence of the horizontal energy flux $\mathcal{F}$ (rate that wave energy is being transferred past a given boundary). Since the pressure p can be treated as a form of energy density (i.e., energy per unit volume), p multiplied with the velocity u is simply the speed at which energy is moving past an area. Integrating pu down the entire water column from $z = 0$ to $z = -h$ gives the flux $\mathcal{F}$:

$$\mathcal{F} = \int_{-h}^{0} up\, dz \tag{4.20}$$

which when differentiated with respect to x produces (4.19). We can also rearrange (4.18) into:

$$\frac{\partial}{\partial t}\left(E_k + E_p\right) = -\frac{\partial \mathcal{F}}{\partial x} \tag{4.21}$$

Figure 4.2 provides a physical interpretation of Eq. (4.21). Wave energy (kinetic and potential) is entering the unit area at boundary x_1 at a rate $\mathcal{F}(x_1)$ and exiting the area at boundary x_2 at a rate of $\mathcal{F}(x_2)$. If $\mathcal{F}(x_2) < \mathcal{F}(x_1)$, energy must be "building up" within the system (since energy is entering faster than it is leaving) at a rate equal to $\partial\left(E_k + E_p\right)/\partial t$. The negative sign on the right-hand side of (4.21) thus reflects this notion.

4.3 Derivation of Kinetic and Potential Components

We now must find expressions for E_k and E_p in order to elucidate how these quantities are related. Since they are time dependent, the most logical approach would be to find E_k, E_p, and $\mathcal{F}$ averaged over a single wave period $T = 2\pi/\omega$.

4.3.1 Kinetic Energy

In terms of the kinetic energy, the averaging procedure is mathematically expressed as:

$$\tilde{E}_k = \frac{\omega}{2\pi}\int_0^{2\pi/\omega} E_k dt = \frac{\omega}{2\pi}\int_0^{2\pi/\omega}\left\{\int_{-h}^{0}\frac{1}{2}\rho|\boldsymbol{u}|^2 dz\right\}dt \tag{4.22}$$

where $\tilde{E}_k$ denotes the kinetic energy averaged over a wave period. In 2D, $|\boldsymbol{u}|^2 = u^2 + w^2$, with the horizontal ($u$) and vertical ($w$) velocities from linear wave theory being:

$$u = \frac{\partial\phi}{\partial x} = \frac{gkA}{\omega}\left(\frac{\cosh[k(h+z)]}{\cosh(kh)}\right)\cos(kx-\omega t) \tag{4.23}$$

$$w = \frac{\partial\phi}{\partial z} = \frac{gkA}{\omega}\left(\frac{\sinh[k(h+z)]}{\cosh(kh)}\right)\sin(kx-\omega t) \tag{4.24}$$

Substituting (4.23) and (4.24) into (4.22) yields:

$$\begin{aligned}\tilde{E}_k =& \frac{\rho g^2k^2A^2}{2\omega^2\cosh^2(kh)}\left(\frac{\omega}{2\pi}\right)\int_0^{2\pi/\omega}\left\{\int_{-h}^{0}\left[\cosh^2[k(h+z)]\cos^2(kx-\omega t)\right.\right.\\ &\left.\left.+\sinh^2[k(h+z)]\sin^2(kx-\omega t)\right]dz\right\}dt\end{aligned} \tag{4.25}$$

The terms $\cos^2(kx-\omega t)$ and $\sin^2(kx-\omega t)$ are not dependent on the depth z. Therefore, we can take them out of the dz integral and integrate them only with respect to time. Fortunately, we find that:

$$\frac{\omega}{2\pi}\int_0^{2\pi/\omega}\cos^2(kx-\omega t)dt = \frac{\omega}{2\pi}\int_0^{2\pi/\omega}\sin^2(kx-\omega t)dt = \frac{1}{2} \tag{4.26}$$

Substituting (4.26) into (4.25) then produces:

$$\tilde{E}_k = \frac{\rho g^2k^2A^2}{2\omega^2\cosh^2(kh)}\frac{1}{2}\int_{-h}^{0}\left\{\cosh^2[k(h+z)] + \sinh^2[k(h+z)]\right\}dz \tag{4.27}$$

Also, we find that:

$$\int_{-h}^{0}\left\{\cosh^2[k(h+z)] + \sinh^2[k(h+z)]\right\}dz = \frac{\sinh(2kh)}{2k} = \frac{\sinh(kh)\cosh(kh)}{k} \tag{4.28}$$

which can be substituted back into (4.27) to give:

$$\tilde{E}_k = \frac{\rho g^2 k A^2}{4\omega^2} \tanh(kh) \tag{4.29}$$

However, we know that the dispersion relation is (recall from Chap. 3 on linear wave theory):

$$\omega^2 = gk \tanh(kh) \tag{4.30}$$

Substituting (4.30) into (4.29) finally yields an expression for the averaged kinetic energy:

$$\tilde{E}_k = \frac{\rho g A^2}{4} \tag{4.31}$$

4.3.2 Potential Energy

Just like the kinetic energy, the potential energy averaged over a single wave period ($\tilde{E}_p$) is:

$$\tilde{E}_p = \frac{\omega}{2\pi} \int_0^{2\pi/\omega} E_p dt = \frac{\omega}{2\pi} \int_0^{2\pi/\omega} \frac{\rho g \eta^2}{2} dt \tag{4.32}$$

From linear wave theory, we know that the free surface η follows the form:

$$\eta = A\cos(kx - \omega t) \tag{4.33}$$

Substituting (4.33) into (4.32) gives:

$$\tilde{E}_p = \frac{\rho g A^2}{2} \left(\frac{\omega}{2\pi}\right) \int_0^{2\pi/\omega} \cos^2(kx - \omega t) dt \tag{4.34}$$

Using the identity from (4.26) finally yields an expression for the averaged potential energy:

$$\tilde{E}_p = \frac{\rho g A^2}{4} \tag{4.35}$$

By comparing (4.31) and (4.35), it is seen that the time averaged kinetic and potential energies are identical, i.e., the total energy is equipartitioned.

4.4 Velocity of Energy Propagation

To reveal the speed at which the total wave energy propagates, we must now determine the energy flux averaged over a wave period $(\tilde{\mathcal{F}})$:

$$\tilde{\mathcal{F}} = \frac{\omega}{2\pi} \int_{0}^{2\pi/\omega} \left(\int_{-h}^{0} up\, dz \right) dt \tag{4.36}$$

where u is given by Eq. (4.23). From linear wave theory, the dynamic pressure p is:

$$p = \rho g A \left(\frac{\cosh[k(h+z)]}{\cosh(kh)} \right) \cos(kx - \omega t) \tag{4.37}$$

Substituting (4.23) and (4.37) into (4.36) gives:

$$\tilde{\mathcal{F}} = \frac{\rho g^2 k A^2}{\omega \cosh^2(kh)} \left(\frac{\omega}{2\pi} \right) \int_{0}^{2\pi/\omega} \left(\int_{-h}^{0} \cosh^2[k(h+z)] \cos^2(kx - \omega t) dz \right) dt \tag{4.38}$$

and:

$$\int_{-h}^{0} \cosh^2[k(h+z)] \cos^2(kx - \omega t) dz = \frac{[kh + \sinh(kh)\cosh(kh)] \cos^2(kx - \omega t)}{2k} \tag{4.39}$$

Substituting (4.39) into (4.38) and integrating $\cos^2(kx - \omega t)$ with time:

$$\tilde{\mathcal{F}} = \frac{\rho g^2 k A^2}{\omega \cosh^2(kh)} \frac{kh + \sinh(kh)\cosh(kh)}{4k} \tag{4.40}$$

which simplifies to:

$$\tilde{\mathcal{F}} = \left(\frac{\rho g A^2}{2} \right) \frac{g}{2\omega} \left[\tanh(kh) + \frac{kh}{\cosh^2(kh)} \right] \tag{4.41}$$

We notice that the group velocity of linear waves is:

$$c_g = \frac{\partial \omega}{\partial k} = \frac{g}{2\omega} \left[\tanh(kh) + \frac{kh}{\cosh^2(kh)} \right] \tag{4.42}$$

and:

$$\tilde{E} = \tilde{E}_k + \tilde{E}_p = \frac{\rho g A^2}{4} + \frac{\rho g A^2}{4} = \frac{\rho g A^2}{2} = \frac{\rho g H^2}{8} \tag{4.43}$$

since $A = H/2$. The averaged energy flux (4.41) can thus be rewritten as:

$$\tilde{\mathcal{F}} = (\tilde{E}_k + \tilde{E}_p)c_g \tag{4.44}$$

Equation (4.44) therefore shows that the total wave energy propagates at the group velocity. Equation (4.18) then becomes:

$$\frac{\partial}{\partial t}(\tilde{E}_k + \tilde{E}_p) + \frac{\partial}{\partial x}[(\tilde{E}_k + \tilde{E}_p)c_g] = 0 \tag{4.45}$$

which expresses the conservation of wave energy. If we assume that no external energy is entering the system such that the first term in Eq. (4.45) is zero, we obtain:

$$\frac{\partial}{\partial x}[(\tilde{E}_k + \tilde{E}_p)c_g] = 0 \tag{4.46}$$

which implies that the energy flux must be a constant:

$$(\tilde{E}_k + \tilde{E}_p)c_g = \text{Constant} \tag{4.47}$$

This is a very important observation since it indicates that any change in the group velocity must result in an opposite change in the total energy density such that their product remains constant (i.e., if c_g decreases then $\tilde{E}_k + \tilde{E}_p$ must increase). One application of Eq. (4.47) involves explaining why waves increase in amplitude as they enter shallower waters. In shallow water, we know that the group velocity is:

$$c_g = \sqrt{gh} \tag{4.48}$$

which states that c_g must decrease with decreasing water depth h. In order to satisfy the condition stipulated by Eq. (4.47), the total energy density $\tilde{E}_k + \tilde{E}_p$ must then increase. From (4.43), we see that the total energy density is $\rho g H^2/8$. Since ρ must remain constant, the wave height H must therefore increase to account for the increase in energy density. This is why waves approaching a beach will slow down while becoming progressively higher before finally breaking as their height reaches a critical limit, being approximately 78% of the local water depth [2]. Equation (4.47) is also the reason for why tsunamis, which can travel at speeds exceeding 200 ms^{-1} in the open ocean [1], can reach tens of meters in height as they approach land, thus causing devastating damage to coastal cities comparable to the aftermath of a global conflict (e.g., Fig. 4.3). Tsunami forces on overland structures will be discussed in Chap. 14.

Fig. 4.3 Left: Hiroshima, Japan after the atom bomb in 1945. Right: North Honshu, Japan after the Tohoku earthquake and tsunami in 2011. Photographs courtesy of the US Navy

References

1. Heller, V., Unger, J., Hager, W., 2005. Tsunami run-up - a hydraulic perspective. Journal of Hydraulic Engineering 131.
2. Tanaka, M., Dold, J.W., Lewy, M., Peregrine, D.H., 1987. Instability and breaking of a solitary wave. Journal of Fluid Mechanics 185, 235–248.

5 The Shallow-Water Equations

When the horizontal length scale of fluid flow is much greater than its vertical scale, such as during tsunamis or coastal floods, the Navier–Stokes equations can be depth-averaged to yield the "shallow-water equations" that are applicable only to shallow-water waves (i.e., when $h/L \ll 0.05$ or $kh \ll 1$). These equations rely upon two crucial assumptions:

1. Accelerations in the vertical direction are negligible such that:

$$\frac{Dw}{Dt} = \frac{\partial w}{\partial t} + u\frac{\partial w}{\partial x} + v\frac{\partial w}{\partial y} + w\frac{\partial w}{\partial z} = 0 \tag{5.1}$$

2. Velocities in the horizontal direction are constant throughout the depth such that:

$$\frac{\partial u}{\partial z} = \frac{\partial v}{\partial z} = 0 \tag{5.2}$$

This chapter will employ these two assumptions to modify the Navier–Stokes equations in order to derive specialized continuity and momentum equations for shallow-water wave propagation. Figure 5.1 illustrates the properties of shallow-water waves where the horizontal fluid velocity does not change with depth z. Note that this velocity $u(x, t)$ can be determined by applying the shallow-water limit ($kh \ll 1$) to the generalized horizontal particle velocity of linear waves given by Eq. (3.44).

S. Wang, *Applied Ocean Mechanics for Structural Engineering and Climate Adaptation*, Synthesis Lectures on Ocean Systems Engineering,
https://doi.org/10.1007/978-3-032-18199-2_5

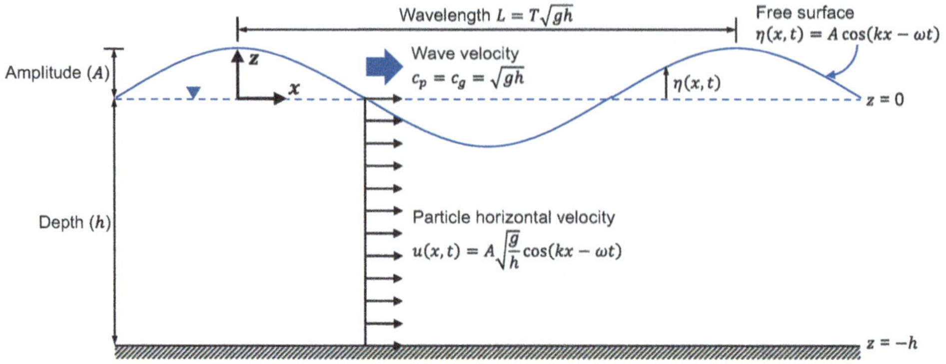

Fig. 5.1 Properties of shallow-water waves

5.1 Derivation of the Shallow-Water Equations

Since we assume that the horizontal velocity field is constant with depth (see Fig. 5.1), we can integrate the Navier–Stokes continuity equation (2.10) over the entire depth from the seabed $z = -h$ to the free surface $z = \eta$:

$$\int_{-h}^{\eta}\left(\frac{\partial u}{\partial x} + \frac{\partial v}{\partial y} + \frac{\partial w}{\partial z}\right)dz = \frac{\partial}{\partial x}\int_{-h}^{\eta} u\,dz + \frac{\partial}{\partial y}\int_{-h}^{\eta} v\,dz + w|_{z=\eta} - w|_{z=-h} = 0 \quad (5.3)$$

In addition, our kinematic and seabed boundary conditions respectively give:

$$w|_{z=\eta} = \frac{\partial \eta}{\partial t} \quad (5.4)$$

$$w|_{z=-h} = 0 \quad (5.5)$$

which when substituted back into (5.3) produces:

$$\frac{\partial \eta}{\partial t} + \frac{\partial}{\partial x}[u(\eta + h)] + \frac{\partial}{\partial y}[v(\eta + h)] = 0 \quad (5.6)$$

Equation (5.6) therefore represents the shallow-water continuity equation. This equation can be further linearized via the elimination of all product terms between η, u, and v to become:

$$\frac{\partial \eta}{\partial t} + \frac{\partial}{\partial x}(hu) + \frac{\partial}{\partial y}(hv) = 0 \quad (5.7)$$

Next, we can apply the shallow-water treatment to the three Navier–Stokes momentum equations, which in their original forms are:

$$\rho\left(\frac{\partial u}{\partial t}+u\frac{\partial u}{\partial x}+v\frac{\partial u}{\partial y}+w\frac{\partial u}{\partial z}\right)=-\frac{\partial p}{\partial x} \tag{5.8}$$

$$\rho\left(\frac{\partial v}{\partial t}+u\frac{\partial v}{\partial x}+v\frac{\partial v}{\partial y}+w\frac{\partial v}{\partial z}\right)=-\frac{\partial p}{\partial y} \tag{5.9}$$

$$\rho\left(\frac{\partial w}{\partial t}+u\frac{\partial w}{\partial x}+v\frac{\partial w}{\partial y}+w\frac{\partial w}{\partial z}\right)=-\frac{\partial p}{\partial z}-\rho g \tag{5.10}$$

By applying the first shallow-water assumption (5.1) to Eq. (5.10) and the second assumption (5.2) to Eqs. (5.8) and (5.9), the three momentum equations are simplified to become:

$$\rho\left(\frac{\partial u}{\partial t}+u\frac{\partial u}{\partial x}+v\frac{\partial u}{\partial y}\right)=-\frac{\partial p}{\partial x} \tag{5.11}$$

$$\rho\left(\frac{\partial v}{\partial t}+u\frac{\partial v}{\partial x}+v\frac{\partial v}{\partial y}\right)=-\frac{\partial p}{\partial y} \tag{5.12}$$

$$\frac{\partial p}{\partial z}=-\rho g \tag{5.13}$$

Equation (5.13) can then be integrated from some arbitrary depth z to the free surface where $z=\eta$:

$$\int_z^\eta \frac{\partial p}{\partial z}dz=\int_z^\eta -\rho g\, dz \tag{5.14}$$

giving:

$$p|_{z=\eta}-p=-\rho g(\eta-z) \tag{5.15}$$

where $p|_{z=\eta}=0$ since the dynamic boundary condition stipulates that the dynamic pressure p at the free surface must equal zero. Differentiating (5.15) with respect to x and y yields:

$$-\frac{\partial p}{\partial x}=-\rho g\frac{\partial \eta}{\partial x} \tag{5.16}$$

$$-\frac{\partial p}{\partial y}=-\rho g\frac{\partial \eta}{\partial y} \tag{5.17}$$

which can now be substituted back into Eqs. (5.11) and (5.12) to produce the shallow-water momentum equations along the horizontal x and y directions:

$$\frac{\partial u}{\partial t} + u\frac{\partial u}{\partial x} + v\frac{\partial u}{\partial y} = -g\frac{\partial \eta}{\partial x} \tag{5.18}$$

$$\frac{\partial v}{\partial t} + u\frac{\partial v}{\partial x} + v\frac{\partial v}{\partial y} = -g\frac{\partial \eta}{\partial y} \tag{5.19}$$

These nonlinear momentum equations can be further linearized then expressed alongside the linearized continuity equation (5.7) to produce the linear shallow-water equations:

$$\frac{\partial u}{\partial t} = -g\frac{\partial \eta}{\partial x} \tag{5.20}$$

$$\frac{\partial v}{\partial t} = -g\frac{\partial \eta}{\partial y} \tag{5.21}$$

$$\frac{\partial \eta}{\partial t} + \frac{\partial}{\partial x}(hu) + \frac{\partial}{\partial y}(hv) = 0 \tag{5.22}$$

We can further combine the three linear shallow-water equations into a single equation by first differentiating the continuity equation (5.22) with respect to time t:

$$\frac{\partial^2 \eta}{\partial t^2} + \frac{\partial}{\partial x}\left(h\frac{\partial u}{\partial t}\right) + \frac{\partial}{\partial y}\left(h\frac{\partial v}{\partial t}\right) = 0 \tag{5.23}$$

The two linear momentum Eqs. (5.20) and (5.21) can now be substituted into (5.23) to yield:

$$\frac{\partial^2 \eta}{\partial t^2} = \frac{\partial}{\partial x}\left(hg\frac{\partial \eta}{\partial x}\right) + \frac{\partial}{\partial y}\left(hg\frac{\partial \eta}{\partial y}\right) \tag{5.24}$$

If we assume wave propagation only along the x direction, all y-terms can be eliminated and Eq. (5.24) becomes:

$$\frac{\partial^2 \eta}{\partial t^2} = c^2\frac{\partial^2 \eta}{\partial x^2} \tag{5.25}$$

where $c^2 = gh$ denotes the wave speed in shallow water (see Table 3.2). This equation is commonly referred to as the "wave equation" and can be used to describe a wide range of wave propagation phenomena (e.g., sound waves traveling through a solid rod) beyond that of shallow-water gravity waves [2].

5.2 Energetics of Shallow-Water Waves

In addition to deriving the shallow-water equations, we can also comment on how the total energy of shallow-water waves is related to its wave properties. As described in Chap. 4, we know that the total energy of a surface wave per unit area is the sum of its kinetic and potential components per Eqs. (4.2) and (4.4):

$$E = E_k + E_p = \int_{-h}^{0} \frac{1}{2}\rho|\boldsymbol{u}|^2 dz + \frac{\rho g \eta^2}{2} \tag{5.26}$$

where $\boldsymbol{u} = \{u, w\}$ for 2D wave propagation. However, for shallow-water waves, we assume that the vertical component of the velocity (w) is negligible when compared with its horizontal component (u). This means that the velocity vector $\boldsymbol{u}$ can be simply replaced by the horizontal velocity u. From Airy wave theory, the horizontal velocity of a surface wave is:

$$u = \frac{\partial \phi}{\partial x} = \frac{gkA}{\omega}\left(\frac{\cosh[k(h+z)]}{\cosh(kh)}\right)\cos(kx - \omega t) \tag{5.27}$$

This equation can be re-expressed for shallow-water waves as:

$$u = \frac{gkA}{k\sqrt{gh}}\cosh(kz)\cos(kx - \omega t) \tag{5.28}$$

since $\cosh(kh) \approx 1$ and $\omega = k\sqrt{gh}$ when $kh \ll 1$. The kinetic energy is therefore:

$$E_k = \int_{-h}^{0} \frac{1}{2}\rho u^2 dz = \rho \int_{-h}^{0} \frac{gA^2}{2h}\cosh^2(kz)\cos^2(kx - \omega t)dz \tag{5.29}$$

where:

$$\int_{-h}^{0} \cosh^2(kz)dz = \frac{2kh + \sinh(2kh)}{4k} \approx h \tag{5.30}$$

since $\sinh(2kh) \approx 2kh$ when $kh \ll 1$. Substituting (5.30) back into (5.29) gives the kinetic energy as:

$$E_k = \frac{\rho g A^2}{2}\cos^2(kx - \omega t) \tag{5.31}$$

Similar to (4.22), we now need to determine the kinetic energy when averaged over a single wave period T:

$$\tilde{E}_k = \frac{\omega}{2\pi}\int_0^{2\pi/\omega} E_k dt = \frac{\omega}{2\pi}\int_0^{2\pi/\omega} \frac{\rho g A^2}{2}\cos^2(kx - \omega t)dt \quad (5.32)$$

which after adopting the identity shown in (4.26), yields:

$$\tilde{E}_k = \frac{\rho g A^2}{4} \quad (5.33)$$

Likewise, the potential energy averaged over a wave period is:

$$\tilde{E}_p = \frac{\omega}{2\pi}\int_0^{2\pi/\omega} \frac{\rho g \eta^2}{2} dt = \frac{\omega}{2\pi}\int_0^{2\pi/\omega} \frac{\rho g A^2}{2}\cos^2(kx - \omega t)dt = \frac{\rho g A^2}{4} \quad (5.34)$$

since $\eta = A\cos(kx - \omega t)$. Equations (5.33) and (5.34) tell us that the energy is equipartitioned between the kinetic and potential components with the total energy per unit area given by:

$$\tilde{E} = \tilde{E}_k + \tilde{E}_p = \frac{\rho g A^2}{4} + \frac{\rho g A^2}{4} = \frac{\rho g A^2}{2} = \frac{\rho g H^2}{8} \quad (5.35)$$

This result demonstrates that the energy of shallow-water waves is proportional to the square of its amplitude A or wave height $H = 2A$. We see that Eq. (5.35) is consistent with (4.43), which is to be expected since (4.43) describes the energy of linear waves across all regimes (i.e., shallow, intermediate, and deep). Similarly, the energy of shallow-water waves propagates at the group velocity $c_g = \sqrt{gh}$.

5.3 Transformation of Shallow-Water Waves

We will now demonstrate how the shallow-water equations can be manipulated to develop analytical solutions that describe the transformation of surface waves over non-uniform water depths following the procedure given by Salmon [3]. In particular, we will describe the spatial and temporal surface profile $\eta(x, t)$ of a shallow-water wave as the water depth suddenly transitions from h_1 to a new depth of h_2 at $x = 0$ (Fig. 5.2). A potential application for such a solution could be to predict the change in wave amplitude of a tsunami as it propagates onto a continental shelf. As shown in Fig. 5.2, η_1 and η_2 denote the free surface corresponding to depths h_1 (where $x < 0$) and h_2 (where $x > 0$), respectively.

An important characteristic of the one-dimensional wave equation given by (5.25) is that it exhibits the following general solution:

$$\eta(x, t) = F(x - ct) + G(x + ct) \quad (5.36)$$

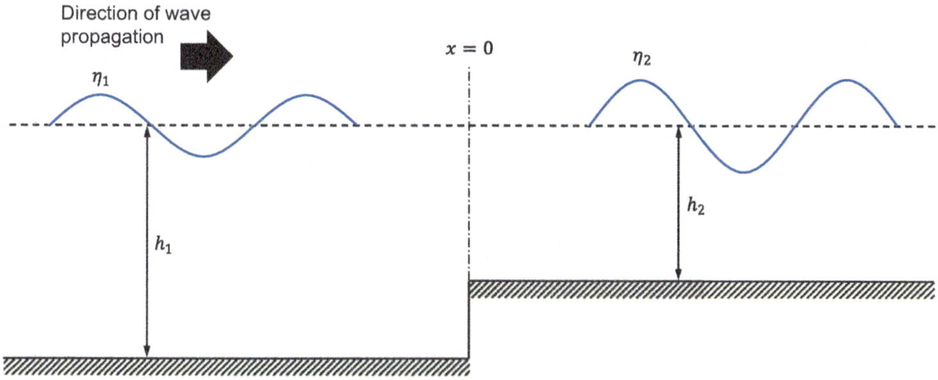

Fig. 5.2 Shallow-water wave transformation by seabed

where F and G are arbituary C^1 functions (i.e., a function whose first derivative, along with itself, is continuous) that must be determined for any given application. For the problem in question, F and G essentially represent waves traveling with velocity c in opposite directions, with F traveling towards the right and G traveling towards the left. Equation (5.36) is also known as "d'Alembert's formula" after the mathematician Jean le Rond d'Alembert who first obtained the solution in the eighteenth century [1].

For a known wave f traveling towards the step, the free surface profile on either side of the step is:

$$\eta(x,t) = \begin{cases} \eta_1 = f(x - c_1 t) + G(x + c_1 t), & x < 0 \\ \eta_2 = F(x - c_2 t), & x > 0 \end{cases} \tag{5.37}$$

where f can represent any incoming wave profile such as $f = A\cos[k(x - c_1 t)]$. As this wave approaches the step at $x = 0$, the discontinuity results in a reflected wave G and a transmitted wave F. To obtain the functions F and G, we first realize that the free surface profiles η_1 and η_2 must exactly match at $x = 0$:

$$f(-c_1 t) + G(c_1 t) = F(-c_2 t) \tag{5.38}$$

After introducing a substitution variable $s = -c_1 t$, Eq. (5.38) can be simplified to:

$$f(s) + G(-s) = F(\beta s) \tag{5.39}$$

where:

$$\beta = \sqrt{\frac{h_2}{h_1}} \tag{5.40}$$

Also recall that in shallow water, the wave velocities are only a function of the water depth (since they are nondispersive) as given by:

$$c_1 = \sqrt{gh_1}, \quad c_2 = \sqrt{gh_2} \tag{5.41}$$

which infers that $c_2 = \beta c_1$. In addition, the rate at which mass is being transported past the vertical boundary at $x = 0$ must also be identical on either side of the step. This condition is expressed as:

$$h_1 u_1 = h_2 u_2 \text{ at } x = 0 \tag{5.42}$$

where u_1 and u_2 denote the horizontal fluid velocity down the entire water depth on the left and right sides of the discontinuity, respectively. After taking the time derivative of (5.42), the shallow-water momentum equation along x as shown in the first equation of (5.20) can be substituted to yield:

$$-h_1 g \frac{\partial \eta_1}{\partial x} = -h_2 g \frac{\partial \eta_2}{\partial x} \text{ at } x = 0 \tag{5.43}$$

The wave profiles η_1 and η_2 from (5.37) can now be substituted into (5.43):

$$-h_1 g\left[f'(-c_1 t) + G'(c_1 t)\right] = -h_2 g\left[F'(-c_2 t)\right] \tag{5.44}$$

where f', G', and F' represent derivatives of their respective functions. After adopting the same substitution variable used for (5.39), Eq. (5.44) becomes:

$$f'(s) + G'(-s) = \beta^2 F'(\beta s) \tag{5.45}$$

We can now integrate (5.45) with respect to s to yield:

$$f(s) - G(-s) = \beta F(\beta s) \tag{5.46}$$

with the constant of integration set to zero. To obtain the function $F(s)$, we sum Eqs. (5.39) and (5.46) to produce:

$$2f(s) = F(\beta s)(1 + \beta) \tag{5.47}$$

We can then eliminate β from the function F by employing the substitution method to finally arrive at a solution for $F(s)$:

$$F(s) = \frac{2}{1+\beta} f\left(\frac{s}{\beta}\right) \tag{5.48}$$

The function $G(s)$ can likewise be determined by subtracting (5.47) from (5.39):

$$2G(-s) = F(\beta s)(1 - \beta) \tag{5.49}$$

After substituting in $F(\beta s)$ from (5.47) then applying the substitution method for $-s$, the solution for $G(s)$ is:

$$G(s) = \frac{1-\beta}{1+\beta} f(-s) \tag{5.50}$$

Substituting $F(s)$ and $G(s)$ back into (5.37) therefore gives the solution only in terms of the known wave $f(s)$:

$$\eta(x,t) = \begin{cases} \eta_1 = f(x - c_1 t) + \dfrac{1-\beta}{1+\beta} f(-x - c_1 t), & x < 0 \\ \eta_2 = \dfrac{2}{1+\beta} f\left(\dfrac{x - c_2 t}{\beta}\right), & x > 0 \end{cases} \tag{5.51}$$

Assuming that the incoming wave $f(s)$ takes the form of a basic linear wave $\eta = A\cos(kx - \omega t)$, the function $f(s)$ must be expressed as:

$$f(s) = A\cos(k_1 s) \tag{5.52}$$

since:

$$A\cos(k_1 s) = A\cos[k_1(x - c_1 t)] = A\cos(k_1 x - k_1 c_1 t) = A\cos(k_1 x - \omega t) \tag{5.53}$$

after recognizing that:

$$\omega = k_1\sqrt{gh_1} = k_2\sqrt{gh_2} \tag{5.54}$$

holds for shallow-water waves. Equation (5.52) can now be substituted into (5.51) to arrive at the complete expression for the wave profile on both sides of the step:

$$\eta(x,t) = \begin{cases} \eta_1 = A\cos(k_1 x - \omega t) + \dfrac{1-\beta}{1+\beta} A\cos(k_1 x + \omega t), & x < 0 \\ \eta_2 = \dfrac{2}{1+\beta} A\cos(k_2 x - \omega t), & x > 0 \end{cases} \tag{5.55}$$

This solution tells us that the amplitude of the reflected (A_1) and transmitted (A_2) wave as a function of the incoming wave amplitude A is:

$$\begin{aligned} A_1 &= \frac{1-\beta}{1+\beta} A \\ A_2 &= \frac{2A}{1+\beta} \end{aligned} \tag{5.56}$$

From (5.56), it is immediately clear that in the absence of a step when $h_1 = h_2$, $\beta = 1$ and the transmitted wave is identical to the incoming wave ($A_2 = A$) with no reflected wave ($A_1 = 0$). On the other hand, if the step is very large such that $h_2 \ll h_1$, $\beta \approx 0$

and the transmitted wave amplitude would be approximately twice the incoming wave's amplitude ($A_2 \approx 2A$). The reflected wave in this instance would then have a similar amplitude to the incoming wave ($A_1 \approx A$).

5.4 Example: Tsunami Propagation onto a Continental Shelf

We will now present an application of the shallow-water wave equations given by (5.55) by modeling the transformation of a hypothetical tsunami as it approaches a continental shelf. The properties of the tsunami wave will be based on the 2011 Tōhoku tsunami which occurred off the east coast of Japan (near Miyagi Prefecture) on 11 March 2011. One of the first observations of this tsunami was made around 30 min after the earthquake by a monitoring station maintained by the National Data Buoy Center (NDBC) located in the Pacific Ocean approximately 500 km east of the epicenter [4]. This station measured a wave height of $H = 2.52\,\text{m}$ and a period of $T = 22\,\text{min}$ (1320 s). Figure 5.3 shows the measured change in free surface elevation as the tsunami propagated past this station.

To illustrate the use of Eq. (5.55), we will consider a hypothetical tsunami with the same H and T as that shown in Fig. 5.3 approaching the Southeast Australian continental shelf approximately halfway between the cities of Brisbane and Sydney. The bathymetry along the direction of wave propagation is visualized in Fig. 5.4. For this example, we will idealize the bathymetric profile such that a discontinuous step exists at the continental shelf. The water depth on the left (deep) and right (shallow) sides of this step are approximated as $h_1 = 4600$ and $h_2 = 110\,\text{m}$, respectively, as shown in Fig. 5.4.

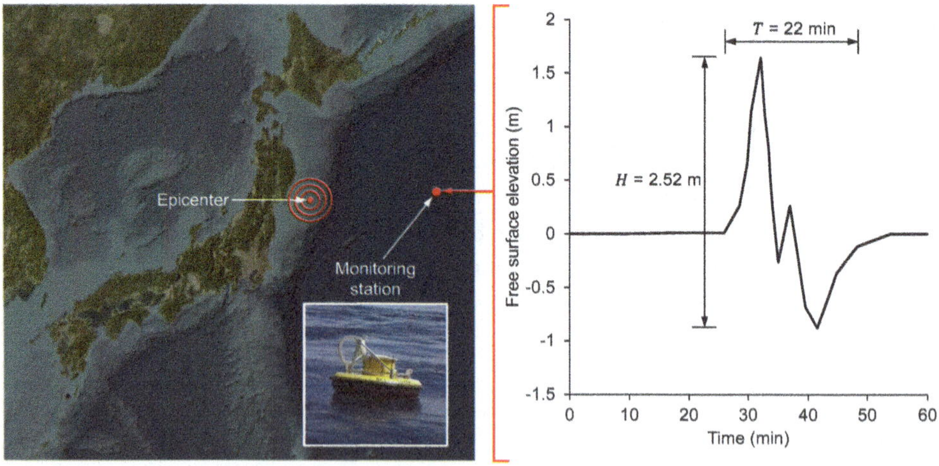

Fig. 5.3 Measurement of the 2011 Tōhoku tsunami reported by Tang et al. [4]. Satellite image and photograph of buoy courtesy of USGS and NOAA

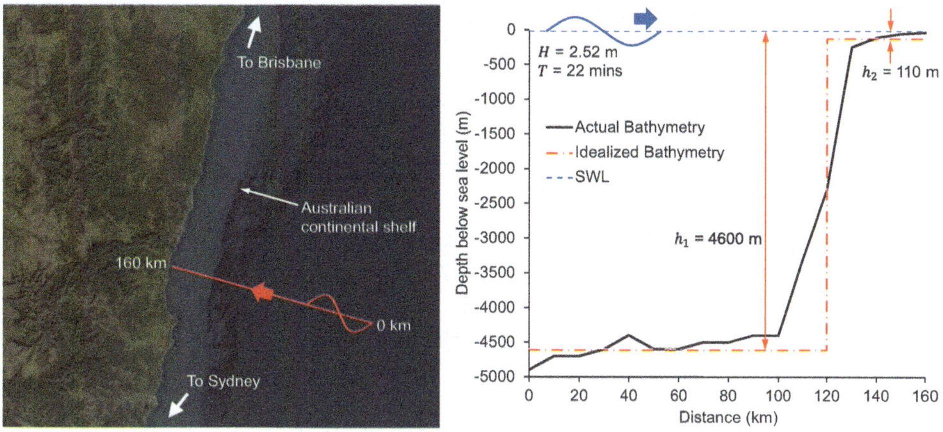

Fig. 5.4 Propagation of a hypothetical tsunami onto the Australian continental shelf. Satellite image courtesy of USGS

We begin this exercise by first confirming that the tsunami can indeed be treated as a shallow-water wave by obtaining its water depth-to-wavelength ratio h/L. By numerically solving the dispersion relation given by Eq. (3.40), the wavenumber k can be determined via:

$$\omega = \sqrt{gk\tanh(kh)}$$

Here, our wave frequency is:

$$\omega = \frac{2\pi}{T} = \frac{2\pi}{1320} = 4.76 \times 10^{-3}\,\text{rad s}^{-1}$$

The wavenumber corresponding to the two water depths $h_1 = 4600$ m and $h_2 = 110$ m are then:

$$k_1 = 2.24 \times 10^{-5}\,\text{m}^{-1}$$

$$k_2 = 1.45 \times 10^{-4}\,\text{m}^{-1}$$

which correspond to wavelengths of:

$$L_1 = \frac{2\pi}{k_1} = \frac{2\pi}{2.24 \times 10^{-5}} = 280{,}000\,\text{m}$$

$$L_2 = \frac{2\pi}{k_2} = \frac{2\pi}{1.45 \times 10^{-4}} = 43{,}000\,\text{m}$$

Instead of solving the dispersion relation numerically to obtain k_1 and k_2, we can alternatively use the approximate solution given by Eq. (3.41) to directly estimate the two wavelengths:

$$L_1 \approx \frac{gT^2}{2\pi}\sqrt{\tanh\left(\frac{4\pi^2 h_1}{T^2 g}\right)} = \frac{9.81 \times 1320^2}{2\pi}\sqrt{\tanh\left(\frac{4\pi^2 \times 4600}{1320^2 \times 9.81}\right)} = 280{,}000\,\text{m}$$

$$L_2 \approx \frac{gT^2}{2\pi}\sqrt{\tanh\left(\frac{4\pi^2 h_2}{T^2 g}\right)} = \frac{9.81 \times 1320^2}{2\pi}\sqrt{\tanh\left(\frac{4\pi^2 \times 110}{1320^2 \times 9.81}\right)} = 43{,}000\,\text{m}$$

which match our exact solutions quite well. The h/L ratios are then:

$$\frac{h_1}{L_1} = \frac{4600}{280{,}000} = 0.016 \quad , \quad \frac{h_2}{L_2} = \frac{110}{43{,}000} = 0.0026$$

which both satisfy the requirement of $h/L < 0.05$ in order to be classified as shallow-water waves. By satisfying this condition, we can now use Eq. (5.55) to describe the profile of this wave as it propagates onto the continental shelf. The parameter β is:

$$\beta = \sqrt{\frac{h_2}{h_1}} = \sqrt{\frac{110}{4600}} = 0.155$$

This parameter can now be used to determine the reflected and transmitted wave amplitudes:

$$A_1 = \frac{1-\beta}{1+\beta}A = \frac{1-0.155}{1+0.155} \times 1.26 = 0.922\,\text{m}$$

$$A_2 = \frac{2A}{1+\beta} = \frac{2 \times 1.26}{1+0.155} = 2.18\,\text{m}$$

where $A = H/2$ is the amplitude of the incident tsunami wave. Hence, we see that the tsunami amplitude has increased from 1.26 to 2.18 m as it travels onto the continental shelf. In addition, the discontinuous step also results in a reflected wave that travels in the opposite direction with an amplitude of 0.922 m. We can additionally use Eq. (5.35) to determine the energy per unit area of the incident (E) and transmitted (E_2) waves via:

$$E = \frac{\rho g A^2}{2} = \frac{1025 \times 9.81 \times 1.26^2}{2} = 7980\,\text{J}\text{m}^{-2}$$

$$E_2 = \frac{\rho g A_2^2}{2} = \frac{1025 \times 9.81 \times 2.18^2}{2} = 23{,}900\,\text{J}\text{m}^{-2}$$

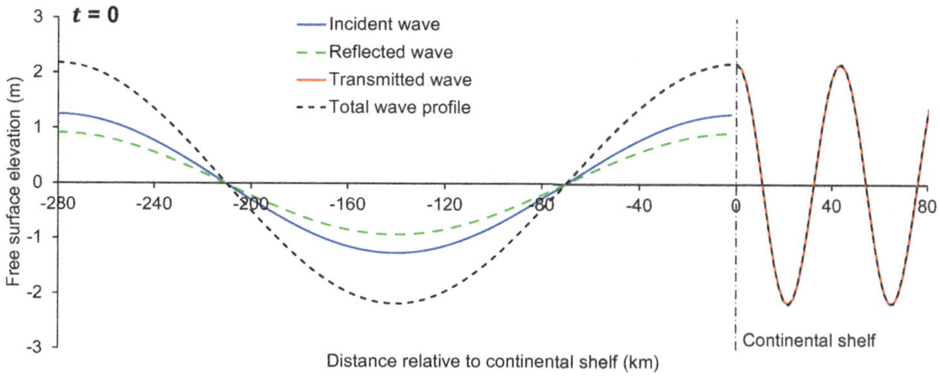

Fig. 5.5 Wave profiles at $t = 0$

where $\rho = 1025\,\text{kg m}^{-3}$ is the density of seawater. These results signify that the energy density of the tsunami has increased by three times after propagating onto the continental shelf. Figures 5.5 and 5.6 illustrates the spatial profile of the incident, reflected, and transmitted waves, in addition to the total wave profile, at $t = 0$ and $t = T/4$, respectively. When $t = 0$, the incident and reflected waves are in-phase, which results in them constructively interfering to produce a wave with a greater amplitude than the original incident wave. Conversely, the incident and reflected waves are out-of-phase at $t = T/4$, resulting in their destructive interference as shown in Fig. 5.6. We also observe that the tsunami has significantly reduced in wavelength as it travels from h_1 to h_2. This is to be expected since the wavelength of shallow-water waves is only a function of the water depth for a constant wave period with:

$$L_1 = T\sqrt{gh_1} = 1320\sqrt{9.81 \times 4600} = 280{,}000\,\text{m}$$

$$L_2 = T\sqrt{gh_2} = 1320\sqrt{9.81 \times 110} = 43{,}000\,\text{m}$$

As expected, these wavelengths are consistent with that previously obtained using the dispersion relation. The reason shallow-water waves maintain the same wave period T despite the reduction in wavelength is due to the reduction in wave speed as they propagate into shallower waters. This wave speed is only a function of the water depth giving:

$$c_1 = \sqrt{gh_1} = \sqrt{9.81 \times 4600} = 212\,\text{ms}^{-1}$$

$$c_2 = \sqrt{gh_2} = \sqrt{9.81 \times 110} = 32.8\,\text{ms}^{-1}$$

which shows a significant reduction in speed as the wave propagates onto the continental shelf.

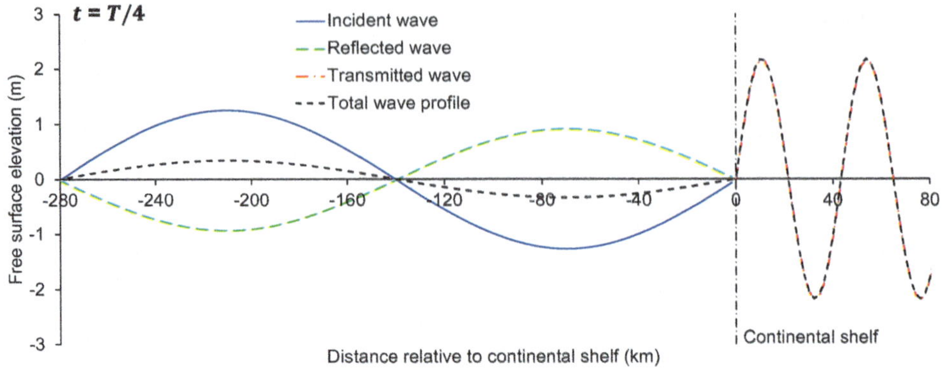

Fig. 5.6 Wave profiles at $t = T/4$

References

1. D'Alembert, J., 1747. Recherches sur la courbe que forme une corde tenduë mise en vibration. Histoire de l'académie royale des sciences et belles lettres de Berlin 3, 214–219.
2. Georgi, H., 1993. The physics of waves. Prentice Hall, New Jersey.
3. Salmon, R., 2015. Introduction to ocean waves. Scripps Institution of Oceanography, University of California, San Diego.
4. Tang, L., Titov, V.V., Bernard, E.N., Wei, Y., Chamberlin, C.D., Newman, J.C., Mofjeld, H.O., Arcas, D., Eble, M.C., Moore, C., Uslu, B., Pells, C., Spillane, M., Wright, L., Gica, E., 2012. Direct energy estimation of the 2011 Japan tsunami using deep-ocean pressure measurements. Journal of Geophysical Research 117.

Irregular Waves and the Wave Spectrum

6

Thus far, we have discussed the propagation of regular waves which exhibit only a single amplitude and frequency. Fortunately, our real world is not quite so boring and purely regular waves rarely occur in nature. As an example, when gazing upon the waves in Sydney Harbor (Fig. 6.1), we can clearly observe the seemingly random patterns that characterize a more realistic sea state. Irregular waves are produced from the superposition (i.e., linear summation) of individual sinusoidal wave components across a wide range of amplitudes and frequencies. The resulting irregular wave profile can then be measured using devices such as the wave buoy pictured in Fig. 6.1, which depicts a monitoring station in Massachusetts Bay maintained by the National Oceanic and Atmospheric Administration (NOAA). This chapter will introduce the statistics of irregular waves and their treatment in both time and frequency domains.

6.1 The Wave Record

As a buoy samples the change in sea surface elevation as a function of time, a wave record is produced of which an example is given in Fig. 6.2. In order to convert this continuous wave record into a discrete number of waves, each with a well-defined wave height and period, two very similar techniques are commonly adopted:

- Zero-upcrossing method
- Zero-downcrossing method.

S. Wang, *Applied Ocean Mechanics for Structural Engineering and Climate Adaptation*,
Synthesis Lectures on Ocean Systems Engineering,
https://doi.org/10.1007/978-3-032-18199-2_6

Fig. 6.1 Left: Irregular waves in Sydney Harbor, Australia. Right: A monitoring station in Massachusetts Bay (courtesy of NOAA)

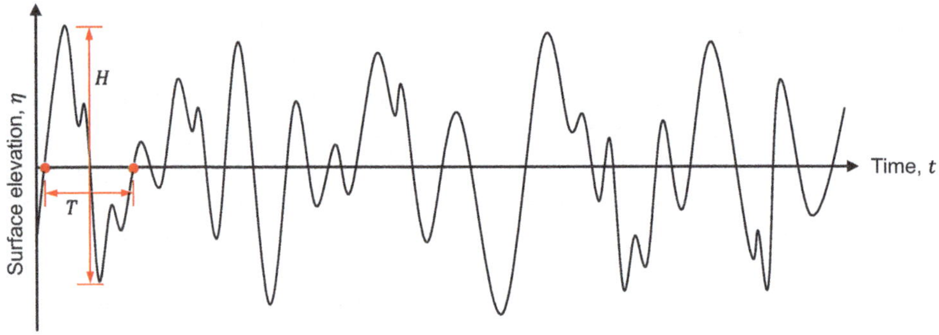

Fig. 6.2 A wave record showing application of the zero-upcrossing method

Before the application of either method, the mean surface elevation (referred to as the "zero line") is first identified (this is simply the x-axis in Fig. 6.2). The zero-upcrossing method works by identifying a point where the free surface crosses the zero line via an upwards trajectory. This point marks the beginning of a single wave. Next, a search is made for the next instant where the free surface upcrosses the zero line after having first gone below the x-axis (minor fluctuations that do not cross the zero line are ignored). This point marks the end of the first wave and the beginning of the second wave. The vertical difference between the highest and lowest points is the wave height while the time difference between the two points signifies the wave period. Figure 6.2 demonstrates an application of the zero-upcrossing method to define a single wave. As the name implies, the zero-downcrossing method works the same way except that the beginning of a wave is defined as the point where the surface profile crosses the zero line in a downwards trajectory. Statistically speaking, the two methods are largely equivalent for most applications.

When using either the zero-upcrossing or the zero-downcrossing method on a wave record of a given duration, a list of individual waves such as that shown in Table 6.1 is produced. Note that while only 30 waves are shown here for demonstrative purposes, the standard practice is to take a record of at least 100 waves [7]. From this list of waves, the following properties can be defined which can be used to characterize a particular sea state:

Highest wave (H_{max} and T_{max}): As the name suggests, this is simply the wave height and period of the highest wave recorded. For the present example, $H_{max} = 5.08$ m and $T_{max} = 8.3$ s.

Mean wave(H_m and T_m): This is the average height and period of all waves within the record. For the present example, $H_m = 2.29$ m and $T_m = 6.82$ s.

Significant wave (H_s and T_s): The significant wave height and period are also sometimes expressed as $H_{1/3}$ and $T_{1/3}$, respectively, which represent the average height and period of the highest 1/3 of all waves in the record. Since there are 30 waves in the present example, the significant wave is the average of the highest 10 waves (bolded cells), yielding $H_s = 3.36$ m and $T_s = 7.77$ s.

The significant wave is the most often used parameter when characterizing a given sea state. The concept of the significant wave was developed in the 1940s by researchers attempting to correlate the wave heights reported by sailors to what was actually being measured [14]. Hence, it is a purely empirical definition driven by historical precedent.

6.2 Irregular Wave Statistics

The statistics of irregular waves are next explored. When plotting the distribution of wave heights for our 30 waves, the left plot in Fig. 6.3 is produced. Since we wish to evaluate the statistical distribution of wave heights, the individual wave heights in Table 6.1 are subsequently normalized against the mean height H_m and the relative frequency is depicted on the y-axis. The relative frequency is called the probability density and is defined as:

$$P(x) = \frac{n}{N\Delta x}, \quad x = \frac{H}{H_m} \geq 0 \tag{6.1}$$

where n represents the total number of waves within a certain band of width Δx and N is the total number of waves in the record. For this example, $\Delta x = 0.5$ and $N = 30$. The normalized wave height histogram is the right plot of Fig. 6.3 to which its total area sums

Table 6.1 Wave heights and periods from a wave record (spanning 30 waves) and their arrangement by wave height in descending order

Wave count	Wave height, H (m)	Wave period, T (s)	Wave ranking by height H	Wave height, H (m)	Wave period, T (s)
1	0.58	4.5	**1**	**5.08**[a]	**8.3**[a]
2	2.16	8.4	**2**	**4.23**	**6.5**
3	4.23	6.5	**3**	**3.95**	**7.9**
4	2.47	11.4	**4**	**3.52**	**8.0**
5	3.52	8.0	**5**	**3.18**	**7.4**
6	1.87	5.4	**6**	**3.02**	**6.9**
7	1.84	4.3	**7**	**2.83**	**7.6**
8	0.93	4.8	**8**	**2.78**	**9.0**
9	2.06	6.3	**9**	**2.57**	**4.7**
10	2.57	4.7	**10**	**2.47**	**11.4**
11	0.99	5.9	11	2.43	9.0
12	2.08	8.5	12	2.35	10.8
13	2.09	8.1	13	2.21	4.0
14	1.53	6.6	14	2.16	8.4
15	3.95	7.9	15	2.15	5.1
16	5.08[a]	8.3[a]	16	2.11	8.7
17	2.43	9.0	17	2.09	8.1
18	2.78	9.0	18	2.08	8.5
19	2.83	7.6	19	2.06	6.3
20	2.15	5.1	20	1.87	5.4
21	3.18	7.4	21	1.85	5.7
22	2.35	10.8	22	1.85	5.3
23	3.02	6.9	23	1.84	4.3
24	1.85	5.3	24	1.84	4.3
25	1.84	4.3	25	1.53	6.6
26	1.02	5.3	26	1.02	5.3
27	1.85	5.7	27	0.99	5.9
28	2.21	4.0	28	0.99	5.9
29	0.99	5.9	29	0.93	4.8
30	2.11	8.7	30	0.58	4.5

[a] Highest wave (wave 16); average of bold = significant wave

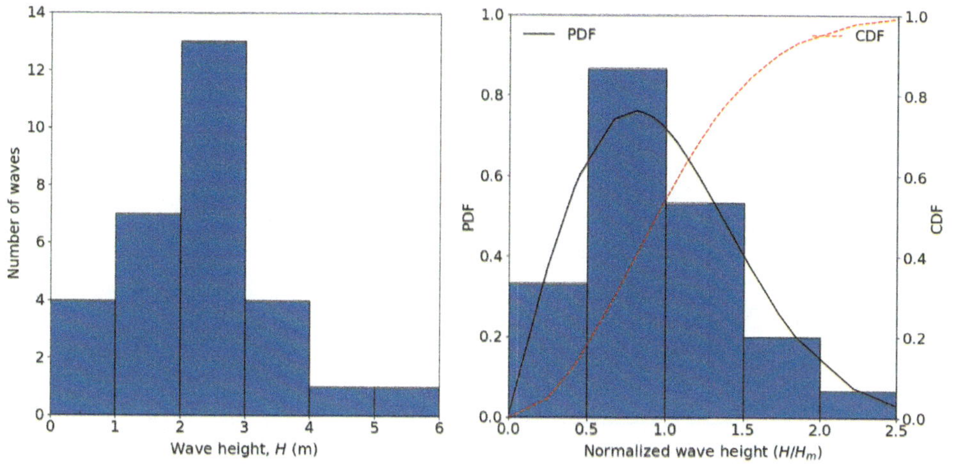

Fig. 6.3 Left: Histogram of wave heights. Right: Normalized wave heights alongside the Rayleigh PDF and CDF distributions

to 1. In deep water, the distribution of normalized wave heights follows the Rayleigh distribution. Originally conceived by Lord Rayleigh to model the distribution of sounds, it was subsequently shown by Longuet-Higgins [12] to also fit the distribution of wave heights in the sea. The probability density function (PDF) and cumulative distribution function (CDF) of the Rayleigh distribution are respectively given by:

$$f(x) = \frac{x}{\sigma^2} \exp\left(-\frac{x^2}{2\sigma^2}\right), \quad \text{PDF} \tag{6.2}$$

$$F(x) = 1 - \exp\left(-\frac{x^2}{2\sigma^2}\right), \quad \text{CDF} \tag{6.3}$$

where $x = H/H_m$ is the normalized wave height and σ is termed the "scale parameter" which gives the spread of the distribution as shown in Fig. 6.4. To adapt Eqs. (6.2) and (6.3) to that of irregular waves, we must first acknowledge that the mean, $E[X]$, of a Rayleigh distribution is related to its scale parameter via:

$$E[X] = \sigma\sqrt{\frac{\pi}{2}} \tag{6.4}$$

Since we are normalizing the wave height H by the mean wave height H_m, $E[X]$ must take a value of unity thus resulting in:

$$\sigma = \sqrt{\frac{2}{\pi}} \tag{6.5}$$

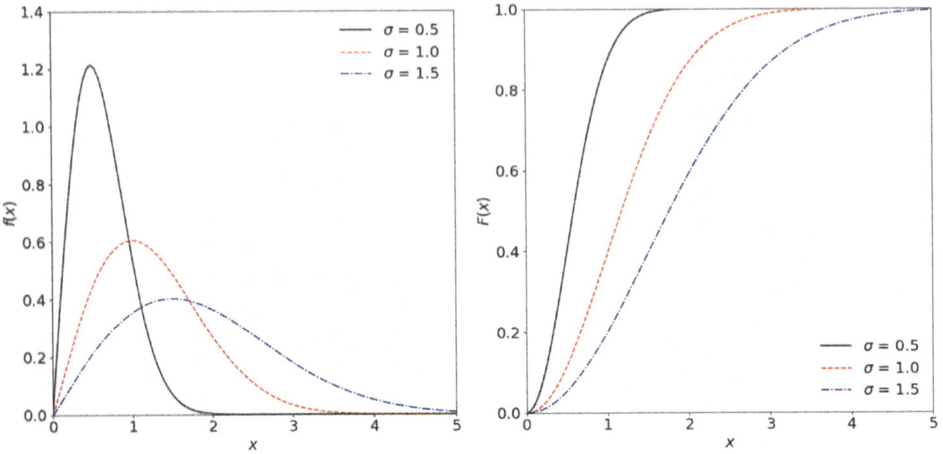

Fig. 6.4 PDF (left) and CDF (right) plots of a Rayleigh distribution for various scale parameters

which can be substituted into Eq. (6.2) to yield the PDF for irregular waves:

$$f(x) = \frac{\pi}{2} x \exp\left(-\frac{\pi}{4} x^2\right), \quad x = \frac{H}{H_m} \tag{6.6}$$

where $f(x)$ is the probability density. In addition to the PDF, the CDF can also be defined via substitution of Eq. (6.5) into (6.3):

$$P(X \leq x) = F(x) = 1 - \exp\left(-\frac{\pi}{4} x^2\right), \quad x = \frac{H}{H_m} \tag{6.7}$$

with $P(X \leq x)$ denoting the probability that a wave with normalized height X will be less than or equal to $x = H/H_m$. On the other hand, the area below the PDF curve between two points, say, x_1 and x_2, gives the probability that a randomly sampled wave will exhibit a normalized wave height falling somewhere between x_1 and x_2. For our example, the PDF and CDF determined via Eqs. (6.6) and (6.7), respectively, are plotted in Fig. 6.3. In addition, if we wish to generate random waves with height H from a Rayleigh distribution with H_m as the mean wave height, the following formulation may be employed:

$$X = \frac{H}{H_m} = \sigma\sqrt{-2\ln U} \tag{6.8}$$

where U is a random number between 0 and 1 and σ is from Eq. (6.5).

We can now discuss some relationships between the various wave heights H_{max}, H_s, and H_m when assuming a Rayleigh distribution. After collecting and analyzing a large amount of wave data [7], it was observed that the significant and mean wave heights are approximately related via:

$$H_s \approx 1.60H_m \tag{6.9}$$

However, the estimation of H_{max} is not so straightforward since wave records having the same value of H_s can all exhibit different H_{max}. This is because H_{max} refers only to a single wave that just happens to be the highest within any given wave record. Through observations, it was identified that the H_{max}/H_s ratio is largely within the range of 1.1–2.4 and is also dependent upon the total number of waves within the record. As expected, records with a larger number of waves will generally have a larger H_{max}. For a wave train with N total waves, a wave with height H_{max} having a probability of exceedance equal to μ is related to H_s via the following:

$$\frac{(H_{max})_\mu}{H_s} \approx 0.706\sqrt{\ln\left[\frac{N}{\ln\left(\frac{1}{1-\mu}\right)}\right]} \tag{6.10}$$

For example, let's say we have a wave record with $N = 100$ waves and $H_s = 5$ m and we wish to determine the maximum wave height such that there is only a 1% chance it will be exceeded. Using Eq. (6.10):

$$(H_{max})_{0.01} \approx 5 \times 0.706\sqrt{\ln\left[\frac{100}{\ln\left(\frac{1}{1-0.01}\right)}\right]} = 10.7\,\text{m}$$

In practice however, the following relations between H_{max} and H_s are commonly adopted as recommended by Goda [7]:

$$H_{max} = \begin{cases} 1.8H_s, & \text{Coastal structures} \\ 2.0H_s, & \text{Offshore structures} \end{cases} \tag{6.11}$$

It is important to remember that the aforementioned relationships between H_{max}, H_s, and H_m are based on the Rayleigh distribution of wave heights (Fig. 6.3) which are applicable only in deep water. In shallow water where extensive wave breaking may be expected due to shallow bathymetry, there is no universally accepted distribution of wave heights, and the Rayleigh distribution will tend to overpredict larger waves. Regardless, the Rayleigh distribution is commonly used even in shallow water but with the recognition that an overestimation of the wave height may occur.

In terms of the wave period, there is also no universal distribution law regardless of deep or shallow water. However, using empirical wave data from field observations, the following relationships were deduced by Goda and Nagai [9]:

$$T_{max} \approx T_s \approx 1.2T_m \tag{6.12}$$

In general, waves with smaller heights tend to have shorter periods whereas waves with heights above the mean height H_m are not correlated with the period. The significant wave period T_s may also be estimated based on H_s using:

$$T_s \approx 3.3H_s^{0.63} \tag{6.13}$$

which was recommended by Goda [6].

6.3 The Wave Spectrum

Up until now, our analysis of irregular waves was in the time domain since statistical wave properties such as H_s were directly determined via a wave record where the x-axis denotes time. In practice, however, the treatment of irregular waves that characterize a particular sea state is usually done in the frequency domain via a wave spectrum. To understand this concept, we first consider the notion that a continuous wave record, such as that shown in Fig. 6.2, can be created from the linear superposition of an infinite number of monochromatic (i.e., sinusoidal) wave trains, all with different frequencies and amplitudes. Figure 6.6 gives an example of five separate monochromatic waves with frequencies ranging from 0.1 to 0.3 s^{-1} combining to produce an irregular wave record (only five waves are used in this example for simplicity). From linear wave theory, we know that the free surface (η) of a monochromatic wave as a function of position and time is given by:

$$\eta = A\cos(kx - \omega t) \tag{6.14}$$

However, we cannot directly sum an infinite number of waves characterized by (6.14) since it assumes that all waves are in phase which is unlikely to be the case. Instead, a phase shift is introduced for each monochromatic wave:

$$\eta = A\cos\left(kx - \omega t + \tilde{\phi}\right) \tag{6.15}$$

Here, the phase shift $\tilde{\phi}$ takes a random value between $-\pi$ and π and follows a uniform distribution (i.e., every value between $-\pi$ and π is equally likely). Figure 6.5 illustrates the application of Eq. (6.15) for a wave with amplitude $A = 1$ m and period $T = 10$ s ($f = 0.1$ Hz, $\omega = 0.628$ rad s^{-1}). In this example the location is taken at $x = 0$ and three phase shifts of $\tilde{\phi} = 0$, $\pi/2$, and $-\pi/2$ were applied. A phase shift therefore serves to "move" the wave in time by $\tilde{\phi}T/2\pi$ seconds with $\tilde{\phi} > 0$ representing a forward time shift and $\tilde{\phi} < 0$ representing a backwards shift. Note that 2π represents a full wave cycle.

Figure 6.6 demonstrates the linear summation of five waves (all at position $x = 0$) adhering to Eq. (6.15) in order to produce the irregular wave train shown in the bottom

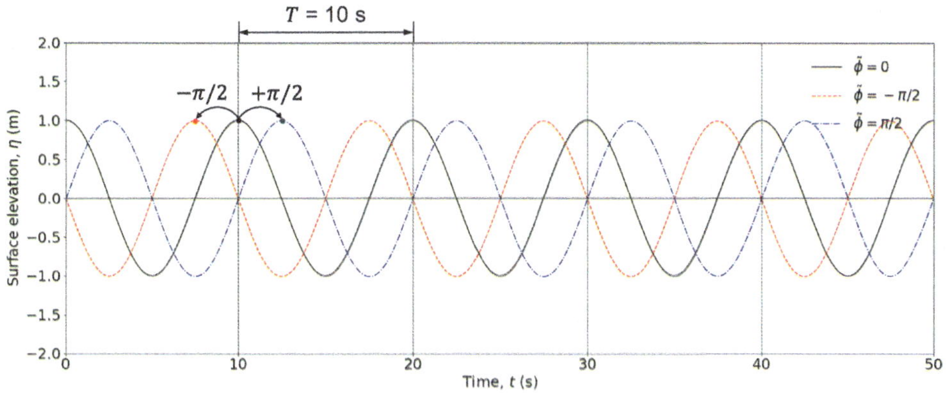

Fig. 6.5 A monochromatic wave train showing phase shifts of $\tilde{\phi} = 0$, $\pi/2$, and $-\pi/2$

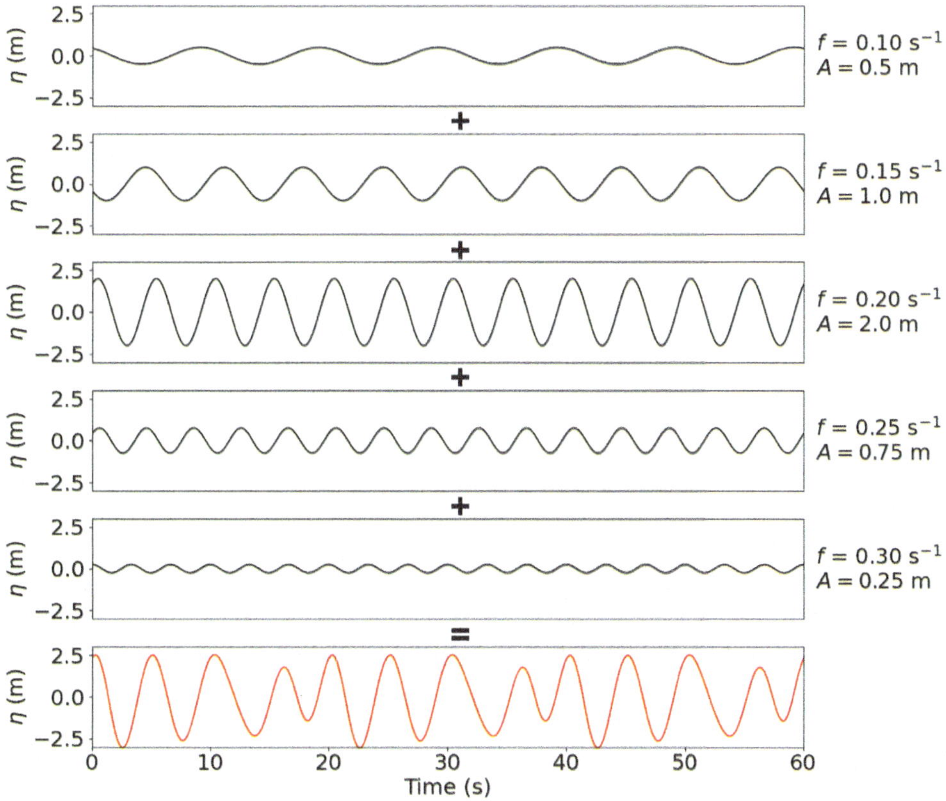

Fig. 6.6 An irregular wave train produced from the linear summation of five monochromatic waves

plot. In practice, an irregular wave record, such as that shown in Fig. 6.2, collected from a buoy is decomposed into its monochromatic constituents via the Fourier Transform.

From Fig. 6.6, we notice that not only are the monochromatic components propagating at different frequencies, but they also have different amplitudes. This relationship between wave frequency and amplitude is given by the wave spectrum. Technically speaking, a wave spectrum tells us the distribution of wave energy with respect to its frequency. However, from our chapter on wave energetics, we know that the total wave energy per unit area is proportional to the square of the amplitude:

$$\tilde{E} = \tilde{E}_k + \tilde{E}_p = \frac{\rho g A^2}{2} \tag{6.16}$$

where $\tilde{E}$ is the energy density from the sum of the kinetic and potential components. Since ρg is a constant, we can redefine the energy density as:

$$E = \frac{\tilde{E}}{\rho g} = \frac{A^2}{2} \tag{6.17}$$

which has units of m^2. We can now plot $E = A^2/2$ for each frequency shown in Fig. 6.6 based on its amplitude. This effectively produces the wave spectrum for the irregular wave train in Fig. 6.6 and is shown in the left plot in Fig. 6.7. In reality, however, since there are an infinite number of monochromatic wave trains, the wave spectrum should take the form of a continuous curve due to the continuous nature of the frequency (right plot in Fig. 6.7). As such, the y-axis denoting the energy instead has units of m^2 s and is called the "spectral density" (S). The total area under the spectral density curve therefore represents the total wave energy associated with the irregular wave record. This area is commonly denoted m_0 and is referred to as the *zeroth-moment* of the wave spectrum:

$$m_0 = \int_0^{\infty} S(f)df \approx \sum_{i=1}^{N} S(f_i)\Delta f \tag{6.18}$$

where Δf is the width of each frequency component f_i, as shown in Fig. 6.7, and N is the total number of monochromatic waves being superimposed to create the spectrum ($N = 5$ in our example). Furthermore, the frequency corresponding to the peak of the spectrum is called the peak frequency (f_p) and its associated period is called the peak period ($T_p = 1/f_p$).

A useful property of the wave spectrum is the ease by which we can determine the significant wave height. When assuming waves follow a Rayleigh distribution, the significant wave height is simply:

$$H_s = 4\sqrt{m_0} \tag{6.19}$$

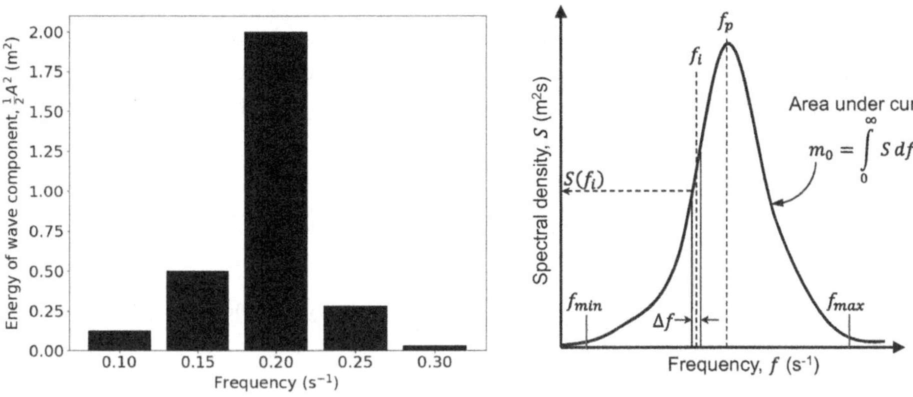

Fig. 6.7 Wave spectrum showing the distribution of wave energy as a function of frequency

The mean and maximum wave heights can then be estimated via Eqs. (6.9) and (6.11), respectively, and the mean wave period is:

$$T_m = \sqrt{\frac{m_0}{m_2}} \tag{6.20}$$

where m_2 is the *second-moment* of the wave spectrum:

$$m_2 = \int_0^\infty S(f)f^2 df \approx \sum_{i=1}^{N} S(f_i)f_i^2 \Delta f \tag{6.21}$$

Likewise, T_{max}. and T_s can then be estimated through Eq. (6.12). We can also use the wave spectrum to determine the amplitude of a wave with frequency f_i. By taking a narrow frequency band of width Δf around f_i, the wave amplitude A_i is:

$$A_i = \sqrt{2S(f_i)\Delta f} \tag{6.22}$$

where $S(f_i)$ is the spectral density S at frequency f_i. The width of the frequency band Δf is selected based on how many N discrete monochromatic waves we wish to split the continuous spectrum into between some minimum (f_{min}) and maximum (f_{max}) frequency threshold (see Fig. 6.7). For example, if we wish to decompose an irregular wave train into only five monochromatic waves with frequencies distributed between 0.1 and 0.3 s^{-1}, similar to that shown in Fig. 6.7, we would set Δf to 0.05 s^{-1}. Therefore, the width of each frequency band is:

$$\Delta f = \frac{f_{max} - f_{min}}{N - 1} \tag{6.23}$$

To adequately capture the randomness of an irregular sea state, at least $N = 200$ monochromatic wave trains are recommended by the Coastal Engineering Manual issued by the US Army Corps of Engineers [17]. The sea surface elevation η for irregular waves then arises from the linear superposition (summation) of each frequency component f_i within the wave spectrum [1]:

$$\eta(x, t) = \sum_{i=1}^{N} A_i \cos\left(k_i x - \omega_i t + \tilde{\phi}_i\right) \tag{6.24}$$

where k_i is the wavenumber associated with frequency f_i determined using the dispersion relation per linear wave theory:

$$2\pi f_i = \sqrt{g k_i \tanh(k_i h)} \tag{6.25}$$

and $\tilde{\phi}_i$ is the random phase of f_i which follows a uniform distribution between $-\pi$ and π.

6.4 Parametric Wave Spectrum for Deep Water

We previously discussed how a wave spectrum (e.g., Fig. 6.7) can be used to obtain statistical wave parameters such as the significant wave height and period. Since waves are ultimately produced by wind blowing over the ocean surface, it is naturally of great interest to develop a mathematical expression that relates wind conditions to the spectral density curve. This will allow us to predict the distribution of wave energy (and heights) for a given storm intensity to inform the design of nearshore or offshore structures under various weather conditions.

In deep water, the distribution of wave energy at a given location on the sea is related to the wind speed, wind duration, and distance the wind has traveled to reach the location of interest (fetch). Figure 6.8 illustrates the concept of fetch F for wind blowing with a constant speed U. As wind blows over a body of water, the waves will become progressively larger due to the transfer of energy from air into the water via friction. This sea state is termed "developing sea" since waves are still in the process of growing. In Chap. 3, we discussed that waves cannot increase in height indefinitely and will eventually break once their height reaches a critical value. Hence, at some point, waves within a developing sea will stop growing and reach an equilibrium point when the energy supplied by the wind is balanced by the energy lost due to breaking. This equilibrium sea state is termed "fully-developed sea".

To devise an empirical spectrum, a large number of wave and wind observations were utilized to formulate parametric equations that best model the distribution of wave energy with frequency based on the wind speed taken at some reference height above sea level. The most well-known and utilized parametric spectrum was developed from the Joint

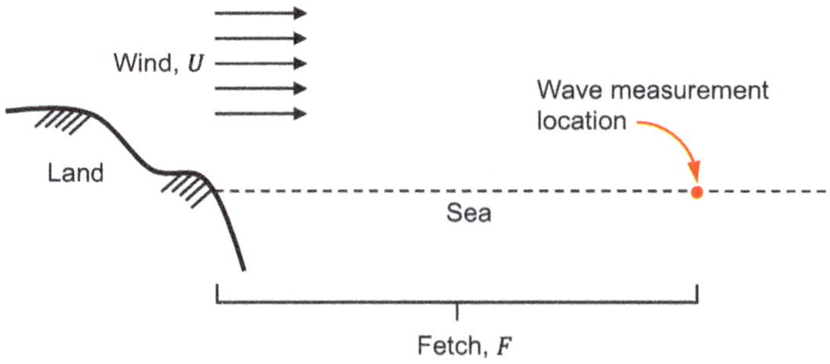

Fig. 6.8 Illustration of fetch distance, F

North Sea Wave Project [10] where wave measurements were taken in the North Sea approximately 160 km west of Denmark in the late 1960s. Referred to as the JONSWAP spectrum, the spectral density S as a function of frequency f follows the form:

$$S(f) = \frac{\alpha g^2}{(2\pi)^4 f^5} \exp\left[-1.25\left(\frac{f_p}{f}\right)^4\right] \gamma^{\exp\left[-\frac{(f-f_p)^2}{2\sigma^2 f_p^2}\right]} \tag{6.26}$$

where $g = 9.81$ m s^{-2} is the acceleration due to gravity and γ is known as the peak enhancement parameter which dictates the shape of the frequency peak. This value is generally between 1 and 7, but $\gamma = 3.3$ is commonly employed for the JONSWAP spectrum. Both the peak frequency f_p and the α parameter are a function of the fetch distance (F) and wind speed 10 m above sea level (U_{10}):

$$f_p = \frac{3.5 g \overline{F}^{-0.33}}{U_{10}} \tag{6.27}$$

$$\alpha = 0.076 \overline{F}^{-0.22} \tag{6.28}$$

where:

$$\overline{F} = \frac{gF}{U_{10}^2} \tag{6.29}$$

is the nondimensionalized fetch and:

$$\sigma = \begin{cases} 0.07, f \leq f_p \\ 0.09, f > f_p \end{cases} \tag{6.30}$$

As an example, suppose we wish to create the wave spectrum produced by wind with $U_{10} = 20$ m s^{-1} traveling over a distance of $F = 200$ km (i.e., 200,000 m). The peak frequency of this spectrum would be:

$$f_p = \frac{3.5g}{U_{10}}\left(\frac{gF}{U_{10}^2}\right)^{-0.33} = \frac{3.5 \times 9.81}{20}\left(\frac{9.81 \times 200{,}000}{20^2}\right)^{-0.33} = 0.104\,\text{Hz}$$

while the α parameter is:

$$\alpha = 0.076\left(\frac{gF}{U_{10}^2}\right)^{-0.22} = 0.076\left(\frac{9.81 \times 200{,}000}{20^2}\right)^{-0.22} = 0.0117$$

The values of f_p and α can now be substituted into Eq. (6.26), in addition to $\gamma = 3.3$ and σ given by (6.30), to plot the relationship between the spectral density S and frequency f. Figure 6.9 presents the JONSWAP wave spectrum between the frequencies of 0 and 0.3 Hz. By computing the area under the spectrum using Eq. (6.18), we find that $m_0 = 1.85\ \text{m}^2$. The significant wave height is therefore:

$$H_s = 4\sqrt{m_0} = 4\sqrt{1.85} = 5.44\,\text{m}$$

which represents the average of the highest 1/3 of waves produced from this particular sea state.

We can also examine how the wave spectrum changes with different fetch distance while maintaining a constant wind speed. As an example, Fig. 6.10 presents the JONSWAP spectrum for $F = 100$, 200, and 300 km for $U_{10} = 20\ \text{m s}^{-1}$. We notice a

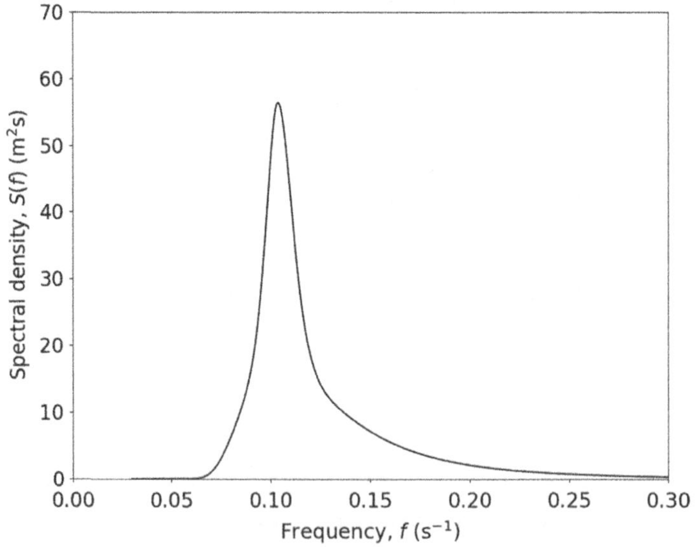

Fig. 6.9 JONSWAP wave spectrum for $U_{10} = 20\ \text{m s}^{-1}$ and $F = 200$ km

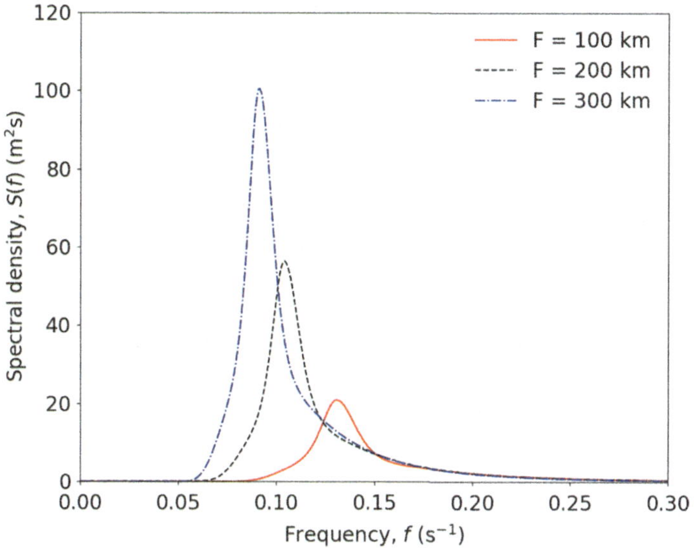

Fig. 6.10 JONSWAP wave spectrum for $U_{10} = 20$ m s^{-1} across $F = 100$, 200, and 300 km

decrease in the peak frequency f_p with increasing fetch, a process referred to as "downshifting". This phenomenon occurs because winds traveling over a longer distance F will transfer more energy to the waves, thereby promoting the formation of longer waves with lower frequencies [5]. Nonlinear wave-wave interactions between individual wave components of different frequencies will also serve to redistribute energy within the spectrum that leads to downshifting of the spectral peak. Young and Verhagen [19] further demonstrated that the extent to which downshifting occurs is reduced as the water depth decreases from deep to shallow water.

The JONSWAP formulation given by (6.26) requires knowledge of the fetch F and wind speed U_{10} to determine the wave spectrum, and thus the significant wave height. Conversely, the spectral density can also be estimated using only the significant wave height H_s as presented by Goda [8]:

$$S(f) = \beta_J H_s^2 \left(\frac{f_p^4}{f^5}\right) \exp\left[-1.25\left(\frac{f_p}{f}\right)^4\right] \gamma^{\exp\left[\frac{-\left(\frac{f}{f_p}-1\right)^2}{2\sigma^2}\right]} \tag{6.31}$$

where the peak frequency f_p can be estimated via:

$$f_p = \frac{1}{T_p} \approx \frac{1 - 0.132(\gamma + 0.2)^{-0.559}}{T_s} \tag{6.32}$$

When adopting Eq. (6.13), the peak frequency can now be reformulated in terms of the significant wave height H_s:

$$f_p = \frac{1}{T_p} = \frac{1 - 0.132(\gamma + 0.2)^{-0.559}}{3.3H_s^{0.63}} \tag{6.33}$$

The β_J parameter is given by:

$$\beta_J = \frac{0.0624\left[1.094 - 0.01915\ln(\gamma)\right]}{0.230 + 0.0336\gamma - 0.185(1.9 + \gamma)^{-1}} \tag{6.34}$$

and σ is given by Eq. (6.30). When adopting $\gamma = 3.3$, the parameters f_p and β_J become:

$$f_p = \frac{0.283}{H_s^{0.63}}, \quad \beta_J = 0.219 \tag{6.35}$$

Suppose now we wish to recreate the JONSWAP spectrum using Eq. (6.31) while adopting the significant wave height of $H_s = 5.44$ m determined from Fig. 6.9. The peak frequency is calculated to be:

$$f_p = \frac{0.283}{H_s^{0.63}} = \frac{0.283}{5.44^{0.63}} = 0.0974\,\text{Hz}$$

which slightly deviates from the previously computed value of 0.104 Hz. Figure 6.11 compares the JONSWAP spectra determined using the wind fetch per Eq. (6.26) and from a specified significant wave height per (6.31), showing some discrepancies between the two approaches. It is also important to realize that the JONSWAP spectrum may not always be internally consistent. This means that the magnitude of the initial H_s used to create the spectrum might not be returned exactly when attempting to re-calculate H_s via Eqs. (6.18) and (6.19). We can demonstrate this by first computing the area under the spectrum created using $H_s = 5.44$ m via Eq. (6.31), giving $m_0 = 1.96$ m^2. The significant wave height is then:

$$H_s = 4\sqrt{m_0} = 4\sqrt{1.96} = 5.60\,\text{m}$$

which is approximately 3% larger than the H_s we have used to create this spectrum. A more internally consistent formulation was later developed by Donelan et al. [4]. However, the Donelan spectrum requires knowledge of the wind speed which makes it less accessible than Eq. (6.31) and has not seen frequent use in engineering practice.

The JONSWAP spectrum in the form of Eq. (6.31) remains the most widely used by coastal and offshore engineers to model sea states at any stage of development, from developing to fully-developed. Yet, many alternative spectral forms have also been proposed that are applicable only to fully-developed sea states (i.e., when waves have stopped growing after reaching their equilibrium point). Examples include Pierson and Moskowitz

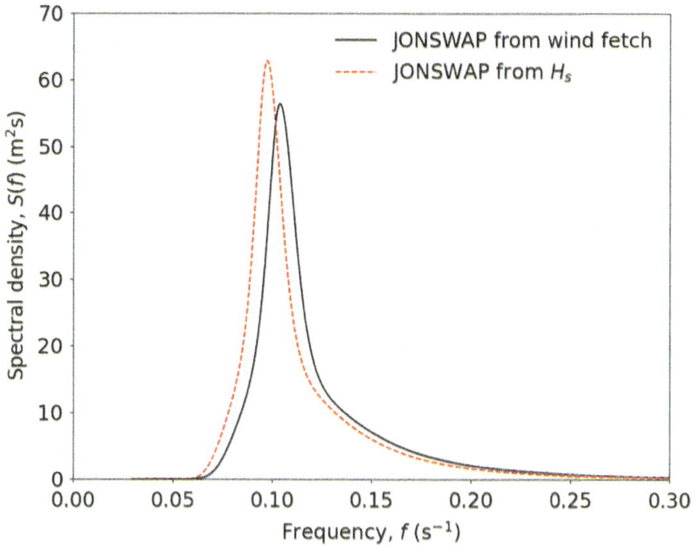

Fig. 6.11 JONSWAP spectra constructed using wind fetch and significant wave height

[15], Bretschneider [3] and Mitsuyasu [13], and Goda [8]. Table 6.2 summarizes these spectral forms expressed in terms of the significant wave height H_s and significant wave period T_s. Note that the Pierson–Moskowitz spectrum was developed under the condition of fully developed seas with unlimited fetch, where the form shown in Table 6.2 is as presented by Reeve et al. [16].

A comparison between the spectral forms shown in Table 6.2 alongside the JONSWAP spectrum from Eq. (6.31) is presented in Fig. 6.12. Here, the significant wave height and significant wave period were fixed at $H_s = 5$ m and $T_s = 10$ s, respectively. We notice that unlike the wave spectra from Table 6.2, the JONSWAP spectrum exhibits a much sharper peak. This behavior results from the inclusion of a peak enhancement parameter

Table 6.2 Alternative spectral forms for fully developed seas

Spectrum	Spectral form
Pierson–Moskowitz[a]	$S(f) = \frac{0.0081g^2}{(2\pi)^4 f^5}\exp\left[-1.25\left(\frac{f_p}{f}\right)^4\right]$
Bretschneider–Mitsuyasu	$S(f) = \frac{0.257H_s^2}{T_s^4 f^5}\exp\left[-\frac{1.03}{(T_s f)^4}\right]$
Goda	$S(f) = \frac{0.205H_s^2}{T_s^4 f^5}\exp\left[-\frac{0.75}{(T_s f)^4}\right]$

[a] The peak frequency f_p can be approximated via H_s using Eq. (6.33)

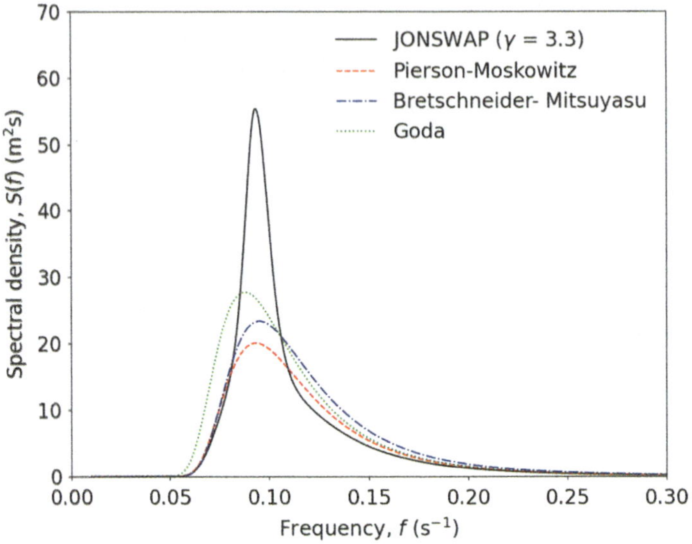

Fig. 6.12 Comparison between different spectral forms for $H_s = 5$ m and $T_s = 10$ s

γ within the JONSWAP spectrum to account for the increased concentration of energy around the peak frequency f_p observed in developing seas. Conversely, this γ parameter is not featured within any of the spectral forms in Table 6.2 since these forms were developed to describe fully developed seas. When $\gamma = 1$, the JONSWAP spectrum from Eq. (6.31) becomes identical to the Goda spectrum.

6.5 Parametric Wave Spectrum for Shallow Water

Our investigation into wave spectra thus far has only focused on deep-water conditions since the JONSWAP spectrum does not include a term for the water depth (h). This is because the spectral formulation was developed from wave observations in deep water where the depth can effectively be considered infinite. In order to adapt the JONSWAP spectrum for nearshore conditions where wave heights may be limited by the bathymetry, the TMA spectrum [2] was developed:

$$S_{TMA}(f, h) = S_{JONSWAP}(f)\Phi(f, h) \tag{6.36}$$

Here, $S_{JONSWAP}(f)$ is the JONSWAP spectrum defined by (6.26) or (6.31), and $\Phi(f, h)$ is a scaling term that accounts for the finite water depth h given by:

$$\Phi(f, h) = \frac{k^{-3}\frac{\partial k}{\partial \omega}}{k_\infty^{-3}\frac{\partial k_\infty}{\partial \omega}} \tag{6.37}$$

where k and k_∞ denote the wavenumber for waves traveling over water with finite and infinite depths, respectively. However, Kitaigordskii et al. [11] identified an approximate solution to $\Phi(f, h)$ in the form of:

$$\Phi(f, h) = \begin{cases} \frac{\omega_h^2}{2}, & \omega_h \leq 1 \\ 1 - \frac{1}{2}(2 - \omega_h)^2, & \omega_h > 1 \end{cases} \tag{6.38}$$

where:

$$\omega_h = 2\pi f \sqrt{\frac{h}{g}} \tag{6.39}$$

When adopting a wind speed of $U_{10} = 20$ m s^{-1} and fetch $F = 200$ km, Fig. 6.13 shows the TMA spectrum for a water depth h of 10 m. The JONSWAP spectrum from Fig. 6.9 is also shown for comparison which uses the same wind speed and fetch. After integrating the area under the TMA curve, we obtain

$$m_0 = \int_0^\infty S_{TMA} df = 0.665\,\text{m}^2$$

The significant wave height is therefore:

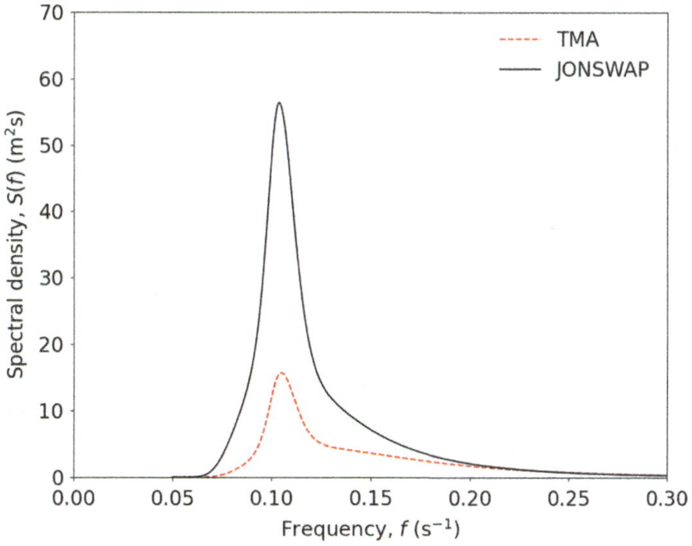

Fig. 6.13 TMA wave spectrum for $U_{10} = 20$ m s^{-1}, $F = 200$ km, and $h = 10$ m. JONSWAP spectrum shown for comparison

$$H_s = 4\sqrt{m_0} = 4\sqrt{0.665} = 3.26\,\text{m}$$

which is equivalent to 60% of the deep-water JONSWAP significant wave height.

A rule of thumb to ascertain whether water of a certain depth h calls for the JONSWAP or TMA spectrum would be to first determine f_p via Eqs. (6.27) or (6.32) then compute the dimensionless shallow water peak frequency:

$$\omega_{hp} = 2\pi f_p \sqrt{\frac{h}{g}} \tag{6.40}$$

If $\omega_{hp} > 0.9$ then the water is of sufficient depth such that the JONSWAP spectrum should be adopted. If $\omega_{hp} \leq 0.9$, the TMA spectrum can be used. This accounts for the observation by Vincent and Hughes [18] that in shallow water, the dimensionless shallow water peak frequency ω_{hp} has an upper limit of approximately 0.9. Hence, when using $f_p = 0.104$ Hz (the peak frequency in Fig. 6.13), we find that:

$$\omega_{hp} = 2\pi f_p \sqrt{\frac{h}{g}} = 2\pi \times 0.104\sqrt{\frac{10}{9.81}} = 0.66$$

Since $\omega_{hp} < 0.9$, our decision to adopt the TMA spectrum is validated.

6.6 Example: Synthesis of Artificial Wave Record

In Sect. 6.3, we explained how a wave record in the time domain can be used to create a wave spectrum to model the distribution of wave energy with frequency. This example will demonstrate how a wave spectrum (e.g., JONSWAP) can be leveraged to recreate a synthetic wave record, showing the change in surface elevation as a function of time, by utilizing Eq. (6.24). Since we are only interested in the change in elevation as a function of time t, the spatial term x in Eq. (6.24) can be set to zero which gives:

$$\eta(t) = \sum_{i=1}^{N} A_i \cos\left(-\omega_i t + \tilde{\phi}_i\right)$$

In this example, we shall use the JONSWAP spectrum shown in Fig. 6.9 produced by wind with $U_{10} = 20$ m s^{-1} blowing over a fetch of $F = 200$ km as described using Eq. (6.26). The first step would be to discretize this spectrum into a fixed number of frequency bands N, where each band represents a single monochromatic wave, between an upper (f_{max}) and lower (f_{min}) frequency threshold. These frequency thresholds should be selected such that they capture most of the spectrum's total energy (i.e., area under the curve). Based on the shape of our wave spectrum, we select $f_{min} = 0.07$ Hz and $f_{max} = 0.20$ Hz. While it is recommended that at least $N = 200$ wave components be considered

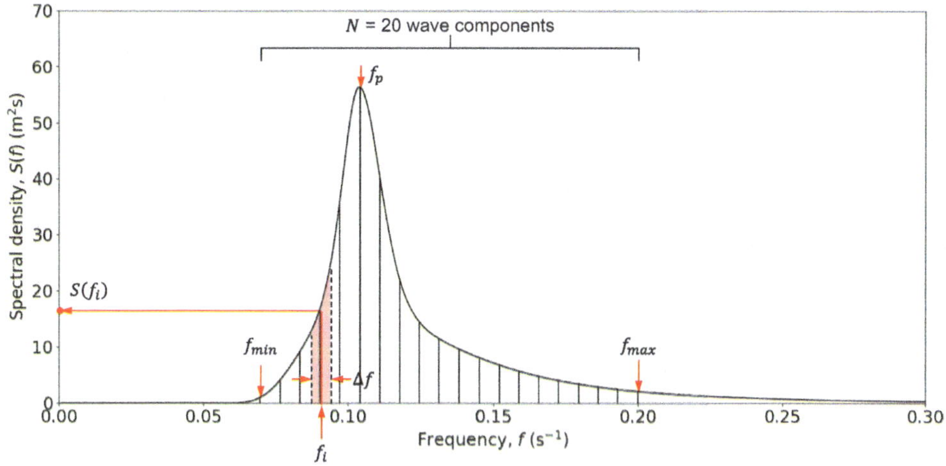

Fig. 6.14 Discretization of a wave spectrum into $N = 20$ wave components

to fully capture the randomness of realistic sea states, we shall only adopt $N = 20$ for demonstration purposes. This results in the width of each frequency band being:

$$\Delta f = \frac{f_{max} - f_{min}}{N - 1} = \frac{0.20 - 0.07}{20 - 1} = 0.00684\,\text{Hz}$$

The discretization of the JONSWAP wave spectrum from Fig. 6.9 into 20 wave components between 0.07 and 0.20 Hz is shown in Fig. 6.14.

Our goal is to determine η at any given point in time t. To do so, we must first calculate the properties of each monochromatic wave corresponding to frequency component i, where $i = 1, 2, 3, \ldots, 20$. These components consist of the wave amplitude A_i and frequency ω_i, in addition to a random phase shift $\tilde{\phi}_i$ that takes on a random value between $-\pi$ and π. The amplitude A_i is determined via:

$$A_i = \sqrt{2S(f_i)\Delta f}$$

where $S(f_i)$ is the spectral density for frequency f_i calculated using Eq. (6.26). For our first wave component where $i = 1$ and $f_{i=1} = 0.07$ Hz, the spectral density is found to be:

$$S(f_{i=1}) = \frac{\alpha g^2}{(2\pi)^4 f_{i=1}^5} \exp\left[-1.25\left(\frac{f_p}{f_{i=1}}\right)^4\right] \gamma^{\exp\left[-\frac{(f_{i=1}-f_p)^2}{2\sigma^2 f_p^2}\right]}$$

$$= \frac{0.0117 \times 9.81^2}{(2\pi)^4 \times 0.07^5} \exp\left[-1.25\left(\frac{0.104}{0.07}\right)^4\right] 3.3^{\exp\left[-\frac{(0.07-0.104)^2}{2\times 0.07^2 \times 0.104^2}\right]}$$

Table 6.3 Properties of individual wave components within the spectrum

i	f_i (Hz)	ω_i (rad s^{-1})	S_i (m^2 s)	A_i (m)	$\tilde{\phi}_i$[a]	$\eta_i(t)$ (m)
1	0.07	0.440	0.988	0.116	− 0.146	$0.116\cos(-0.440t - 0.146)$
2	0.0768	0.483	4.116	0.237	2.957	$0.237\cos(-0.483t + 2.957)$
3	0.0837	0.526	9.223	0.355	− 2.308	$0.355\cos(-0.526t - 2.308)$
…	…	…	…	…	…	…
20	0.20	1.257	2.064	0.168	− 2.621	$0.168\cos(1.257t - 2.621)$

[a] Values in this column were randomly generated between a range of $-\pi$ and π

$$= 0.988\ \mathrm{m^2\,s}$$

which can be used to find the amplitude of the first wave component:

$$A_{i=1} = \sqrt{2 \times 0.988 \times 0.00684} = 0.116\,\mathrm{m}$$

The profile of this first wave component at time t is then:

$$\eta_{i=1}(t) = A_{i=1}\cos\left(-\omega_{i=1}t + \tilde{\phi}_{i=1}\right) = 0.116\cos(-0.440t - 0.146)$$

where:

$$\omega_{i=1} = 2\pi f_{i=1} = 2\pi \times 0.07 = 0.440\,\mathrm{rad\,s^{-1}}$$

in addition to assuming a random phase shift of $\tilde{\phi}_{i=1} = -0.146$. This process is next repeated for all values of i up to $N = 20$ as summarized in Table 6.3.

The free surface elevation $\eta(t)$ at any given time t is then created from the linear superposition of each individual wave $\eta_i(t)$. This is achieved by simply summing $\eta_i(t)$ shown in the final column of Table 6.3 from $i = 1$ to 20:

$$\eta(t) = \sum_{i=1}^{N=20} \eta_i(t)$$

giving:

$$\begin{aligned}\eta(t) = {} & 0.116\cos(-0.440t - 0.146) \\ & + 0.237\cos(-0.483t + 2.957) \\ & + 0.355\cos(-0.526t - 2.308) \\ & + \cdots + 0.168\cos(1.257t - 2.621)\end{aligned}$$

This summation procedure and the resulting wave record produced by the spectrum is shown in Fig. 6.15 over a total duration of 600 s. However, if we increase the total

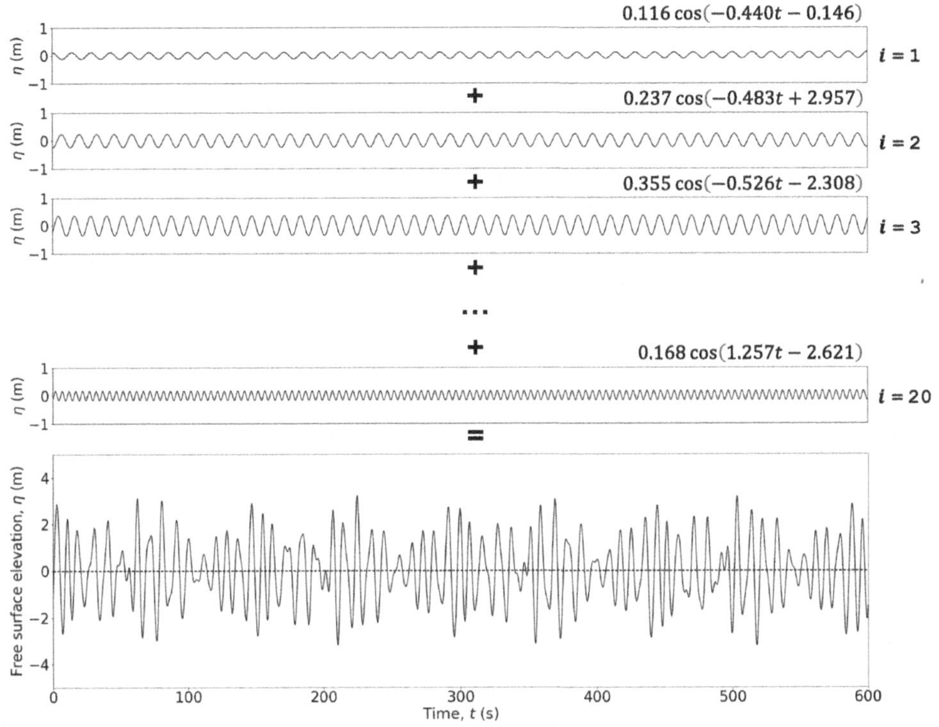

Fig. 6.15 Synthetic wave record produced from a wave spectrum with $N = 20$

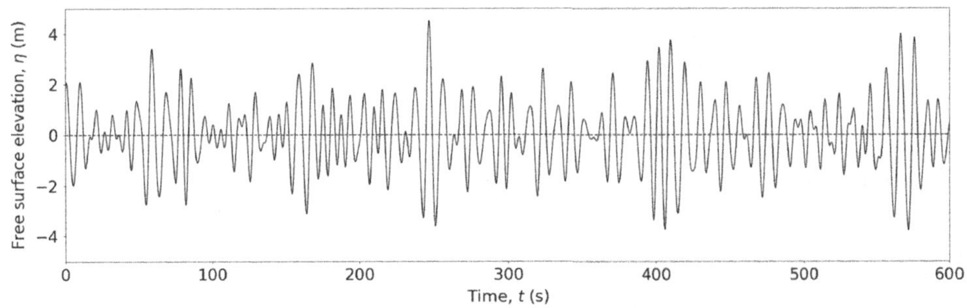

Fig. 6.16 Synthetic wave record produced from a wave spectrum with $N = 200$

number of wave components to $N = 200$, the profile shown in Fig. 6.16 is produced. We notice that as the number of wave components increase, the wave record exhibits more randomness and will be more representative of sea states occurring in nature.

References

1. Borgman, L.E., 1969. Ocean wave simulation for engineering design. Journal of Waterway, Port, Coastal, and Ocean Engineering 95, 557–583.
2. Bouws, E., Gunther, H., Rosenthal, W., Vincent, C.L., 1985. Similarity of the wind wave spectrum in finite water. Journal of Geophysical Research 90, 975–986.
3. Bretschneider, C.L., 1957. Hurricane design wave practices. Journal of the Waterways and Harbors Division 83, 1–33.
4. Donelan, M.A., Hamilton, J., Hui, W.H., 1985. Directional spectra of wind-generated ocean waves. Philosophical Transactions of the Royal Society A 315, 509–562.
5. Fontaine, E., 2013. A Theoretical Explanation of the Fetch- and Duration-Limited Laws. Journal of Physical Oceanography 43, 233–247.
6. Goda, Y., 2003. Revisiting Wilson's Formulas for Simplified Wind-Wave Prediction. Journal of Waterway, Port, Coastal, and Ocean Engineering 129, 93–95.
7. Goda, Y., 2000. Random seas and design of maritime structures. World Scientific Publishing Co., Singapore.
8. Goda, Y., 1988. Statistical variability of sea state parameters as a function of a wave spectrum. Coastal Engineering in Japan 31, 39–52.
9. Goda, Y., Nagai, K., 1974. Investigation of the statistical properties of sea waves with field and simulation data. Report of the Port and Harbour Technical Research Institute 13, 3–37.
10. Hasselmann, K., Barnett, T.P., Bouws, E., Carlson, H., Cartwright, D.E., Enke, K., Ewing, J.A., Gienapp, H., Hasselmann, D.E., Kruseman, P., Meerburg, A., Muller, P., Olbers, D.J., Richter, K., Sell, W., Walden, H., 1973. Measurements of wind-wave growth and swell decay during the joint North Sea wave project (JONSWAP). Deutches Hydrographisches Institut 8, 1–95.
11. Kitaigordskii, S.A., Krasitskii, V.P., Zaslavskii, M.M., 1975. On Phillips' theory of equilibrium range in the spectra of wind-generated gravity waves. Journal of Physical Oceanography 5, 410–420.
12. Longuet-Higgins, M.S., 1953. On the statistical distribution of the heights of sea waves. Journal of Marine Research 11, 245–266.
13. Mitsuyasu, H., 1970. On the growth of spectrum of wind-generated waves (2) - spectral shape of wind waves at finite fetch, in: Proceedings of the Japanese Conference on Coastal Engineering. pp. 1–7.
14. Munk, W.H., 1944. Proposed uniform procedure for observing waves and interpreting instrument records. Scripps Institution of Oceanography, La Jolla, California.
15. Pierson, W.J., Moskowitz, L., 1964. A proposed spectral form for fully developed wind seas based on the similarity theory of S. A. Kitaigorodskii. Journal of Geophysical Research 69, 5181–5190.
16. Reeve, D., Chadwick, A., Fleming, C., 2018. Coastal engineering: processes, theory & design practice. Taylor and Francis Group, Boca Raton, FL.
17. USACE, 2015. Coastal Engineering Manual. U.S. Army Corps of Engineers, Washington D.C.
18. Vincent, C.L., Hughes, S.A., 1985. A note on wind wave growth in shallow water. Journal of Waterway, Port, Coastal, and Ocean Engineering 3.
19. Young, I.R., Verhagen, L.A., 1996. The growth of fetch limited waves in water of finite depth. Part 1. Total energy and peak frequency. Coastal Engineering 29, 47–78.

Experimental Generation of Surface Waves 7

The reproduction of surface waves within hydraulic laboratories or numerical simulations is crucial when attempting to understand how different types of coastal and offshore structures perform under various sea states. These waves are usually produced within a wave flume via the forward and backward motion of a piston wavemaker. As an example, Fig. 7.1 shows the wavemaker within the 104 m long Large Wave Flume housed within the O.H. Hinsdale Wave Research Laboratory at Oregon State University. Likewise, Fig. 7.2 depicts a piston wavemaker within a numerical simulation used to study the response of a floating breakwater excited by waves in 2D. By controlling the translational motion of these physical or numerical wavemakers, regular waves with a constant wave height and wavelength, or irregular waves corresponding to different wave spectra, can be created. Such waves can then be used to study the behavior of structures placed within the flume that may be susceptible to similar wave conditions in the field. This chapter will introduce the methods by which different types of waves can be generated.

7.1 Generation of Regular Waves

To understand how waves trains of any arbitrary profile (e.g., Fig. 6.16) can be generated within a wave flume, we must first explore the relationship between the translational motion of a piston wavemaker (referred to as the "stroke") and the height of a wave produced by this motion. When assuming an incompressible and irrotational fluid with constant pressure at the free surface, Biesel and Suquet [2] developed a function for the wave height (H) to piston stroke (e_0) ratio in terms of the wavenumber k and water depth h:

S. Wang, *Applied Ocean Mechanics for Structural Engineering and Climate Adaptation*,
Synthesis Lectures on Ocean Systems Engineering,
https://doi.org/10.1007/978-3-032-18199-2_7

Fig. 7.1 Left: Large Wave Flume (LWF) in the O.H. Hinsdale Wave Research Laboratory at Oregon State University. Right: Close-up of piston wavemaker within the flume

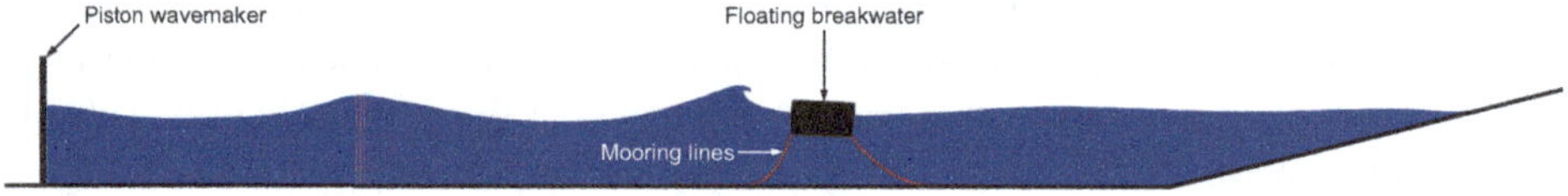

Fig. 7.2 Numerical simulation of waves generated using a piston wavemaker impacting a floating breakwater

$$\chi = \frac{H}{e_0} = \frac{2[\cosh(2kh) - 1]}{\sinh(2kh) + 2kh} \tag{7.1}$$

which can also be expressed as:

$$\chi = \frac{H}{e_0} = \frac{2\sinh^2(kh)}{\sinh(kh)\cosh(kh) + kh} \tag{7.2}$$

and is often referred to as the "Biesel transfer function". Figure 7.3 plots this function in terms of kh which shows χ asymptotically converging to a value of 2. This behavior implies that for large kh, the piston stroke e_0 is equal to half of the desired wave height. In addition, recall that the free surface η of a linear wave is:

$$\eta(t) = A\cos(\omega t) \tag{7.3}$$

where $A = H/2$ is the wave amplitude and $\omega = 2\pi/T$ is the frequency. From Eqs. (7.1) and (7.3), we can deduce that the motion of a piston wavemaker $e(t)$ for the generation of linear waves with a fixed wave height of H (or amplitude A) is:

$$e(t) = \frac{e_0}{2}\sin(\omega t) = \frac{A}{\chi}\sin(\omega t) \tag{7.4}$$

Here, we have adopted $\sin(\omega t)$ in lieu of $\cos(\omega t)$ to ensure that the wavemaker remains in its neutral position (i.e., $e = 0$) when $t = 0$.

Strictly speaking, Eq. (7.4) pertains only the generation of waves falling within the linear region of Le Méhauté's abacus (i.e., region "A" in Fig. 3.9). To generate nonlinear waves with greater amplitude, additional higher order terms must be added to the first-order piston motion given by Eq. (7.4). Madsen [7] proposed the following stroke for the generation of regular waves classified as linear or Stokes second-order (i.e., region "B" in Fig. 3.9):

$$e(t) = \frac{A}{\chi}\sin(\omega t) + e_2(t) \tag{7.5}$$

where:

$$e_2(t) = \frac{H^2}{32h}\left(\frac{3\cosh(kh)}{\sinh^3(kh)} - \frac{2}{\chi}\right)\sin(2\omega t) \tag{7.6}$$

is the second-order term. In addition, Eq. (7.5) is valid only when the following condition is satisfied:

$$\frac{HL^2}{h^3} < \frac{8\pi^2}{3} \tag{7.7}$$

where L is the desired wavelength. Note that the HL^2/h^3 term in Eq. (7.7) is known as the Ursell number which describes the nonlinearity of surface waves [11]. While the generation of regular waves beyond Stokes second-order is beyond the scope of this text, an in-depth discussion into higher order waves is presented by Fenton [4].

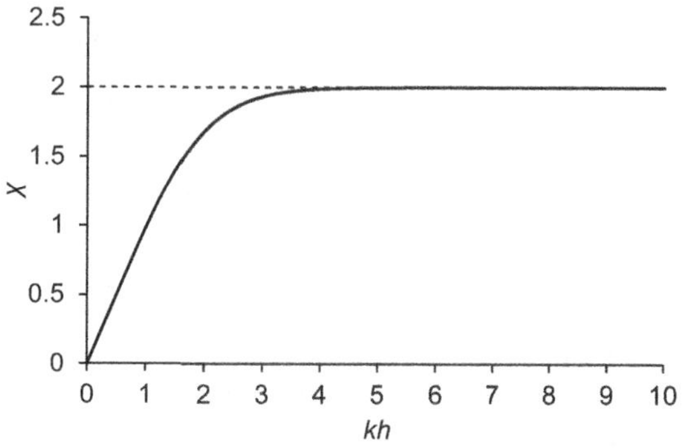

Fig. 7.3 Behavior of the Biesel transfer function

7.2 Generation of Irregular Waves

We can now extend the concept given by Eq. (7.4) towards the generation of irregular waves via the wave superposition principle previously demonstrated in Sect. 6.6. By discretizing a wave spectrum $S(f)$ into N frequency bands of width Δf, the motion of a piston wavemaker as a function of time (see Fig. 7.4) is given by the linear summation of Eq. (7.4) over all frequencies $f_i = f_1, f_2, \ldots, f_N$ within the spectrum:

$$e(t) = r_t \sum_{i=1}^{N} \frac{A_i}{\chi_i} \sin\left(\omega_i t + \tilde{\phi}_i\right) \tag{7.8}$$

where A_i is the amplitude associated with frequency f_i:

$$A_i = \sqrt{2S(f_i)\Delta f} \tag{7.9}$$

and $\omega_i = 2\pi f_i$. Likewise, the Biesel transfer function is now expressed in terms of each frequency component i:

$$\chi_i = \frac{2[\cosh(2k_i h) - 1]}{\sinh(2k_i h) + 2k_i h} \tag{7.10}$$

where the wavenumber k_i is obtained from each f_i via the dispersion relation:

$$\omega_i = \sqrt{g k_i \tanh(k_i h)} \tag{7.11}$$

Similar to Eq. (7.4), we wish to ensure that the piston remains at its neutral position ($e = 0$) when $t = 0$. We can achieve this condition through the introduction of an artificial

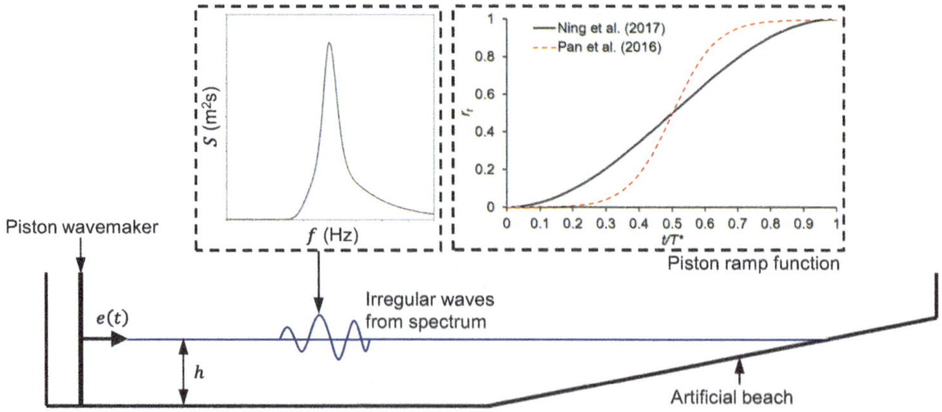

Fig. 7.4 Generation of irregular waves via a piston wavemaker within a wave flume

ramp function r_t into (7.8) that gradually increases from 0 to 1 over a set period of time T^* from the beginning of the piston's motion. The selection of a ramp function is largely arbitrary and is given by Ning et al. [8] as:

$$r_t(t) = \begin{cases} \frac{1}{2}\left[1 - \cos\left(\frac{\pi t}{T^*}\right)\right], & t < T^* \\ 1, & t \geq T^* \end{cases} \tag{7.12}$$

while Pan et al. [9] suggests the form of:

$$r_t(t) = \frac{1}{2}\left\{1 + \tanh\left[7.6\left(\frac{t}{T^*} - \frac{1}{2}\right)\right]\right\} \tag{7.13}$$

to which both are plotted in Fig. 7.4 in terms of t/T^*. In Eqs. (7.12) and (7.13), T^* is the smoothing time scale which can be set to an arbitrary duration or simply taken as the mean period of the wave spectrum T_m. Finally, a phase shift $\tilde{\phi}_i$ is introduced into Eq. (7.8) for each frequency component which takes a random value between $-\pi$ and π.

Figure 7.5 illustrates an example of a piston wavemaker generating a train of irregular waves in water $h = 0.8$ m deep from the JONSWAP wave spectrum per Eq. (6.31) with $H_s = 0.3$ m and $T_p = 1.3$ s. Here, the spectrum was discretized into $N = 100$ frequency bands between the upper and lower frequencies of 2.0 and 0.5 Hz, respectively. The ramp function given by Eq. (7.12) was adopted where T^* was set equal to T_p. In Fig. 7.5, the piston stroke $e(t)$ and corresponding free surface profile $\eta(t)$ produced by the piston's motion is shown over a period of 60 s.

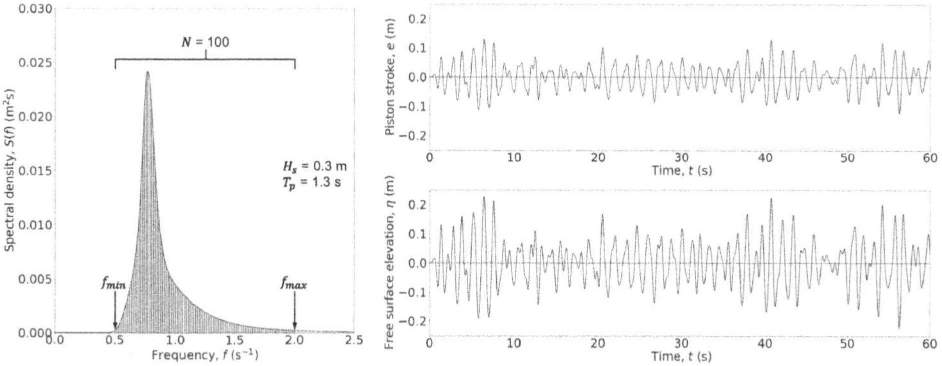

Fig. 7.5 Motion of a piston wavemaker and profile of waves generated from a JONSWAP wave spectrum with $H_s = 0.3$ m and $T_p = 1.3$ s

7.3 Generation of Focused Waves

Wave focusing is a popular technique used to generate a highly nonlinear extreme wave at a predefined point in space and time from an underlying wave spectrum, thus describing the most probable shape of a large wave produced by the given sea state [12]. By superimposing a series of small amplitude waves at a specified location, a large wave can be reproduced with great repeatability. The most common approach towards the generation of a focused wave is via "NewWave" theory proposed by Tromans et al. [10], with the wavemaker stroke given by:

$$e(t) = \sum_{i=1}^{N} \frac{a_i}{\chi_i} \cos\left[k_i\left(x - x_f\right) - \omega_i\left(t - t_f\right)\right] \tag{7.14}$$

Here, x_f and t_f denote the position (relative to the wavemaker) and time of the focused wave, respectively, and x represents the rest position of the wavemaker which is often taken as $x = 0$. The Biesel transfer function χ_i for each component of the wave spectrum $S(f)$ bounded by f_{min} and f_{max} (see Fig. 7.5) is given by Eq. (7.10), while the frequency ω_i and wavenumber k_i are related through Eq. (7.11). The wave amplitude a_i of each frequency component is expressed as:

$$a_i = A \frac{S(f_i)\Delta f}{\sum_{i=1}^{N} S(f_i)\Delta f} \tag{7.15}$$

where A denotes the target theoretical linear wave amplitude of the focused wave. Note that for a desired focusing position x_f, the time t_f needs to be calibrated. Alternatively, Ning et al. [8] suggests that the following relationship can be used to define t_f for a given x_f:

$$t_f = \frac{2x_f T_{min}}{L_{min}} \tag{7.16}$$

where $T_{min} = 1/f_{max}$ is the minimum wave period within the spectrum and L_{min} is the wavelength associated with T_{min}. Extreme wave focusing is also commonly used to produce breaking waves within laboratory wave flumes as demonstrated by Chuang et al. [3].

7.4 Generation of Solitary Waves

Solitary waves are a special type of cnoidal wave (see Fig. 3.9) that consists of a single wave crest with theoretically infinite wavelength. As such, solitary waves are commonly used to approximate the behavior of long waves observed in nature such as tsunamis [13]. The generation of a single solitary wave with height H in water of depth h via a piston wavemaker was described by Goring [5], to which its stroke is given as:

$$e(t) = s(t) + \frac{H}{k_s h} \tag{7.17}$$

where:

$$s(t) = \frac{H}{k_s h} \tanh\left\{k_s\left[c_s\left(t - \frac{T_f}{2}\right) - s(t)\right]\right\} \tag{7.18}$$

Here, c_s represents the speed of the solitary wave:

$$c_s = \sqrt{g(H + h)} \tag{7.19}$$

and:

$$k_s = \sqrt{\frac{3H}{4h^3}} \tag{7.20}$$

is known as the "outskirt coefficient" which describes how the free surface tends towards the still water level (SWL) at infinity. Note that while solitary waves theoretically have infinite wavelength (i.e., zero wavenumber), the wave profile will rapidly decay towards zero as we move further away from the crest. Hence, the outskirt coefficient can be treated as a wavenumber used to approximate the "effective wavelength" between two points on either side of the crest where the wave profile η is very close to the SWL (i.e., $\eta \approx 0$). Similar to regular waves, this effective wavelength is determined via:

$$L_s = \frac{2\pi}{k_s} \tag{7.21}$$

Knowledge of both L_s and c_s thus allows us to assign a representative time scale T_r to the wave per:

$$T_r = \frac{L_s}{c_s} \tag{7.22}$$

which denotes the time taken for a solitary wave bounded by L_s to travel past a given point in space.

Unlike the cyclical movement of a piston wavemaker used to generate regular or irregular waves (e.g., Fig. 7.5), the wavemaker's motion is unidirectional when producing a solitary wave, with $e(t)$ monotonically increasing with time (e.g., Fig. 7.6). The duration of wavemaker motion is given by T_f in Eq. (7.18) and is expressed as:

$$T_f = \frac{2}{k_s c_s}\left(3.8 + \frac{H}{h}\right) \tag{7.23}$$

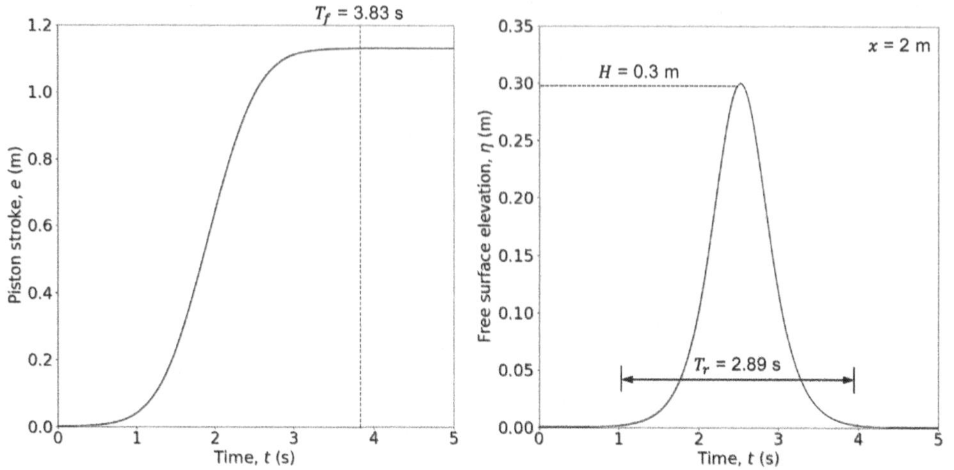

Fig. 7.6 Left: Motion of a piston wavemaker. Right: Profile of a solitary wave with $H = 0.3$ m and $h = 0.8$ m at $x = 2$ m

Finally, Eq. (7.18) must be solved iteratively for $s(t)$ for all values of time t, which can then be substituted into (7.17) in order to obtain the piston stroke $e(t)$. The free surface profile η at a distance x downstream from the wavemaker is then:

$$\eta(x, t) = \frac{H}{\cosh^2\left\{k_s\left[c_s\left(t - \frac{T_f}{2}\right) - x\right]\right\}} \tag{7.24}$$

As an example, Fig. 7.6 illustrates the piston stroke obtained using Eq. (7.17) used to generate a solitary wave with $H = 0.3$ m in water $h = 0.8$ m deep. The wave speed and outskirt coefficient from Eqs. (7.19) and (7.20) were determined to be $c_s = 3.28\ \mathrm{ms}^{-1}$ and $k_s = 0.663\ \mathrm{m}^{-1}$, respectively. These properties give rise to a wavelength of $L_s = 9.48$ m and a representative time scale of $T_r = 2.89$ s as shown in the right plot of Fig. 7.6. The wave profile at a distance of $x = 2$ m from the wavemaker as described using Eq. (7.24) is also illustrated. To produce this wave, the wavemaker must travel forwards for a duration of $T_f = 3.83$ s until it reaches a maximum stroke of approximately 1.13 m.

7.5 Scaling Relationships

It is important to realize that when generating waves within laboratory wave flumes (i.e., model scale), all wave properties such as the significant wave height (H_s) and peak period (T_p) must be scaled down from their corresponding observations in the real world (i.e., prototype scale). Froude scaling is commonly used to achieve this, with the prototype-to-model scaling relationships being:

Table 7.1 Froude scaling relationships for different properties [1]

Property	Symbol	Units	Model scale[a]	Prototype scale[a]
Length	L	m	$L_m = L_p/\lambda$	$L_p = \lambda L_m$
Area	A	m^2	$A_m = A_p/\lambda^2$	$A_p = \lambda^2 A_m$
Volume	V	m^3	$V_m = V_p/\lambda^3$	$V_p = \lambda^3 V_m$
Mass	m	kg	$m_m = m_p/\lambda^3$	$m_p = \lambda^3 m_m$
Velocity	v	ms^{-1}	$v_m = v_p/\sqrt{\lambda}$	$v_p = v_m\sqrt{\lambda}$
Acceleration	a	ms^{-2}	$a_m = a_p$	$a_p = a_m$
Force	F	N	$F_m = F_p/\lambda^3$	$F_p = \lambda^3 F_m$
Moment	M	Nm	$M_m = M_p/\lambda^4$	$M_p = \lambda^4 M_m$
Stress	σ	Nm^{-2}	$\sigma_m = \sigma_p/\lambda$	$\sigma_p = \lambda\sigma_m$
Strain	ε	Unitless	$\varepsilon_m = \varepsilon_p/\lambda$	$\varepsilon_p = \lambda\varepsilon_m$
Time	t	s	$t_m = t_p/\sqrt{\lambda}$	$t_p = t_m\sqrt{\lambda}$
Frequency	f	Hz (s^{-1})	$f_m = f_p\sqrt{\lambda}$	$f_p = f_m/\sqrt{\lambda}$

[a] Subscript (m) = model; (p) = prototype

$$H_s(\text{model}) = \frac{H_s(\text{prototype})}{\lambda}, \quad T_p(\text{model}) = \frac{T_p(\text{prototype})}{\sqrt{\lambda}} \tag{7.25}$$

where λ refers to the scale factor. As an example, suppose we wish to model real-world waves with $H_s = 5$ m and $T_p = 10$ s within a wave flume at 1:20 scale. The significant wave height and peak period used to define the piston's motion in Eq. (7.8) would be $H_s = 5/20 = 0.25$ m and $T_p = 10/\sqrt{20} = 2.24$ s, respectively. Scaling of the water depth would also follow the same relationship as the wave height (i.e., $h_{model} = h_{prototype}/\lambda$). The piston stroke would then be determined using these model-scale properties.

Froude scaling relationships between various model and prototype properties as a function of λ are further summarized in Table 7.1. The reader is encouraged to consult Heller [6] for a more detailed discussion into appropriate scaling strategies for hydraulic experiments.

References

1. Bairrao, R., Vaz, C.T., 2000. Shaking table testing of civil engineering structures - the LNEC 3D simulator experience, in: Proceedings 12th World Conference on Earthquake Engineering. Auckland, New Zealand.
2. Biesel, F., Suquet, F., 1951. Les Appareils Générateurs de Houle en Laboratoire. La Houille Blanche 37, 147–165.
3. Chuang, W., Chang, K., Mercier, R., 2017. Impact pressure and void fraction due to plunging breaking wave impact on a 2D TLP structure. Experiments in Fluids 58.

4. Fenton, J.D., 1985. A fifth-order Stokes theory for steady waves. Journal of Waterway, Port, Coastal, and Ocean Engineering 111, 216–234.
5. Goring, D.G., 1979. Tsunamis - the propagation of long waves onto a shelf (Ph.D. Thesis). California Institute of Technology, Pasadena, California.
6. Heller, V., 2011. Scale effects in physical hydraulic engineering models. Journal of Hydraulic Research 49, 293–306.
7. Madsen, O.S., 1971. On the generation of long waves. Journal of Geophysical Research 76, 8672–8683.
8. Ning, D., Wang, R., Chen, L., Li, J., Zang, J., Cheng, L., 2017. Extreme wave run-up and pressure on a vertical seawall. Applied Ocean Research 67, 188–200.
9. Pan, K., Ijzermans, R.H.A., Jones, B.D., Thyagarajan, A., van Beest, B.W.H., Williams, J.R., 2016. Application of the SPH method to solitary wave impact on an offshore platform. Computational Particle Mechanics 3, 155–166.
10. Tromans, P.S., Anaturk, A.R., Hagemeijer, P., 1991. A new model for the kinematics of large ocean waves-application as a design wave. Presented at the The First International Offshore and Polar Engineering Conference, Edinburgh.
11. Ursell, F., 1953. The long-wave paradox in the theory of gravity waves. Mathematical Proceedings of the Cambridge Philosophical Society 49, 685–694.
12. Wang, S., Chuang, W.-L., 2025. Dynamic analysis of breaking wave impact on a floating offshore wind turbine via smoothed particle hydrodynamics. Marine Structures 100.
13. Yao, Y., Huang, Z., Lo, E.Y.M., Shen, H., 2014. A Preliminary Laboratory Study of Motion of Floating Debris Generated by Solitary Waves Running up a Beach. Journal of Earthquake and Tsunami 8.

Wind and Waves in Tropical Cyclones

8

Tropical cyclones (also called hurricanes or typhoons) produce devastating wind and wave conditions that often result in significant loss of life and damage to coastal infrastructure. With ongoing climate change and sea level rise, the frequency and intensity of tropical cyclones occurring in the Atlantic, Pacific, and Indian Oceans are expected to increase [9]. Storm surge flooding caused by tropical cyclones represents some of the deadliest and costliest natural disasters in the world (e.g., Fig. 8.1). As of 2025 (Table 8.1), nine out of the ten most expensive natural disasters impacting the United States were attributed to hurricane-induced flooding, all of which having occurred after 1992 [6]. This chapter will explore the properties of tropical cyclones, as well as methods used to estimate their wind speed and the sea states they produce.

8.1 Terminology

It is important to realize that the terms "hurricane", "cyclone", and "typhoon" all refer to the same natural phenomena, i.e., a rotating storm system with a low-pressure central region. The terminology used for these storms is differentiated only by their geographic point of origin as follows (Fig. 8.2):

- Storms that form over the North Atlantic and Northeast Pacific are termed **hurricanes**
- Storms that form over the Northwest Pacific are termed **typhoons**
- Storms that form over the South Pacific and Indian Ocean are termed **cyclones**.

S. Wang, *Applied Ocean Mechanics for Structural Engineering and Climate Adaptation*,
Synthesis Lectures on Ocean Systems Engineering,
https://doi.org/10.1007/978-3-032-18199-2_8

Fig. 8.1 Left: Storm surge flooding caused by Hurricane Ike in Texas (2008). Right: Flooding from Hurricane Sandy in New York (2012). Photos courtesy of NOAA and the New Jersey National Guard

Table 8.1 Ten most expensive natural disasters in United States history as of 2025

Rank	Year	Natural disaster	Cost of damages[a] (USD)	Fatalities
1	2005	Hurricane Katrina	$201.3 billion	1833
2	2017	Hurricane Harvey	$160 billion	89
3	2022	Hurricane Ian	$119.6 billion	152
4	2017	Hurricane Maria	$115.2 billion	2981
5	2012	Hurricane Sandy	$88.5 billion	159
6	2021	Hurricane Ida	$84.6 billion	96
7	2024	Hurricane Helene	$78.7 billion	219
8	2017	Hurricane Irma	$64 billion	97
9	1992	Hurricane Andrew	$60.5 billion	61
10	1988	North American drought	$54.6 billion	454

Data courtesy of NCEI [7]
[a] Costs have been CPI-adjusted to 2025 US dollars

Note that these names will be used interchangeably throughout this chapter. In addition, beyond a few exceptions, tropical cyclones do not usually form within 5° of latitude north or south of the equator as shown in Fig. 8.2. Furthermore, severe storms that form in the Mediterranean Sea are sometimes informally referred to as "medicanes" [1].

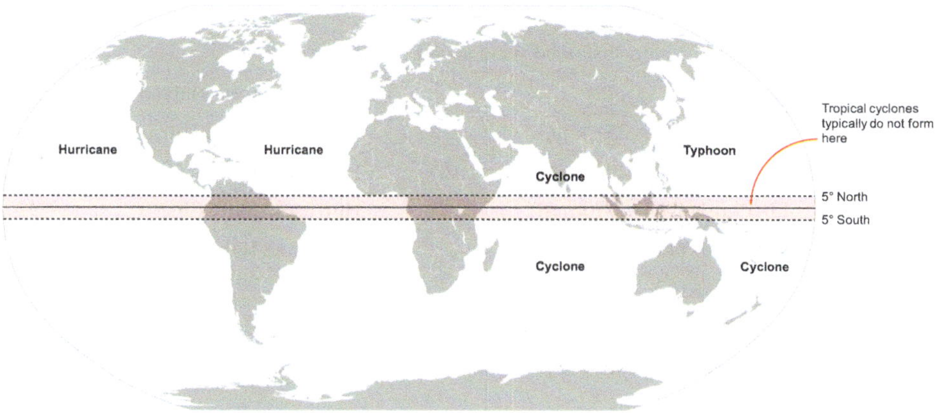

Fig. 8.2 Terminology used for tropical cyclones based on their geographic origin

8.2 Hurricane Intensity

Tropical cyclones are characterized by high winds rotating around a central region of low pressure called the "eye". The direction of rotation is anticlockwise in the Northern Hemisphere and clockwise in the Southern Hemisphere due to anticlockwise rotation of the earth when viewed from the north pole. Broadly speaking, the intensity of tropical cyclones (i.e., maximum wind speeds) depends on the difference between the central pressure p_0 in the eye and the ambient atmospheric pressure p_n outside the storm. This difference in pressure is termed the "pressure deficit" and tropical cyclones with larger pressure deficits tend to exhibit greater wind speeds. For example, Hurricane Patricia (making landfall in Mexico in 2015) recorded a pressure deficit of 141 hPa (1 hectopascal = 100 Pa) which was the largest ever recorded in the Western Hemisphere. This storm exhibited a maximum sustained wind speed of 345 km/h (faster than many propeller aircraft) and was the highest hurricane wind speed on record [10].

The intensity of tropical cyclones is commonly quantified using the Saffir-Simpson hurricane wind scale (SSHWS) shown in Table 8.2. This scale classifies hurricanes into five distinct categories (Category 1 to 5) based on the maximum wind speed at 10 m above sea level that has been sustained for at least one minute. To qualify as a cyclone, this wind speed must be at least 33 m/s (119 km/h). Winds below this threshold are classified as "tropical storms" (TS) or "tropical depressions" (TD).

Table 8.2 Saffir-Simpson hurricane wind scale (SSHWS)

Category	Maximum 1-min sustained wind speed	Example
5	> 70 m/s (252 km/h)	Super Typhoon Haiyan (2013)
4	58–70 m/s (209–251 km/h)	Great Galveston Hurricane (1900)
3	50–58 m/s (178–208 km/h)	Hurricane Eloise (1975)
2	43–49 m/s (154–177 km/h)	Hurricane Sally (2020)
1	33–42 m/s (119–153 km/h)	Hurricane Isaias (2020)
TS[a]	18–32 m/s (63–118 km/h)	
TD[a]	< 17 m/s (62 km/h)	

[a] TS = tropical storm; TD = tropical depression

8.3 Hurricane Wind Field

The wind speed inside a tropical cyclone varies as a function of the distance from the eye. Figure 8.3 shows an image of Hurricane Katrina one day before making landfall in Louisiana. The wind speed U_{10}' increases sharply from the calm region inside the eye up to a certain radius termed the "radius to maximum wind", R_m, before decreasing again beyond this radius. There are many empirical approaches to estimating R_m such as the one proposed by Vickery et al. [13] for storms in the North Atlantic:

$$R_m = \begin{cases} \exp\left(2.097 + 0.0187793\Delta p - 0.00018672\Delta p^2 + 0.0381328\phi_0\right), & \phi_0 \leq 30^\circ \\ \exp\left(2.636 - 5.086 \times 10^{-5}\Delta p^2 + 0.0394899\phi_0\right), & \phi_0 > 30^\circ \end{cases} \tag{8.1}$$

where ϕ_0 is the latitude (i.e., north–south coordinate) of the eye in degrees and $\Delta p = p_n - p_0$ with p_n and p_0 denoting the ambient (1013 hPa) and central pressure, respectively. Δp is therefore the pressure deficit and must be expressed in hPa, while R_m determined using (8.1) will have units of km. Note that many other models also exist for the prediction of R_m such as those proposed by Chavas and Knaff [2], Hsu and Yan [4], and Takagi and Wu [12].

Naturally, it is of great interest to predict the distribution of wind speed within a tropical cyclone for a given pressure deficit. This information will enable engineers to not only design for the anticipated wind loads but also allow us to use wave spectra to estimate the expected wave heights coastal structures may be subjected to (see Chap. 6 on irregular waves). The most direct and common approach to determine hurricane wind speeds (as a function of distance from the eye) is through the use of parametric wind field models that takes as input only the pressure deficit. One of the most popular is the Holland model [3] first published in 1980 that expresses the wind speed U_{10}' measured 10 m above sea level at radius r from the center of the storm:

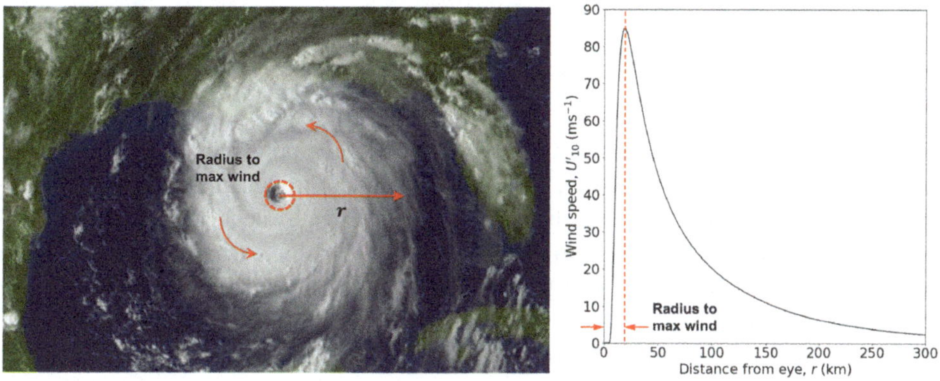

Fig. 8.3 Left: Satellite image of Hurricane Katrina one day before landfall (courtesy of NOAA). Right: Relationship between wind speed and distance from the eye predicted using a parametric model

$$U'_{10} = \left[\frac{100AB(p_n - p_0)\exp\left(-\frac{A}{r^B}\right)}{\rho_a r^B} + \frac{r^2 f_c^2}{4} \right]^{1/2} - \frac{rf_c}{2} \tag{8.2}$$

Here, $\rho_a = 1.2$ kg/m^3 is the air density and p_n and p_0 are expressed in hPa. The parameter A depends on the radius to maximum wind:

$$A = R_m^B \tag{8.3}$$

where R_m is expressed in meters. Likewise, B determines the overall shape of the wind profile with distance from the eye and is governed by the central pressure:

$$B = 1.5 + \frac{980 - p_0}{120} \tag{8.4}$$

The parameter f_c is called the "Coriolis parameter" and serves to quantify the extent to which the rotation of the earth influences the wind speed inside a tropical cyclone. The Coriolis parameter is only dependent on the latitude ϕ and is determined via:

$$f_c = 2\Omega \sin\phi, \quad \Omega = \frac{2\pi}{86{,}400} \tag{8.5}$$

Hence, f_c is shown to be largest at the poles ($\phi = \pm 90°$) while being zero at the equator ($\phi = 0°$). In Eq. (8.5), Ω denotes the rotation rate of the earth as there are 86,400 s in a day. Broadly speaking, f_c only plays a noticeable role when r is large. In the context of tropical cyclones, it is common to take $f_c \approx 0$ when evaluating wind speeds within the radius to maximum wind (i.e., $r \leq R_m$). Hence, if we are only interested in determining the maximum wind speed within a given tropical cyclone, we can substitute $r = R_m$ into

Eq. (6.6) and set $f_c = 0$, yielding:

$$U'_{10,max} = \left[\frac{100B(p_n - p_0)}{\rho_a e}\right]^{1/2} \tag{8.6}$$

where $e = 2.718$ is Euler's number. However, for design applications, we often require the average wind speed sustained for at least one minute. Through empirical observations, it has been shown that $U'_{10,max}$ can be converted into an equivalent 1-min averaged wind speed at 10 m above sea level by multiplying with 0.8. Finally, we can include the contribution of the cyclone forward speed V_{fm} (the speed at which the entire storm is translating on the surface of the earth) to obtain the maximum 1-min averaged wind speed at 10 m above sea level as proposed by Young [15]:

$$U_{10,max} = 0.8\left[\frac{100B(p_n - p_0)}{\rho_a e}\right]^{1/2} + \frac{V_{fm}}{2} \tag{8.7}$$

Note that V_{fm} is divided by 2 to account for various asymmetrical effects within the wind field of the cyclone.

8.4 Example: Parametric Wind Field of Hurricane Katrina (2005)

We will use the Holland model to estimate the wind field of Hurricane Katrina during its most intense phase over the Gulf of Mexico as a practical demonstration of Eq. (8.2). Details about this storm were comprehensively documented in a report by Knabb et al. [5]. Hurricane Katrina made landfall near New Orleans, Louisiana, on 29 August 2005 and resulted in approximately 1833 fatalities (Table 8.1). The entire track of the hurricane along with the evolution of its storm category is shown in Fig. 8.4. We see that Hurricane Katrina originated as a tropical depression in the Caribbean Sea (near the Lesser Antilles) on 23 August 2005 before evolving into a Category 5 Hurricane over the Gulf of Mexico.

The lowest central pressure was recorded as $p_0 = 902\,\text{hPa}$ which took place on 28 August at 1800 UTC. The coordinates of the eye during this instant were at latitude $\phi_0 = 26.3°$ and longitude $\lambda_0 = -88.6°$. We can therefore estimate the radius to maximum wind using the first equation in (8.1) since $\phi_0 \leq 30°$:

$$\begin{aligned} R_m &= \exp\left(2.097 + 0.0187793\Delta p - 0.00018672\Delta p^2 + 0.0381328\phi_0\right) \\ &= \exp\left(2.097 + 0.0187793 \times 111 - 0.00018672 \times 111^2 + 0.0381328 \times 26.3\right) \\ &= 17.88\,\text{km} \end{aligned}$$

where $\Delta p = 1013 - 902 = 111\,\text{hPa}$ and $\phi_0 = 26.3°$. The parameter B can next be found using Eq. (8.4):

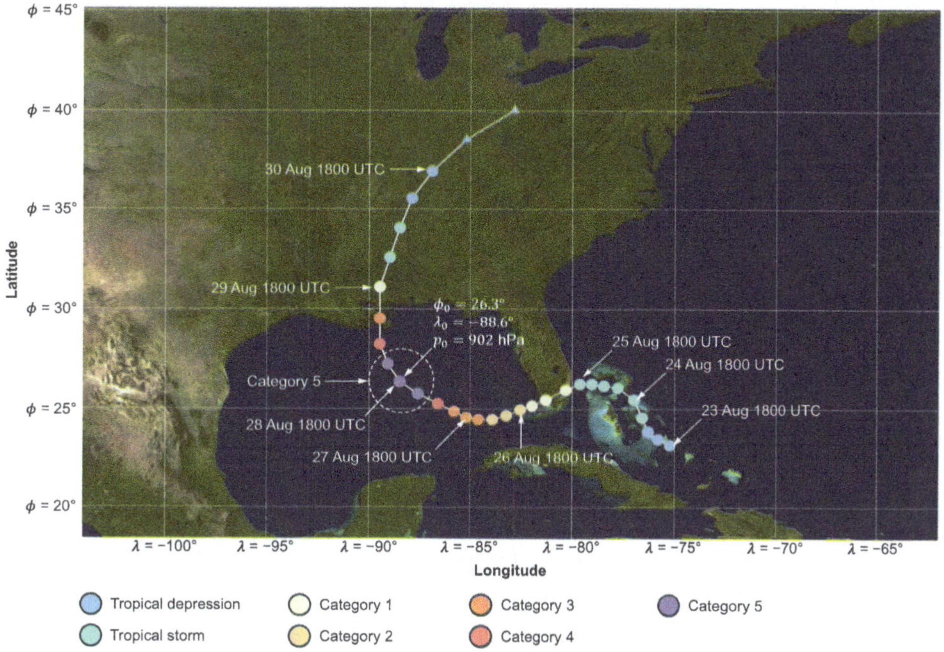

Fig. 8.4 Track of Hurricane Katrina over the southern United States in late August 2005 (courtesy of NASA)

$$B = 1.5 + \frac{980 - p_0}{120} = 1.5 + \frac{980 - 902}{120} = 2.15$$

leading to:

$$A = R_m^B = 17{,}880^{2.15} = 1.39 \times 10^9$$

Note that we must express R_m in meters when calculating A. With the A and B parameters determined, we can now calculate the wind speed U'_{10} using Eq. (8.2) as a function of the distance from the eye (r). Since we are interested in modeling the entire storm, as opposed to only winds at R_m, the Coriolis parameter f_c will also need to be calculated using (8.5) at every location within the wind field as a function of its latitude ϕ. We can determine the distance r between any given set of latitude ϕ and longitude λ coordinates within the wind field and the eye at ϕ_0 and λ_0 via:

$$r = R_E \sqrt{x_r^2 + y_r^2}$$

where $R_E = 6371 \times 10^3$ m is the radius of the earth and:

$$x_r = (\lambda - \lambda_0)\cos\left(\frac{\phi_0 + \phi}{2}\right), \quad y_r = \phi - \phi_0$$

with all values of latitude and longitude expressed in radians.

As an example, suppose we wish to find the local wind speed U'_{10} at coordinates $\phi = 27° = 0.471$ rad and $\lambda = -88° = -1.536$ rad. The distance of this location from the eye of Hurricane Katrina ($\phi_0 = 0.459$ rad, $\lambda_0 = -1.546$ rad) would first need to be calculated:

$$x_r = [-1.536 - (-1.546)]\cos\left(\frac{0.459 + 0.471}{2}\right) = 0.00894$$

$$y_r = 0.471 - 0.459 = 0.012$$

giving:

$$r = 6371 \times 10^3\sqrt{0.00894^2 + 0.012^2} = 95{,}336\,\text{m}$$

At this location, the Coriolis parameter would be:

$$f_c = 2\Omega\sin\phi = 2\left(\frac{2\pi}{86{,}400}\right)\sin 27° = 6.60 \times 10^{-5}\,\text{rad s}^{-1}$$

and the wind speed from Eq. (8.2) is:

$$U'_{10} = \left[\frac{100AB(p_n - p_0)\exp\left(-\frac{A}{r^B}\right)}{\rho_a r^B} + \frac{r^2 f_c^2}{4}\right]^{\frac{1}{2}} - \frac{rf_c}{2}$$

$$= \left[\frac{100 \times \left(1.39 \times 10^9\right) \times 2.15 \times (1013 - 902)\exp\left(-\frac{1.39\times10^9}{95{,}336^{2.15}}\right)}{1.2 \times 95{,}336^{2.15}}\right.$$

$$\left. + \frac{95{,}336^2 \times \left(6.60 \times 10^{-5}\right)^2}{4}\right]^{\frac{1}{2}}$$

$$- \frac{95{,}336 \times 6.60 \times 10^{-5}}{2} = 20\,\text{ms}^{-1}$$

Using the above procedure, the wind speed of Hurricane Katrina can be computed at all coordinates ϕ and λ within the region of interest to construct the entire wind field. However, the forward velocity of the entire storm must also be taken into account when determining the wind speed at any given location within the storm. At the instant considered in this example (28 August at 1800 UTC), the storm was translating in a northwest direction towards New Orleans with a bearing of 330° (measured clockwise from north) at a velocity of approximately V_{fm}= 5.4 ms^{-1}. The wind field shown in Fig. 8.5 was thus

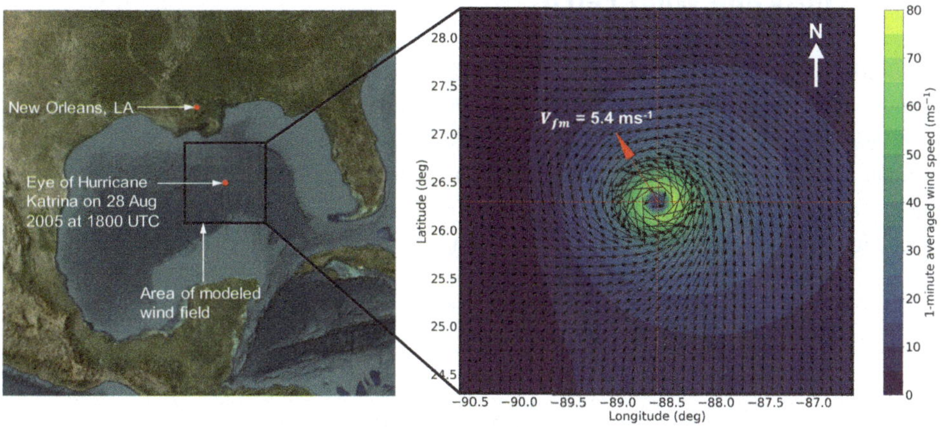

Fig. 8.5 Parametric wind field of Hurricane Katrina within the Gulf of Mexico predicted via the Holland model. Satellite image courtesy of USGS

produced by superimposing the 1-min wind velocity given by $0.8U'_{10}$ over a constant velocity field oriented along the direction of storm propagation with magnitude V_{fm}. A constant inflow angle of 25°, as suggested by Shea and Gray [11], was also adopted for the wind vectors every location, since hurricane winds typically exhibit a slight inwards trajectory and are not perfectly tangential to the radius of rotation.

From Fig. 8.5, the radius to maximum wind can be clearly visualized. Furthermore, we notice that the wind field is not symmetrical since the largest wind speeds are occurring on the right side of the storm. This is because hurricanes in the Northern Hemisphere translate towards the north while rotating in a counterclockwise direction. Hence, winds on the right (eastern) side of the eye that are blowing towards the north will match the direction of storm translation, thus amplifying its velocity (i.e., the effects of the storm's rotation and forward motion are combined). Conversely, winds on the left (western) side of the eye that are blowing towards the south will tend to oppose the direction of forward motion, thereby reducing its velocity. The maximum velocity within the entire wind field was predicted to be 73 ms^{-1}, which deviates from the actual recorded maximum wind speed of 77 ms^{-1} by only 5%.

Evidently, calculating the maximum wind speed by constructing the entire wind field can be a rather laborious process. Instead, we can attempt to directly estimate the maximum 1-min averaged wind speed at R_m using Eq. (8.7):

$$U_{10,max} = 0.8\left[\frac{100B(p_n - p_0)}{\rho_a e}\right]^{1/2} + \frac{V_{fm}}{2} = 0.8\left[\frac{100 \times 2.15(1013 - 902)}{1.2 \times 2.718}\right]^{1/2} + \frac{5.4}{2}$$
$$= 71\,\mathrm{ms}^{-1}$$

which yields a result that is still very close to the measured wind speed.

8.5 Hurricane Wind Fetch

In addition to the maximum 1-min sustained wind speed given by Eq. (8.7), it is also of interest to obtain the fetch distance F_m associated with this maximum wind. This will allow us to use parametric wave spectra such as JONSWAP or TMA (see Chap. 6) to estimate the wave heights for a given storm. Using a large number of storm observations, Young [15] developed an empirical formulation to estimate the equivalent fetch associated with winds at the radius to maximum wind R_m:

$$F_m = R'\psi\left(aU_{10}^2 + bU_{10}V_{fm} + cV_{fm}^2 + dU_{10} + eV_{fm} + f\right) \tag{8.8}$$

where:

$$\begin{aligned} a &= -2.175 \times 10^{-3} \\ b &= 1.506 \times 10^{-2} \\ c &= -1.223 \times 10^{-1} \\ d &= 2.190 \times 10^{-1} \\ e &= 6.737 \times 10^{-1} \\ f &= 7.980 \times 10^{-1} \end{aligned} \tag{8.9}$$

and:

$$\begin{aligned} R' &= 22.5 \times 10^3 \log_{10} R_m - 70.8 \times 10^3 \\ \psi &= -0.015U_{10} + 0.0431V_{fm} + 1.30 \end{aligned} \tag{8.10}$$

Here, U_{10} is taken as the maximum 1-min averaged wind speed from Eq. (8.7). Using Hurricane Katrina as an example, Eq. (8.8) gives the fetch associated with $U_{10} = 71\,\text{ms}^{-1}$ predicted using Eq. (8.7) and $V_{fm} = 5.4\,\text{ms}^{-1}$ as $F_m = 130.7\,\text{km}$. We can now adopt the JONSWAP formulation (see Chap. 6) to plot the wave spectrum of the hurricane in deep water:

$$S(f) = \frac{\alpha g^2}{(2\pi)^4 f^5}\exp\left[-1.25\left(\frac{f_p}{f}\right)^4\right]\gamma^{\exp\left[-\frac{(f-f_p)^2}{2\sigma^2 f_p^2}\right]}, \quad \begin{aligned} f_p &= \frac{3.5g}{U_{10}}\left(\frac{gF}{U_{10}^2}\right)^{-0.33} \\ \alpha &= 0.076\left(\frac{gF}{U_{10}^2}\right)^{-0.22} \\ \sigma &= \begin{cases} 0.07, & f \le f_p \\ 0.09, & f > f_p \end{cases} \end{aligned} \tag{8.11}$$

when adopting $U_{10} = 71\,\text{ms}^{-1}$ and $F = 130{,}700\,\text{m}$, Eq. (8.11) produces the wave spectrum shown in Fig. 8.6. The area below this spectrum is then:

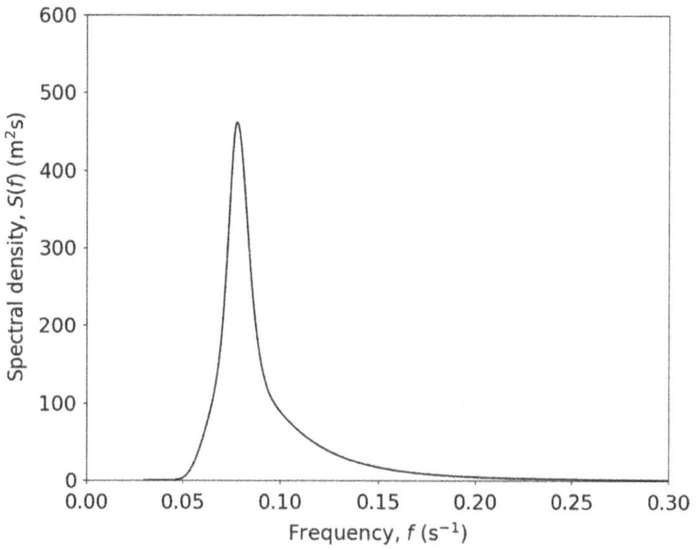

Fig. 8.6 JONSWAP spectrum for Hurricane Katrina at radius to maximum wind based on equivalent fetch

$$m_0 = \int_0^{\infty} S(f)df \tag{8.12}$$

which yields $m_0 = 11.4\,\text{m}^2$. The significant wave height can next be found using:

$$H_s = 4\sqrt{m_0} \tag{8.13}$$

giving $H_s = 13.5\,\text{m}$.

8.6 Hurricane Wave Spectrum

The JONSWAP wave spectrum in Fig. 8.6 was estimated based on the equivalent fetch of hurricane winds. However, the wave spectrum under hurricane conditions in deep water may also be directly determined via its wind speed U_{10} without the need to first calculate F_m. Empirical observations of hurricanes show that waves closely follow the JONSWAP form:

$$S(f) = \frac{\alpha g^2}{(2\pi)^4 f^5} \exp\left[-1.25\left(\frac{f_p}{f}\right)^4\right] \gamma^{\exp\left[-\frac{(f-f_p)^2}{2\sigma^2 f_p^2}\right]} \tag{8.14}$$

with $\sigma = 0.1$ and α and γ parameters are modified to become:

$$\alpha = 4.5H_s^2 f_p^4 \tag{8.15}$$

$$\gamma = 9.5H_s^{0.34} f_p \tag{8.16}$$

which were determined by Ochi [8] based on observations from NOAA wave buoys for hurricanes in close proximity to the United States. In addition, the significant wave height H_s was found to increase linearly with the wind speed U_{10}:

$$H_s = 0.235U_{10} \tag{8.17}$$

We now only require the peak frequency f_p associated with hurricane waves. When adopting the JONSWAP formation, a relationship exists between the peak frequency and the total wave energy:

$$\frac{g^2 m_0}{U_{10}^4} = 7.13 \times 10^{-6} \left(\frac{f_p U_{10}}{g} \right)^{-3.03} \tag{8.18}$$

where the zeroth moment of the wave spectrum m_0 can be directly calculated via the significant wave height determined using (8.17):

$$m_0 = \left(\frac{H_s}{4} \right)^2 \tag{8.19}$$

Substituting (8.19) into (8.18) and rearranging for f_p finally gives:

$$f_p = \frac{g}{U_{10}} \left(\frac{484g^2}{U_{10}^2} \right)^{-\frac{1}{3.03}} = 0.282U_{10}^{-0.34} \tag{8.20}$$

Hence, we may now directly determine the hurricane wave spectrum based on only the wind speed U_{10}. Figure 8.7 presents the JONSWAP spectrum obtained using Eq. (8.14) for Hurricane Katrina at the radius to maximum wind with the wind speed $U_{10} = 71\ \text{ms}^{-1}$ predicted via Eq. (8.7). A comparison is also shown with the spectrum predicted based on the equivalent hurricane fetch (shown in Fig. 8.6). Discrepancies are seen between the two methods because the equivalent fetch is sensitive to the forward speed V_{fm} and the radius to maximum wind R_m, while Eq. (8.14) is only a function of the wind speed U_{10}. In addition, instead of determining the significant wave height by integrating the wave spectrum for m_0 as done previously via (8.12), H_s is directly calculated from Eq. (8.17). When adopting $U_{10} = 71\ \text{ms}^{-1}$, we obtain $H_s = 16.7$ m. This result is around 24% higher than our H_s estimated using the equivalent fetch method. In addition, hurricane wave spectra in shallow water may also be obtained by multiplying the JONSWAP spectrum

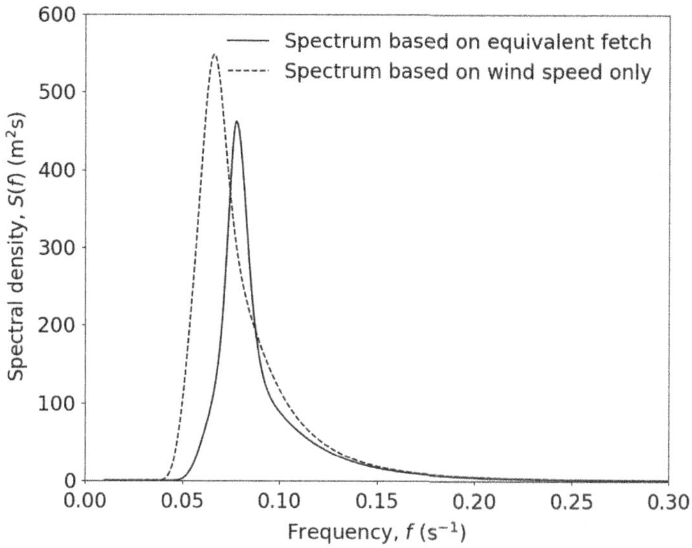

Fig. 8.7 Predicted JONSWAP spectrum for Hurricane Katrina at radius to maximum wind based on equivalent fetch versus spectrum obtained directly from wind speed

from (8.14) with the TMA scaling factor given by Eq. (6.38), as demonstrated by Wang et al. [14].

8.7 Summary: Hurricane Wave Spectrum at Radius to Maximum Wind

The procedure to predict the 1-min sustained wind speed $U_{10,max}$ and corresponding wave spectrum $S(f)$ at the radius to maximum wind R_m of a hurricane with known central pressure p_0 and forward velocity V_{fm} are summarized in Table 8.3.

Table 8.3 Calculation of hurricane wave spectrum at radius to maximum wind

Step 1	Determine B via (8.4). Note that p_0 has units of hPa	$B = 1.5 + \frac{980 - p_0}{120}$
Step 2	Determine $U_{10,max}$ via (8.7)	$U_{10,max} = 0.8\left[\frac{100B(p_n - p_0)}{\rho_a e}\right]^{1/2} + \frac{V_{fm}}{2}$ where $\rho_a = 1.2$ kg m^{-3}, $p_n = 1013$ hPa, and $e = 2.718$
Step 3	Determine H_s via (8.17)	$H_s = 0.235U_{10,max}$
Step 4	Determine f_p via (8.20)	$f_p = 0.282U_{10,max}^{-0.34}$
Step 5	Determine α and γ via (8.15) and (8.16)	$\alpha = 4.5H_s^2 f_p^4, \quad \gamma = 9.5H_s^{0.34} f_p$
Step 6	Determine $S(f)$ via (8.14)	$S(f) = \frac{\alpha g^2}{(2\pi)^4 f^5}\exp\left[-1.25\left(\frac{f_p}{f}\right)^4\right]\gamma^{\exp\left[-\frac{(f-f_p)^2}{2\sigma^2 f_p^2}\right]}$ where $\sigma = 0.1$

References

1. Cavicchia, L., von Storch, H., Gualdi, S., 2013. A long-term climatology of medicanes. Climate Dynamics 43, 1183–1195.
2. Chavas, D.R., Knaff, J.A., 2022. A Simple Model for Predicting the Tropical Cyclone Radius of Maximum Wind from Outer Size. Weather and Forecasting 37, 563–579.
3. Holland, G.J., 1980. An analytic model of the wind and pressure profiles in hurricanes. Monthly Weather Review 108, 1212–1218.
4. Hsu, S.A., Yan, Z., 1998. A note on the radius of maximum wind for hurricanes. Journal of Coastal Research 14, 667–668.
5. Knabb, R.D., Rhome, J.R., Brown, D.P., 2023. Tropical cyclone report: Hurricane Katrina. National Hurricane Center.
6. Muhlbaum, D., 2025. The Most Expensive Natural Disasters in US History. Kiplinger. https://www.kiplinger.com/slideshow/business/t019-s001-most-expensive-natural-disasters-in-u-s-history/index.html.
7. NOAA National Centers for Environmental Information (NCEI), 2025. U.S. Billion-Dollar Weather and Climate Disasters (2025). https://www.ncei.noaa.gov/access/billions, https://doi.org/10.25921/stkw-7w73
8. Ochi, M.K., 2003. Hurricane-generated seas. Elsevier, Kidlington, Oxford.
9. Pérez-Alarcón, A., Fernández-Alvarez, J.C., Coll-Hidalgo, P., 2023. Global Increase of the Intensity of Tropical Cyclones under Global Warming Based on their Maximum Potential Intensity and CMIP6 Models. Environmental Processes. 10, 36.

10. Rogers, R.F., Aberson, S., Bell, M.M., Cecil, D.J., Doyle, J.D., Kimberlain, T.B., Morgerman, J., Shay, L.K., Velden, C., 2017. Rewriting the Tropical Record Books: The Extraordinary Intensification of Hurricane Patricia (2015). Bulletin of the American Meteorological Society 98, 2091–2112.
11. Shea, D.J., Gray, W.M., 1973. The hurricane's inner core region. I. Symmetric and asymmetric structure. Journal of Atmospheric Sciences 30, 1544–1564.
12. Takagi, H., Wu, W., 2016. Maximum wind radius estimated by the 50kt radius: improvement of storm surge forecasting over the western North Pacific. Natural Hazards and Earth System Sciences 16, 705–717.
13. Vickery, P.J., Skerlj, P.F., Twisdale, L.A., 2000. Simulation of hurricane risk in the U.S. using empirical track model. Journal of Structural Engineering 126, 1222–1237.
14. Wang, S., Garlock, M., Glisic, B., 2021. Parametric modeling of depth-limited wave spectra under hurricane conditions with applications to Kinetic Umbrellas against storm surge inundation. Water 13.
15. Young, I.R., 2003. A review of the sea state generated by hurricanes. Marine Structures 16, 201–218.

Wave Shoaling and Breaking

9

In our discussion of wave energetics (Chap. 4), we proved that waves slow down as they enter shallower water while increasing their height, a phenomenon referred to as wave shoaling. This increase in wave height is due to the increase in energy density since the energy flux shown in Eq. (4.47) must remain constant. Once this height reaches a critical value, the crest collapses back into the water in a process known as wave breaking. Understanding the extent to which deep-water waves increase their height as they propagate into coastal regions and the conditions by which breaking occurs is important for the design of many types of coastal structures. This is because when designing such structures to resist wave forces or overtopping, local wave heights at the location of the structure must be employed when carrying out the required calculations.

9.1 Wave Shoaling

We wish to determine the significant wave height H_s of waves on a sloping beach at any given water depth h. The input conditions will be the deep-water wave height (H_0') and wavelength (L_0) as shown in Fig. 9.1. To calculate the change in wave height due to shoaling, recall that the energy flux from Eq. (4.47) is:

$$(\tilde{E}_k + \tilde{E}_p)c_g = \text{Constant} \tag{9.1}$$

where c_g is the group velocity and:

$$\tilde{E}_k + \tilde{E}_p = \frac{\rho g H^2}{8} \tag{9.2}$$

S. Wang, *Applied Ocean Mechanics for Structural Engineering and Climate Adaptation*, Synthesis Lectures on Ocean Systems Engineering,
https://doi.org/10.1007/978-3-032-18199-2_9

is the wave energy density (i.e., energy per unit area) which is a function of the wave height H. Since the energy flux must remain constant for all water depths, we can equate the flux for waves in deep and shallow water:

$$\frac{\rho g H_0'^2}{8} c_{g,0} = \frac{\rho g H^2}{8} c_g \tag{9.3}$$

where H_0' and $c_{g,0}$ respectively denote the wave height and group velocity of waves in deep water. This expression can be rearranged to form:

$$\frac{H}{H_0'} = \sqrt{\frac{c_{g,0}}{c_g}} = K_{si} \tag{9.4}$$

where K_{si} is referred to as the "shoaling coefficient" for linear waves. From Table 3.2, we know that the group velocity of deep-water waves is:

$$c_{g,0} = \frac{1}{2}\sqrt{\frac{g}{k}} = \frac{1}{2}\sqrt{\frac{g^2}{\omega^2}} = \frac{g}{2\omega} \tag{9.5}$$

where $k = \omega^2/g$ is the deep-water wavenumber. In addition, the group velocity at any arbitrary depth h is:

$$c_g = \frac{g}{2\omega}\left[\tanh(kh) + \frac{kh}{\cosh^2(kh)}\right] \tag{9.6}$$

Substituting Eqs. (9.5) and (9.6) into (9.4) thus yields the shoaling coefficient K_{si}:

$$K_{si} = \frac{H}{H_0'} = \frac{1}{\sqrt{\tanh(kh) + \frac{kh}{\cosh^2(kh)}}} \tag{9.7}$$

where $k = 2\pi/L_0$. Note that for shoaling calculations, the deep-water wavelength L_0 is determined using the significant wave period T_s:

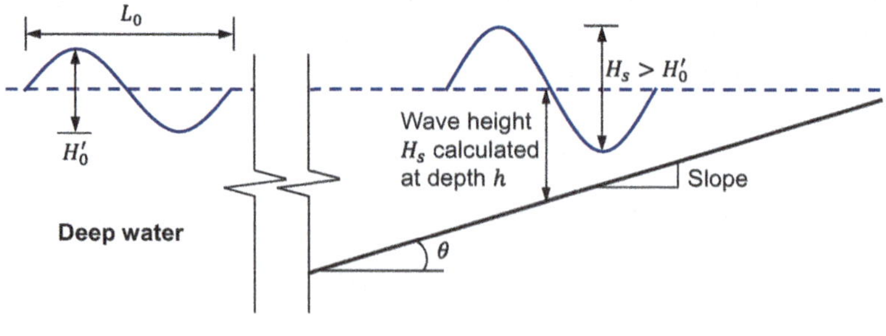

Fig. 9.1 Wave shoaling over a sloping beach

$$L_0 = \frac{gT_s^2}{2\pi} \tag{9.8}$$

The significant wave height (H_s) at any depth h can now be determined by multiplying the deep-water significant wave height ($H_{s,0} = H_0'$ in the absence of refraction or diffraction effects) with the shoaling coefficient. However, it is important to understand that the K_{si} from Eq. (9.7) is valid only for linear waves with small amplitude. The generalized shoaling coefficient K_s that can be used for all waves (linear and nonlinear) should be determined using Fig. 9.2 which was adapted from Goda [3]. Here, K_s is determined by matching the curve for H_0'/L_0 with the h/L_0 value from the x-axis, while making sure that K_s does not exceed the breaking limit for the associated slope. The contribution of shoaling is also influenced by the h/L_0 ratio and the slope of nearshore bathymetry θ. Goda [2] gives the significant wave height after shoaling at a depth of h as:

$$H_s = \begin{cases} K_s H_0', & h/L_0 \geq 0.2 \\ \min\left[\left(\beta_0 H_0' + \beta_1 h\right), \beta_{max} H_0', K_s H_0'\right], & h/L_0 < 0.2 \end{cases} \tag{9.9}$$

where the β coefficients are:

$$\beta_0 = 0.028\left(\frac{H_0'}{L_0}\right)^{-0.38} \exp\left(20 \tan^{1.5}\theta\right) \tag{9.10}$$

$$\beta_1 = 0.52 \exp(4.2 \tan\theta) \tag{9.11}$$

$$\beta_{max} = \max\left[0.92, 0.32\left(\frac{H_0'}{L_0}\right)^{-0.29} \exp(2.4 \tan\theta)\right] \tag{9.12}$$

with θ denoting the slope angle. The procedure used to determine the significant wave height of waves after shoaling is summarized in Table 9.1.

9.2 Wave Breaking

Waves propagating towards the shoreline will continue to increase in height due to shoaling. This growth is not indefinite and at some critical height waves will break, resulting in the release of their energy and momentum. If not properly accounted for, wave breaking can result in the failure of coastal structures, of which many examples have been documented [5]. Given their importance to coastal design, it is of great interest to predict the wave height at which breaking occurs in the proximity of nearshore structures. The region between the shoreline and where waves first begin to break is referred to as the "surf zone", with its width being highly dependent on local wave conditions and nearshore

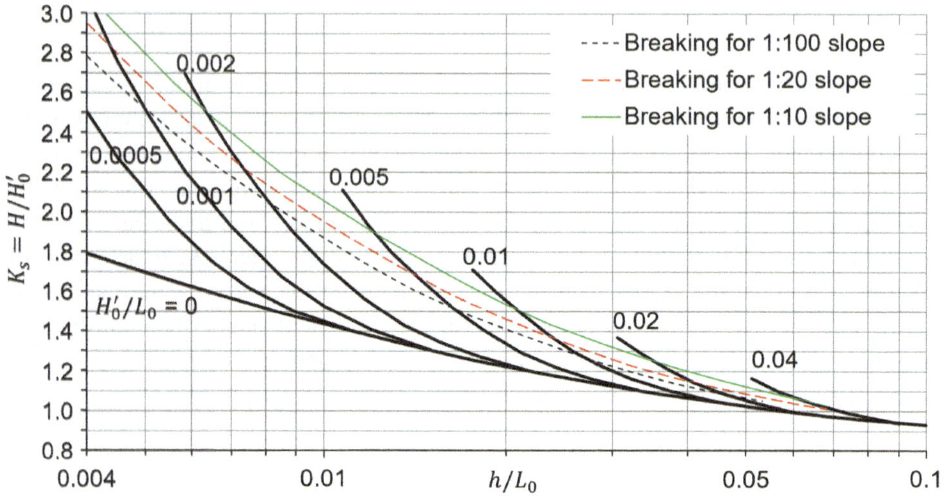

Fig. 9.2 Shoaling coefficient for nonlinear waves

Table 9.1 Determination of wave height due to shoaling at a local water depth h

Step 1	Determine the deep-water wavelength L_0 per (9.8)	$L_0 = \frac{gT_s^2}{2\pi}$
Step 2	Determine β_0, β_1, and β_{max} via (9.10)–(9.12)	$\beta_0 = 0.028\left(\frac{H_0'}{L_0}\right)^{-0.38} \exp\left(20 \tan^{1.5} \theta\right)$ $\beta_1 = 0.52 \exp(4.2 \tan \theta)$ $\beta_{max} = \max\left[0.92, 0.32\left(\frac{H_0'}{L_0}\right)^{-0.29} \exp(2.4 \tan \theta)\right]$
Step 3	Determine K_s via Fig. 9.2	Compute h/L_0 and H_0'/L_0 first, then read off plot to obtain K_s from the y-axis
Step 4	Determine H_s via (9.9)	$H_s = \begin{cases} K_s H_0', & h/L_0 \geq 0.2 \\ \min\left[(\beta_0 H_0' + \beta_1 h), \beta_{max} H_0', K_s H_0'\right], & h/L_0 < 0.2 \end{cases}$

bathymetry. Storms that produce larger waves will cause the surf zone to extend a considerable distance offshore while calmer weather may see this zone resemble a narrow band close to the shoreline (e.g., Fig. 9.3).

In shallow water, the breaking wave height H_b can generally be approximated as $H_b = 0.78h$ if the seabed slope is shallower than 1:50 (i.e., $\theta < 1.15°$), where h is the local water depth. For other slope angles θ, Goda [2] proposed an alternative formula to estimate H_b

Fig. 9.3 Surf zone near a beach in Tauranga, New Zealand

at any given depth h. The breaking wave height within the surf zone is given as:

$$H_b = \begin{cases} 0.78h, & \theta \leq 1.15^\circ \\ A_b L_0 \left\{ 1 - \exp\left[-\frac{1.5\pi h}{L_0} \left(1 + 15 \tan^{4/3} \theta \right) \right] \right\}, & \theta > 1.15^\circ \end{cases} \tag{9.13}$$

where L_0 is the deep-water wavelength and A_b is an empirical parameter that takes a value of 0.17 for regular waves and is between 0.12 and 0.18 for irregular waves in order to account for the wide distribution of wave frequencies. When using Eq. (9.13), the wavelength L_0 can be calculated using linear wave theory assuming deep water conditions (see Table 3.2) via:

$$L_0 = \frac{gT_p^2}{2\pi} \tag{9.14}$$

where T_p is the peak period. If using the JONSWAP spectrum, this peak period can be estimated directly from the significant wave period T_s following Eq. (6.32):

$$T_p \approx \frac{T_s}{1 - 0.132(\gamma + 0.2)^{-0.559}} \tag{9.15}$$

where $\gamma = 3.3$ is often adopted as the JONSWAP peak enhancement factor. If no information regarding T_s is available, recall that the significant wave period can also be estimated using:

$$T_s \approx 3.3 H_s^{0.63} \tag{9.16}$$

where H_s is the deep-water significant wave height. Figure 9.4 illustrates the transformation of a wave in deep water into a breaking wave as it traverses over a sloping shoreline. Here, h_s is used to denote the water depth where breaking occurs.

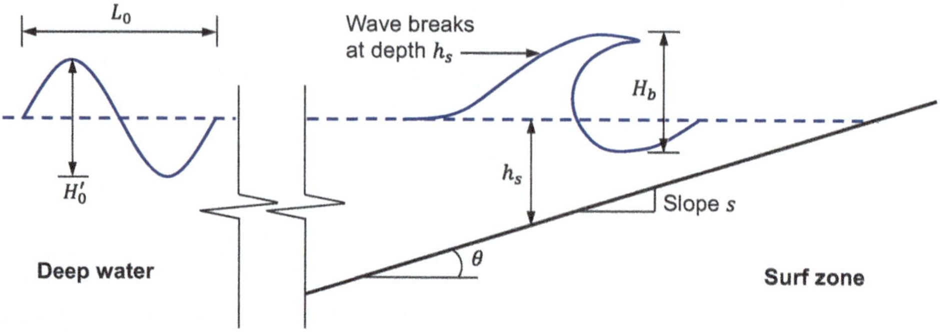

Fig. 9.4 Schematic of wave breaking in the surf zone

The breaking wave height can also be calculated from the equivalent deep water wave height H_0' through the H_b/H_0' ratio, referred to as the "breaker height index". This equivalent deep-water wave height can be determined from the significant wave height H_s via:

$$H_0' = K_r K_d H_s \tag{9.17}$$

where K_r and K_d are coefficients of refraction and diffraction, respectively. Refraction refers to the change in wave direction as they travel from deeper to shallower water (due to the decrease in wave velocity), which occurs when waves approach a sloping beach from a non-perpendicular direction. Diffraction refers to the change in wave direction as they encounter an object in its path, such as a protruding rock. Hence, in the absence of refraction or diffraction effects, the equivalent deep water wave height H_0' can simply be set to the significant wave height H_s. This knowledge of H_0' can then be used to determine the breaking wave height H_b as a function of the slope and peak period T_p via Fig. 9.5, which has been adapted from Goda [4]. Once H_b has been determined, the depth at which breaking occurs can also be estimated by rearranging Eq. (9.13) in terms of h:

$$h_s = -\frac{2L_0}{3\pi\left(1 + 15\tan^{4/3}\theta\right)}\ln\left(1 - \frac{H_b}{A_b L_0}\right) \tag{9.18}$$

where we have replaced h in Eq. (9.13) with h_s to represent the water depth at wave breaking. Recall that for irregular waves, A_b takes a value between 0.12 and 0.18. As such, two water depths will be obtained for irregular waves, with the larger and smaller depths given by $A_b = 0.12$ and 0.18, respectively. The region between these two depths is referred to as the "outer surf zone", with waves experiencing the most rapid change in their shape due to breaking. The region between the outer surf zone and the shoreline is the "inner surf zone" which largely comprises broken waves traveling towards the shore as a turbulent bore. These two regions collectively constitute the entire surf zone (also called

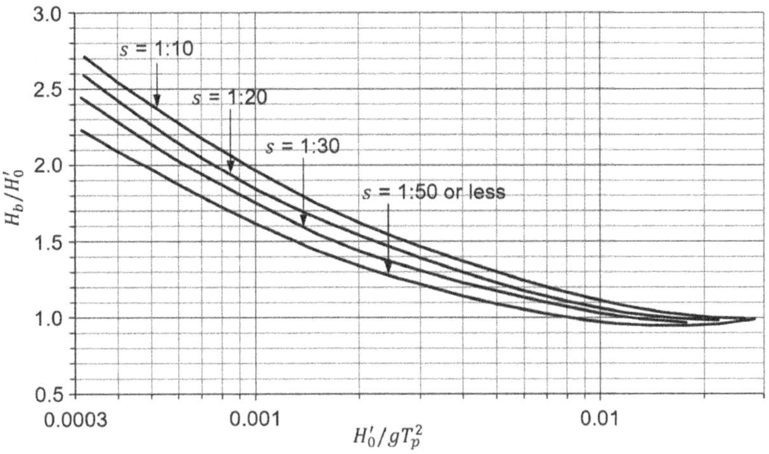

Fig. 9.5 Breaker height index for different slopes

Table 9.2 Determination of breaking wave height at a local water depth h

Step 1	Determine the beach slope angle θ	
Step 2	If $\theta \leq 1.15°$, calculate H_b from the local water depth h	$H_b = 0.78h$
Step 3	If $\theta > 1.15°$, determine the deep-water wavelength L_0 per (9.14)	$L_0 = \frac{gT_p^2}{2\pi}$
Step 4	Next, determine H_b via the second equation in (9.13)	$A_b L_0\left\{1 - \exp\left[-\frac{1.5\pi h}{L_0}\left(1 + 15\tan^{4/3}\theta\right)\right]\right\}$ where: $A_b = \begin{cases} 0.17, & \text{Regular waves} \\ 0.12 \text{ to } 0.18, & \text{Irregular waves} \end{cases}$

the "breaker zone") and is characterized by highly turbulent flows and intense sediment transport [1]. On the other hand, for a given water depth h, the procedure to determine the wave height that breaking will occur (H_b) is summarized in Table 9.2.

9.3 Example: Wave Shoaling and Breaking on a Sloping Beach

Let us consider an example where a storm generates waves in deep water with significant wave height and period of $H_0' = 4.5$ m and $T_s = 12$ s, respectively. The waves are propagating towards a beach with a constant slope of 1:10 ($\theta = 5.71°$). For this exercise, we are interested in the following:

- The significant wave height H_s due to shoaling when the water depth is $h = 9$ m
- The height of breaking waves H_b and the location where they are expected to occur (i.e., the surf zone) relative to the shoreline.

9.3.1 Wave Shoaling

To determine the wave height after shoaling, we must first obtain the deep-water wavelength. Equation (9.8) gives:

$$L_0 = \frac{gT_s^2}{2\pi} = \frac{9.81 \times 12^2}{2\pi} = 225 \text{ m}$$

which corresponds to a deep-water steepness of:

$$\frac{H_0'}{L_0} = \frac{4.5}{225} = 0.02$$

while the h/L_0 ratio is:

$$\frac{h}{L_0} = \frac{9}{225} = 0.04$$

From Fig. 9.2, the shoaling coefficient is approximately $K_s = 1.15$. The β parameters are next determined:

$$\begin{aligned}
\beta_0 &= 0.028\left(\frac{H_0'}{L_0}\right)^{-0.38} \exp\left(20 \tan^{1.5} \theta\right) \\
&= 0.028\left(\frac{4.5}{225}\right)^{-0.38} \exp\left[20 \tan^{1.5}(5.71)\right] = 0.233
\end{aligned}$$

$$\beta_1 = 0.52 \exp(4.2 \tan \theta) = 0.52 \exp\left(4.2 \tan 5.71°\right) = 0.791$$

$$\beta_{max} = \max \begin{cases} 0.92 \\ 0.32\left(\frac{H_0'}{L_0}\right)^{-0.29} \exp(2.4 \tan \theta) = 0.32\left(\frac{4.5}{225}\right)^{-0.29} \exp(2.4 \tan 5.71°) = 1.26 \end{cases}$$

$$\beta_{max} = 1.26$$

Since $h/L_0 = 0.04 < 0.2$, we must use the second equation in (9.9) to determine H_s:

$$H_s = \min \begin{cases} \beta_0 H_0' + \beta_1 h = 0.233 \times 4.5 + 0.791 \times 9 = 8.17 \text{ m} \\ \beta_{max} H_0' = 1.26 \times 4.5 = 5.67 \text{ m} \\ K_s H_0' = 1.15 \times 4.5 = 5.18 \text{ m} \end{cases}$$

The significant wave height after shoaling is therefore $H_s = 5.18$ m at a depth of $h =$ 9 m.

9.3.2 Wave Breaking

We now need to estimate the breaking wave height H_b and the corresponding water depth h_s that wave breaking is expected to occur. To do so, we first need to find the deep-water wavelength L_0 based on the peak period T_p. Since the peak period has not been explicitly provided, we can approximate this value as:

$$\begin{aligned} T_p &\approx \frac{T_s}{1 - 0.132(\gamma + 0.2)^{-0.559}} \\ &= \frac{12}{1 - 0.132(3.3 + 0.2)^{-0.559}} = 12.8 \text{ s} \end{aligned}$$

when using $\gamma = 3.3$ as the JONSWAP peak enhancement factor. The deep-water wavelength is then:

$$L_0 = \frac{gT_p^2}{2\pi} = \frac{9.81 \times 12.8^2}{2\pi} = 256 \text{ m}$$

Figure 9.5 can now be used to determine the breaker height index when assuming $H_0' = H_s = 4.5$ m. First calculating H_0'/gT_p^2:

$$\frac{H_0'}{gT_p^2} = \frac{4.5}{9.81 \times 12.8^2} = 0.0028$$

When using the curve for a 1:10 slope, the breaker height index is approximately 1.5. The height of the breaking wave is thus:

$$H_b = 1.5H_0' = 1.5 \times 4.5 = 6.75 \text{ m}$$

We can now use Eq. (9.18) to estimate the water depth when breaking occurs. Since we are dealing with irregular waves, the parameter A_b will be between 0.12 and 0.18. This means that the water depth at breaking will exhibit an upper bound (when $A_b = 0.12$) of:

$$\begin{aligned} h_s &= -\frac{2L_0}{3\pi\left(1 + 15\tan^{4/3}\theta\right)} \ln\left(1 - \frac{H_b}{A_b L_0}\right) \\ &= -\frac{2 \times 256}{3\pi\left[1 + 15\tan^{4/3}(5.71°)\right]} \ln\left(1 - \frac{6.75}{0.12 \times 256}\right) \\ &= 7.9 \text{ m} \end{aligned}$$

and a lower bound (when $A_b = 0.18$) of:

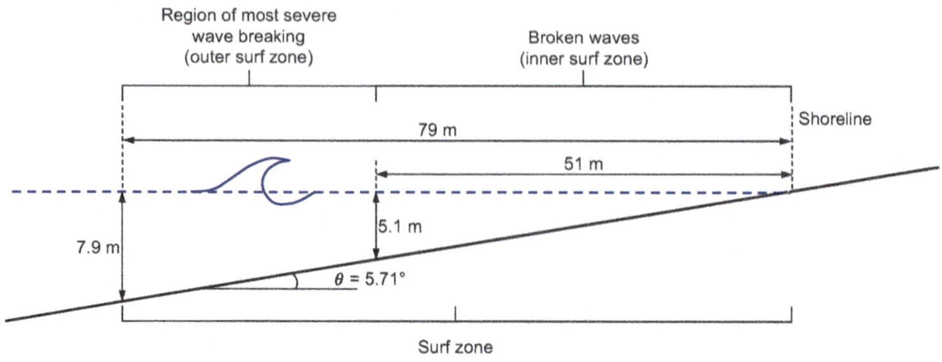

Fig. 9.6 Zone of wave breaking on a 1:10 sloping beach for $H_0' = 4.5$ m and $T_s = 12$ s

$$
\begin{aligned}
h_s &= -\frac{2L_0}{3\pi\left(1 + 15\tan^{4/3}\theta\right)}\ln\left(1 - \frac{H_b}{A_b L_0}\right) \\
&= -\frac{2 \times 256}{3\pi\left[1 + 15\tan^{4/3}(5.71°)\right]}\ln\left(1 - \frac{6.75}{0.18 \times 256}\right) \\
&= 5.1 \text{ m}
\end{aligned}
$$

Since the beach has a constant slope angle of $\theta = 5.71°$, these water depths correspond to a distance of $7.9/\tan(5.71°) = 79$ m and $5.1/\tan(5.71°) = 51$ m away from the shoreline. Waves for this particular sea state would be expected to start breaking within this zone as shown in Fig. 9.6.

To summarize our findings, deep-water waves with a significant wave height of $H_0' = 4.5$ m traveling towards the shore gradually increase in height due to the reduction in water depth h. At a depth of 9 m (or 90 m from the shoreline), the significant wave height has grown to approximately $H_s = 5.18$ m. As these waves approach within 79 m of the shoreline, breaking commences with a height of $H_b = 6.75$ m and waves experience the most rapid changes in their profile. The breaking process continues up until 51 m from the shore within the outer surf zone. As waves move past this region into the inner surf zone, any changes in shape become more gradual as broken waves approach the shoreline in the form of a turbulent bore.

References

1. Aagaard, T., Brinkkemper, J., Christensen, D.F., Hughes, M.G., Ruessink, G., 2021. Surf Zone Turbulence and Suspended Sediment Dynamics—A Review. Journal of Marine Science and Engineering 9, 1300.
2. Goda, Y., 2000. Random seas and design of maritime structures. World Scientific Publishing Co., Singapore.

3. Goda, Y., 1975. Deformation of irregular waves due to depth-controlled wave breaking. Report of the Port and Harbour Research Institute 14, 59–106.
4. Goda, Y., 1970. A synthesis of breaker indices. Transactions of the Japanese Society of Civil Engineers 2, 39–49.
5. Oumeraci, H., 1994. Review and analysis of vertical breakwater failures - lessons learned. Coastal Engineering 22, 3–29.

Part III

Wave Interactions with Fixed Structures

10 Hydrostatic Forces

Our discussions thus far have focused on the fundamental mechanics of surface gravity waves. We are now ready to apply this knowledge towards the analysis of engineering structures interacting with the ocean environment. However, before introducing pressures and forces exerted on structures by fluids in motion (hydrodynamics), this chapter will first provide a brief overview of pressures and forces resulting from fluids at rest (hydrostatics).

10.1 Derivation of Hydrostatic Principles

Recall that the Navier–Stokes momentum equations (along x, y, and z coordinates) for incompressible and inviscid fluids are:

$$\rho\frac{D\boldsymbol{u}}{Dt} = -\nabla p + \rho\boldsymbol{g} \rightarrow \begin{cases} \rho\left(\frac{\partial u}{\partial t} + u\frac{\partial u}{\partial x} + v\frac{\partial u}{\partial y} + w\frac{\partial u}{\partial z}\right) = -\frac{\partial p}{\partial x} \\ \rho\left(\frac{\partial v}{\partial t} + u\frac{\partial v}{\partial x} + v\frac{\partial v}{\partial y} + w\frac{\partial v}{\partial z}\right) = -\frac{\partial p}{\partial y} \\ \rho\left(\frac{\partial w}{\partial t} + u\frac{\partial w}{\partial x} + v\frac{\partial w}{\partial y} + w\frac{\partial w}{\partial z}\right) = -\frac{\partial p}{\partial z} - \rho g \end{cases} \quad (10.1)$$

When a fluid is at rest, its velocity everywhere and at all times must equal zero. Therefore, all velocity terms (u, v, and w) and their derivatives in time and space (e.g., $\partial u/\partial t$ and $\partial u/\partial x$) must equal zero and Eq. (10.1) becomes:

S. Wang, *Applied Ocean Mechanics for Structural Engineering and Climate Adaptation*, Synthesis Lectures on Ocean Systems Engineering,
https://doi.org/10.1007/978-3-032-18199-2_10

$$\frac{\partial p}{\partial x} = 0$$
$$\frac{\partial p}{\partial y} = 0 \tag{10.2}$$
$$\frac{\partial p}{\partial z} = -\rho g$$

The first two equations in (10.2) state that the pressure does not change along the horizontal (x and y) directions at any fixed depth for a body of fluid at rest. If we integrate the third equation with respect to the depth z, then set the constant of integration to zero (since there is no pressure at $z = 0$), we obtain:

$$p = p_h = -\rho g z \tag{10.3}$$

which states the hydrostatic pressure (p_h) varies linearly with depth z. Note that the negative sign exists on the right-hand side of (10.3) since we assign $z = 0$ at the free surface. Since z will be negative at any point below the surface, the hydrostatic pressure p will always be positive. A positive pressure means that it is acting "inward normal", which implies a direction oriented towards the contacted surface (i.e., the pressure is "pushing" on the surface). While pressure is itself a scalar quantity (i.e., it only has a magnitude without any direction), it can be converted into a force vector via the integration over its contacted surface since force = pressure × area. The direction of this pressure force will then be normal to and directed towards this surface.

Technically speaking, the atmospheric pressure (101.3 kPa at sea level) due to the weight of the air at the free surface should also be included as part of the total hydrostatic pressure. However, we will not explicitly account for this term (since it is a constant everywhere) and treat $p_h = 0$ when only the atmospheric pressure is present at the free surface where there is no water pressure. Note that this chapter has used p_h to denote the hydrostatic pressure in order to differentiate it from p, which represents the dynamic pressure induced only by fluids in motion. Therefore, to obtain the total pressure acting on a surface, the hydrostatic pressure $p_h = -\rho g z$ must be added to the dynamic pressure p. This detail will become important when dealing with wave forces acting on coastal structures.

To summarize, hydrostatic pressures p_h always act inward normal to a surface in contact with the fluid with a magnitude equal to ρg multiplied by the vertical distance to the free surface. The resulting pressure force will have a magnitude equal to the pressure multiplied by the contacted surface area.

10.2 Hydrostatics of Submerged Bodies

Figure 10.1 shows the application of hydrostatic pressure on a submerged object. The magnitude of the pressure at any point on the object is simply equal to $p_h = \rho g d$, where d is the depth below the surface. As previously mentioned, this pressure will be acting inward normal (i.e., perpendicular) to any given surface. These pressures will result in an upwards buoyancy force acting on the submerged body.

10.2.1 Buoyancy Force on Rectangular Bodies

Consider an object with dimensions d_x, d_y, and d_z along the x, y, and z directions, respectively, as shown in Fig. 10.1. The pressure force on its top surface located at $z = -d$ will be acting downwards while on the bottom surface $z = -(d + d_z)$ it will be acting upwards. These pressures are respectively expressed as $\rho g d$ and $\rho g(d + d_z)$. A net upwards force will therefore be produced with a magnitude:

$$F_z = \rho g(d + d_z) \times d_x d_y - \rho g d \times d_x d_y \tag{10.4}$$

where $d_x d_y$ is the area of the top and bottom surface (note that d_y is oriented into the page). The equation can be expanded into:

$$F_z = \rho g d \times d_x d_y + \rho g d_z \times d_x d_y - \rho g d \times d_x d_y \tag{10.5}$$

giving:

$$F_z = \rho g d_x d_y d_z = \rho g V \tag{10.6}$$

where $V = d_x d_y d_z$ is the volume of the submerged object. Equation (10.6) is known as Archimedes' principle which states that an object fully or partially submerged in a fluid

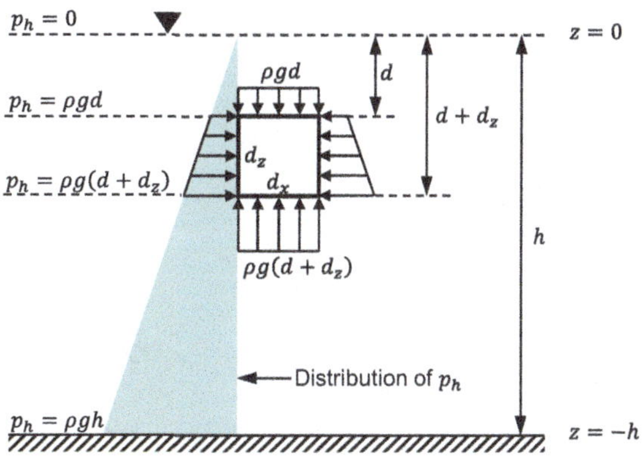

Fig. 10.1 Hydrostatic pressure on a prismatic submerged body with side lengths d_x, d_y, and d_z

will experience an upwards buoyancy force F_z equal to the weight of the displaced fluid $\rho g V$, where ρ is the fluid's density.

10.2.2 Buoyancy Force on Arbitrary Bodies

Strictly speaking, the proof described in Eqs. (10.4)–(10.6) is valid only for rectangular bodies (e.g., Fig. 10.1). However, we can also demonstrate that Archimedes' principle applies for bodies of any arbitrary geometry such as that shown in Fig. 10.2. To prove this, we first acknowledge that the resultant force vector $\boldsymbol{F}$ produced by a fluid in contact with a body in 3D is equal to the inward normal hydrostatic pressure integrated over the entire surface S (i.e., sum of all pressure forces acting on the surface).

The resultant force $\boldsymbol{F}$ can be mathematically expressed as:

$$\boldsymbol{F} = \iint_S \boldsymbol{\sigma} \cdot \boldsymbol{n} dS \tag{10.7}$$

where the double integral represents integration over a surface and $\boldsymbol{n}$ is the unit normal vector (a vector perpendicular to the surface at any given point with a magnitude of 1). Here, $\boldsymbol{\sigma}$ denotes the Cauchy stress tensor, being a 3 × 3 matrix where normal stresses in x, y, and z are represented by the main diagonal and shear stresses are represented by the off-diagonals. Under hydrostatic conditions, only normal stresses exist, and the Cauchy stress tensor is:

$$\boldsymbol{\sigma} = -p_h \begin{bmatrix} 1 & 0 & 0 \\ 0 & 1 & 0 \\ 0 & 0 & 1 \end{bmatrix} = \begin{bmatrix} -p_h & 0 & 0 \\ 0 & -p_h & 0 \\ 0 & 0 & -p_h \end{bmatrix} \tag{10.8}$$

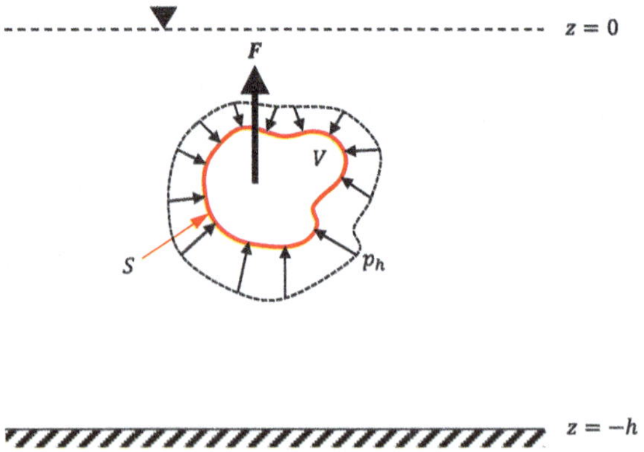

Fig. 10.2 Hydrostatic pressure on an arbitrary submerged body with surface S and volume V

which implies that the hydrostatic pressure from Eq. (10.3) exists as a purely normal stress that acts everywhere on the body. Note that the sign convention for the Cauchy stress tensor is such that positive normal stresses act outwards from the surface (i.e., tension). This is opposite to the convention we have adopted for fluid pressures (positive = inwards); hence we must use $-p_h$ to ensure that hydrostatic stresses are always acting inwards towards the surface (i.e., compression). We can re-express Eq. (10.7) using Gauss's theorem, also known as the divergence theorem, which states that the surface integral of a vector field over a closed surface is equal to the volume integral of the divergence over the region enclosed by the surface [1]. Mathematically, Gauss's theorem states that:

$$\boldsymbol{F} = \iint_S \boldsymbol{\sigma} \cdot \boldsymbol{n} dS = \iiint_V \nabla \cdot \boldsymbol{\sigma} dV \tag{10.9}$$

where V is the volume of the body and the triple integral represents the fact that we are integrating over a volume in 3D. From Eq. (1.30), the divergence of $\boldsymbol{\sigma}$ is:

$$\nabla \cdot \boldsymbol{\sigma} = \begin{Bmatrix} \frac{\partial}{\partial x} \\ \frac{\partial}{\partial y} \\ \frac{\partial}{\partial z} \end{Bmatrix} \cdot \begin{bmatrix} \rho g z & 0 & 0 \\ 0 & \rho g z & 0 \\ 0 & 0 & \rho g z \end{bmatrix} = \begin{Bmatrix} 0 \\ 0 \\ \rho g \end{Bmatrix} \tag{10.10}$$

with $-p_h = \rho g z$ per Eq. (10.3). Substituting (10.10) into (10.9) gives:

$$\boldsymbol{F} = \iiint_V \nabla \cdot \boldsymbol{\sigma} dV = \iiint_V \begin{Bmatrix} 0 \\ 0 \\ \rho g \end{Bmatrix} dV = \begin{Bmatrix} 0 \\ 0 \\ \iiint_V \rho g dV \end{Bmatrix} = \begin{Bmatrix} 0 \\ 0 \\ \rho g V \end{Bmatrix} \tag{10.11}$$

Equation (10.11) states that the resultant force vector $\boldsymbol{F}$ is acting upwards (in the direction of positive z) with magnitude equal to the weight of the displaced fluid $\rho g V$. This result is equivalent to Eq. (10.6), thereby proving that Archimedes' principle applies irrespective of the body's geometry.

10.3 Hydrostatics of Surface-Piercing Bodies

The determination of hydrostatic pressure on surface-piercing bodies is crucial to the design of many types of structures for coastal resilience. Common examples include seawalls or flood barriers which often carry large hydrostatic pressures in addition to hydrodynamic forces due to wave action (to be discussed in Chap. 11). Figure 10.3 presents the distribution of hydrostatic pressure acting on a wall inclined at an angle θ to the horizontal. The area of the triangular pressure distribution, increasing linearly from 0 to $\rho g h$, gives the force per unit length of wall (F) with units of N/m. Multiplying F with the wall's length (L) yields the total hydrostatic force in N. As can be seen, F

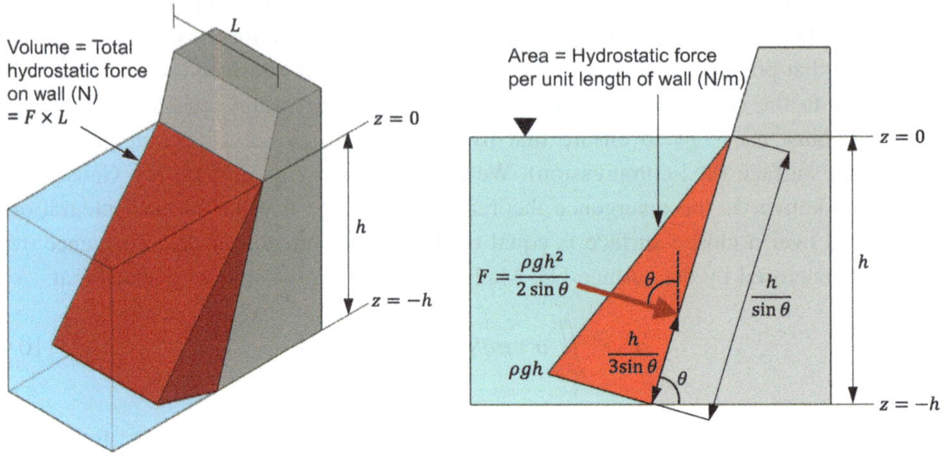

Fig. 10.3 Hydrostatic pressure distribution on an inclined wall

acts perpendicular to the surface though the geometric centroid of the triangular pressure distribution, being 1/3 of its height from the base. The total force per unit length acting inward normal to the wall is thus:

$$F = \rho g h \times \frac{h}{\sin\theta} \times \frac{1}{2} = \frac{\rho g h^2}{2\sin\theta} \tag{10.12}$$

This force resultant acts at a distance of $h/3\sin\theta$ from the base of the wall with an inclination of θ from the vertical as shown in Fig. 10.3.

The hydrostatic force given in Eq. (10.12) can be further discretized into its horizontal and vertical components, respectively denoted F_x and F_z. The horizontal component can be computed from the hydrostatic pressure acting on a surface that is perpendicular to the horizontal direction (i.e., a vertical surface with height h), shown in Fig. 10.4, giving:

$$F_x = \rho g h \times h \times \frac{1}{2} = \frac{\rho g h^2}{2} \tag{10.13}$$

which acts at an elevation of $h/3$ from the seabed. The vertical component is equal to the total weight of the fluid above the toe of the wall:

$$F_z = -\rho g \times h \times \frac{h}{\tan\theta} \times \frac{1}{2} = -\frac{\rho g h^2}{2\tan\theta} \tag{10.14}$$

where the minus sign signifies that it is acting in a downward direction. Since the fluid area above the toe takes the shape of an inverted triangle as illustrated in Fig. 10.4, the line of action of F_z will be 1/3 the distance of the base of the triangle ($h/\tan\theta$) from the left edge, i.e., $h/3\tan\theta$. The force resultant can then be determined using the Pythagorean

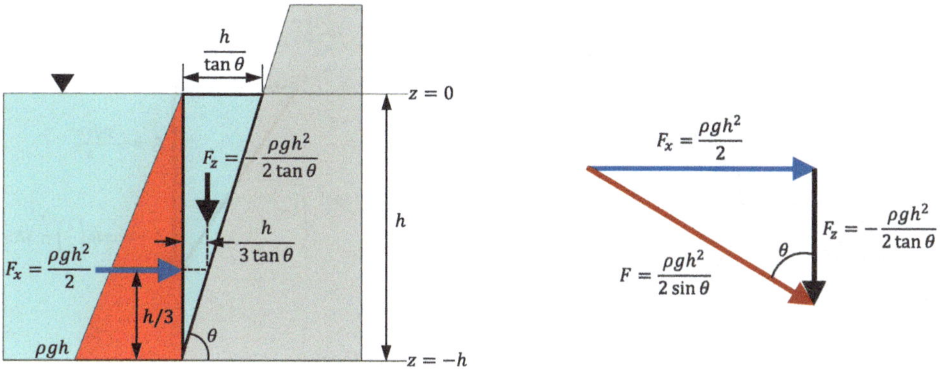

Fig. 10.4 Separation of hydrostatic pressure into horizontal and vertical forces on an inclined surface

theorem:

$$F = \sqrt{F_x^2 + F_z^2} = \sqrt{\left(\frac{\rho g h^2}{2}\right)^2 + \left(-\frac{\rho g h^2}{2\tan\theta}\right)^2}$$
$$= \frac{\rho g h^2}{2}\sqrt{1 + \frac{1}{\tan^2\theta}} = \frac{\rho g h^2}{2\sin\theta} \tag{10.15}$$

which returns Eq. (10.12) as expected. As shown in Fig. 10.4, F is oriented at an angle θ from the vertical axis. When $\theta = 90°$ (i.e., the wall is vertical), $\sin\theta = 1$ and $\tan\theta$ approaches either $+\infty$ or $-\infty$. This means that:

$$F_x = F = \frac{\rho g h^2}{2}, \quad F_z = 0 \tag{10.16}$$

with F acting $h/3$ above the base.

The procedure illustrated in Fig. 10.4 can also be applied to determine hydrostatic forces acting on curved surfaces. Consider the example of a wall with curvature matching that of a quarter circle as shown in Fig. 10.5. The radius of the circular wall matches that of the water depth h. To determine the resultant hydrostatic force, we must first compute the horizontal and vertical components. As before, the horizontal force F_x is equal to the pressure acting over the vertical projection of the wall and is given by Eq. (10.13). Likewise, the vertical force F_z is equal to the weight of the fluid above the toe of the wall. These forces are summarized as:

$$F_x = \frac{\rho g h^2}{2}, \quad F_z = -\frac{\rho g \pi h^2}{4} \tag{10.17}$$

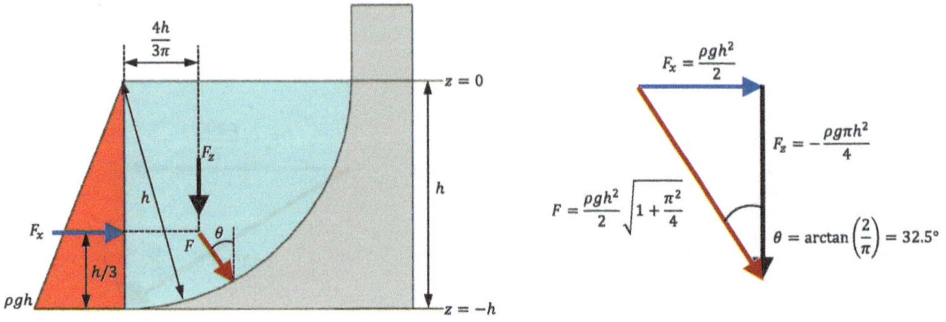

Fig. 10.5 Treatment of hydrostatic pressure on a curved surface

where $\pi h^2/4$ is the area of a quarter circle with radius h. As shown in Fig. 10.5, F_x acts through the centroid of the triangular hydrostatic pressure distribution located $h/3$ above the toe, while F_z acts vertically through the centroid of the fluid area used to determine its weight. For a quarter circle with radius h, the centroid is located $4h/3\pi$ from the edge. The resultant force F is then:

$$F = \sqrt{F_x^2 + F_z^2} = \sqrt{\left(\frac{\rho g h^2}{2}\right)^2 + \left(-\frac{\rho g \pi h^2}{4}\right)^2} = \frac{\rho g h^2}{2}\sqrt{1 + \frac{\pi^2}{4}} \tag{10.18}$$

The line of action of F passes through the intersection of F_x and F_y and is oriented at angle θ relative to the vertical. This angle can be found via:

$$\theta = \arctan\left(\left|\frac{F_x}{F_z}\right|\right) = \arctan\left(\frac{2}{\pi}\right) = 32.5^\circ \tag{10.19}$$

Note that Eqs. (10.17)–(10.19) apply only for a quarter circular wall where the water depth h is exactly equal to its radius. However, the same procedure can be applied to determine the horizontal, vertical, and resultant hydrostatic force vectors on any generic surface.

10.4 Example: Hydrostatic Forces on a Rotating Gate

We will now demonstrate how the principles of hydrostatic pressure can be applied to the case of a rotating flood barrier. Figure 10.6 shows a gate with length l and total mass m per unit length (into the page). The gate is connected to the floor via a frictionless hinge and makes an angle θ to the horizontal. Water with depth h exists on its left side. Suppose we wish to determine the relationship between the water depth h, and the gate's mass m, length l, and angle of inclination θ, under purely hydrostatic conditions.

Fig. 10.6 Hydrostatics of a rotating flood barrier

We begin by computing the total hydrostatic force F acting on the gate. For a given water depth h, the length of gate in contact with the water will be $h/\sin\theta$. Since $p_h = -\rho g z$, the hydrostatic pressure at the hinge ($z = -h$) will be $\rho g h$. This gives:

$$F = \rho g h \times \frac{h}{\sin\theta} \times \frac{1}{2} = \frac{\rho g h^2}{2\sin\theta}$$

as the total hydrostatic force, which acts at a distance of $h/3\sin\theta$ along the gate from the hinge. This force produces a clockwise (negative) moment at the hinge:

$$M_{water} = -F \times \frac{h}{3\sin\theta} = -\frac{\rho g h^3}{6\sin^2\theta}$$

For static equilibrium to occur, this clockwise moment must be balanced by the counterclockwise moment produced by the weight of the gate:

$$M_{gate} = mg \times \frac{l}{2}\cos\theta = \frac{mgl\cos\theta}{2}$$

Considering moment equilibrium at the hinge then gives us:

$$M_{water} + M_{gate} = -\frac{\rho g h^3}{6\sin^2\theta} + \frac{mgl\cos\theta}{2} = 0$$

which can be rearranged in terms of h:

$$h = \left(\frac{3ml\cos\theta\sin^2\theta}{\rho}\right)^{1/3}$$

However, this water depth h must be less than the maximum possible depth capable of being held back by the gate with a total length of l for each angle of θ, given by:

$$h_{max} = l\sin\theta$$

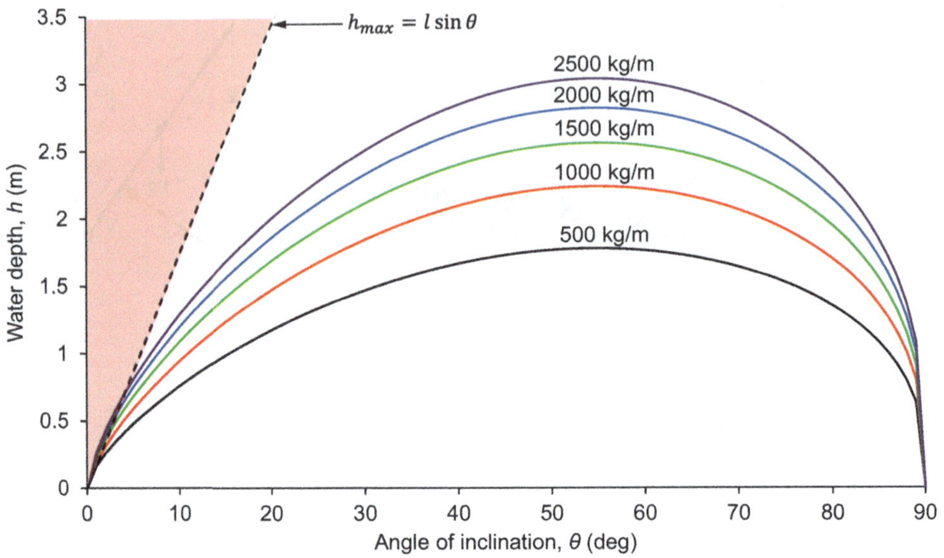

Fig. 10.7 Relationship between water depth and inclination angle for a gate with $l = 10\,\mathrm{m}$

When adopting a gate with length $l = 10\,\mathrm{m}$ holding back seawater ($\rho = 1025\,\mathrm{kg/m^3}$), Fig. 10.7 demonstrates the relationship between water depth and the inclination angle across five different gate masses m from 500 to 2500 kg/m. We notice that for any given depth h below the maximum possible depth, two angles of θ exist for which the system will be in hydrostatic equilibrium. In addition, we find that gates with greater mass are capable of sustaining higher water depths. The upper limit of h_{max} is also shown in Fig. 10.7 and is satisfied for all m at angles θ above approximately 4°.

Reference

1. Kreyszig, E., 2020. Advanced Engineering Mathematics, 10th Edition. Wiley.

11 Wave Forces on Walls

The determination of wave pressures and forces on vertical surfaces such as walls or breakwaters constitute some of the most important problems encountered in coastal and ocean engineering. Naturally, when considering the design of structures for coastal resilience, a method to compute the forces exerted on coastal structures with vertical facades resulting from waves exhibiting different characteristics is paramount. In this context, we are mainly interested in three types of waves:

- **Nonbreaking**: regular or irregular waves that have well defined crests and troughs.
- **Breaking**: waves that have reached a critical steepness such that their crests are in the process of overturning.
- **Broken**: highly turbulent waves that form after wave breaking.

Figure 11.1 highlights the aforementioned types of waves as they approach a rocky beach. This chapter will introduce both historical and modern design formulations that can be used to estimate the pressures and forces produced by such waves on wall-type structures.

11.1 Breaking, Nonbreaking, and Broken Waves

By far the most destructive form of wave loading arises when a collapsing wavefront impinges directly upon a structure in the form of a breaking wave. As shown in Fig. 11.2, such waves typically trap air between the wave front and structure which translates to a sharp impulsive peak in the force history, labeled F_I in Fig. 11.2, as the crest strikes the structure. This impulsive peak is often followed by an oscillatory component produced

S. Wang, *Applied Ocean Mechanics for Structural Engineering and Climate Adaptation*, Synthesis Lectures on Ocean Systems Engineering,
https://doi.org/10.1007/978-3-032-18199-2_11

Fig. 11.1 Waves approaching a beach near Hokitika, New Zealand, at various stages of breaking

from oscillations of the entrained air pocket. These oscillations rapidly dampen as the air cavity is filled with water which leads to a secondary quasi-static force maximum, labeled F_{QS} in Fig. 11.2, as the wave runs up the wall. In contrast, the initial force maximum (F_1) for nonbreaking, slightly breaking, or broken waves will be significantly subdued and are often approximated as quasi-static loads. Since the physics of wave breaking is still not well understood, there are currently no reliable formulations for the prediction of impulsive pressures or forces resulting from breaking waves. However, Kortenhaus and Oumeraci [10] proposed that different types of wave impact on coastal structures can be differentiated based on the ratio between the primary (F_1) and secondary (F_2) force maxima:

$$\begin{array}{ll} F_1/F_2 \approx 1, & \text{Nonbreaking wave} \\ 1 < F_1/F_2 < 2.5, & \text{Slightly breaking or broken wave} \\ F_1/F_2 > 2.5, & \text{Breaking wave} \end{array} \tag{11.1}$$

While a theoretical ratio of F_1/F_2 exists as a lower limit for forces to be classified as "impulsive", it is important to recognize that impact loads from wave breaking can be up to 50 times greater than nonbreaking wave forces [1]. Unsurprisingly, breaking waves have been responsible for the failure of numerous coastal structures (e.g., breakwaters) around the world during the twentieth century [12]. Figure 11.3 illustrates the evolution of a breaking wave as it makes contact with a vertical wall. With ongoing climate change, the increasing frequency and severity of tropical cyclones and storms will significantly increase the susceptibility of coastal infrastructure such as buildings, seawalls, or breakwaters to breaking waves [14].

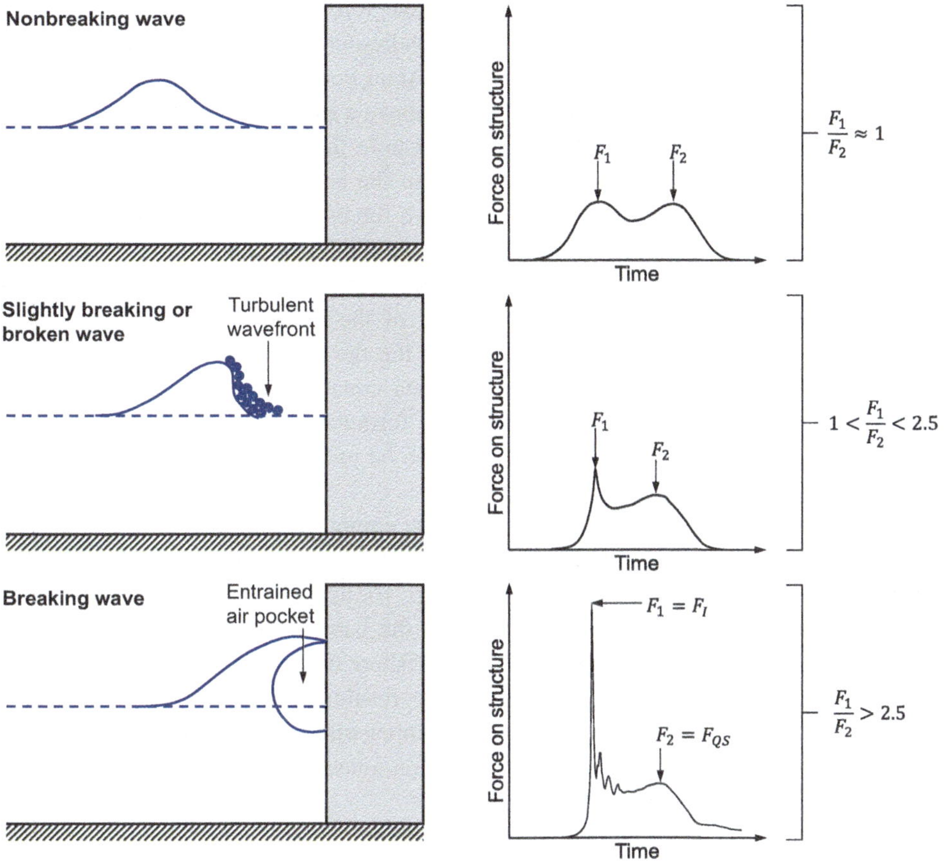

Fig. 11.2 Nonbreaking (top), slightly breaking or broken (middle), and breaking (bottom) waves impacting a coastal structure with typical force history shown

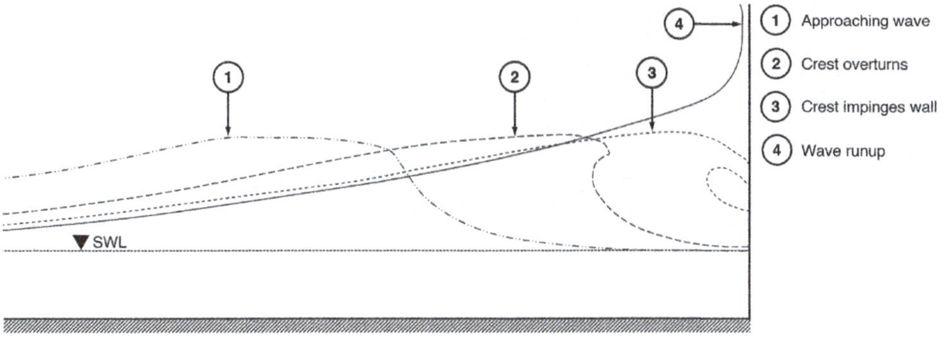

Fig. 11.3 Evolution of a breaking wave before and after impingement on a vertical wall

The development of equations to predict wave forces and pressures on walls was largely driven by the construction of vertical breakwaters throughout Japan in the early twentieth century. As shown in Fig. 11.4, using Akita Port as an example, vertical breakwaters often exist as concrete caissons elevated above a rubble mound (used to distribute the caisson's weight over the seabed) placed offshore that prevent waves from entering a harbor. As such, they are frequently exposed to the action of waves on their seaward side which necessitates the determination of wave forces in order to inform their design for strength and stability. Despite the difficulties in predicting wave forces, many design formulas that seek to address breaking, nonbreaking, and broken waves on vertical breakwaters have been proposed since the beginning of the twentieth century. This chapter will first present a historical overview covering the development of wave pressure formulas before discussing the semi-empirical Goda equation which represents the most widely used design procedure today. While such formulas were originally developed for the design of offshore breakwaters, they may also be utilized for the estimation of wave forces on coastal structures of all types.

Note that the pressures p, as shown in Fig. 11.4, computed using the wave pressure formulas presented in this chapter depict only the dynamic component and does not include the hydrostatic pressure p_h. As discussed in Chap. 10, hydrostatic pressures vary linearly from 0 at the still water level (SWL) to $\rho g h$ at the base of the structure. Nevertheless, this hydrostatic component is often ignored for offshore breakwaters since the same water level usually exists on both sides of the structure, resulting in equal and opposite hydrostatic forces cancelling each other out. For structures where water is present on one side only, such as a flood barrier, hydrostatic pressures must be included in addition to the dynamic pressure distribution p due to waves.

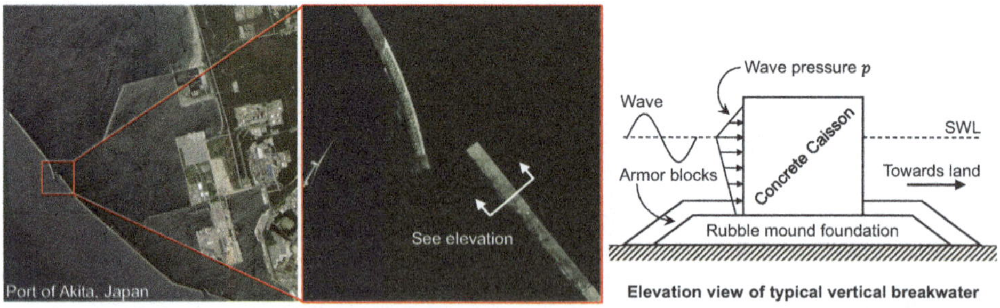

Fig. 11.4 Vertical breakwaters at Akita Port, Japan. Satellite images courtesy of USGS

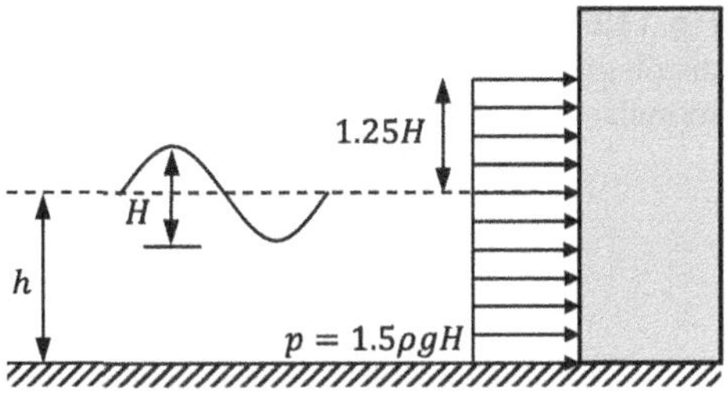

Fig. 11.5 Wave pressure distribution from Hiroi's formula

11.2 Historical Wave Pressure Formulations

11.2.1 Hiroi's Formula

In 1919, the Japanese engineer Hiroi proposed the first widely used formula for the estimation of impulsive wave pressures arising from breaking waves impacting vertical breakwaters [8]. From experimental measurements and assuming that the impulsive pressure is similar to that of a water jet, Hiroi suggested that the wave pressure p can be simply related to the design wave height H via:

$$p = 1.5\rho g H \tag{11.2}$$

As seen in Fig. 11.5, the pressure p was assumed to act uniformly on the structure up to a height of $1.25H$ above the SWL. If H is unknown, Hiroi recommends taking 90% of the local water depth as the design wave height ($H = 0.9h$) for use in Eq. (11.2). Since its introduction, Hiroi's equation has enjoyed use within Japan for over 60 years.

11.2.2 Sainflou's Formula

Evidently, Hiroi's formula serves only to guide the design of structures exposed to breaking waves. In 1928, the French engineer Sainflou, while affiliated with the Ecole Nationale des Ponts et Chaussees, developed equations to predict nonbreaking wave forces on vertical walls [16]. Using trochoidal wave theory, Sainflou proposed the following equations to determine the wave pressures p_1 (at the SWL) and p_2 (at the base) for nonbreaking waves with height H as shown in Fig. 11.6:

$$p_1 = (p_2 + \rho g h)\frac{H + \delta_0}{h + H + \delta_0} \tag{11.3}$$

$$p_2 = \frac{\rho g H}{\cosh\left(\frac{2\pi h}{L}\right)} \tag{11.4}$$

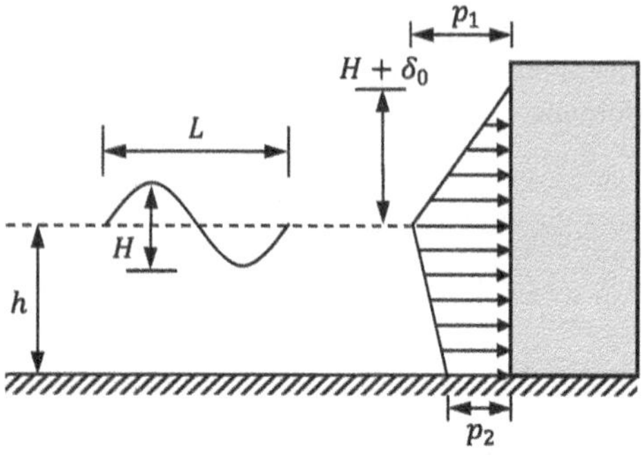

Fig. 11.6 Wave pressure distribution from Sainflou's formula

where:

$$\delta_0 = \frac{\pi H^2}{L} \coth\left(\frac{2\pi h}{L}\right) \tag{11.5}$$

First discovered by Gerstner [4], the trochoidal wave theory used by Sainflou to develop Eqs. (11.3) and (11.4) is a nonlinear theory that gives an exact solution to the Euler equations (Table 2.2) for surface wave propagation over infinite depths. Due to its relatively limited applications when compared with other more common nonlinear theories (e.g., Stokes), and its inability to account for finite water depths, trochoidal wave theory is rarely used in modern coastal engineering. In addition, Eqs. (11.3) and (11.4) cannot be applied for breaking waves or if waves overtop the structure. Therefore, harbor engineers have traditionally employed Sainflou's equations in conjunction with Hiroi's formula to address nonbreaking and breaking waves, respectively.

11.2.3 Minikin's Formula

In addition to the works of Hiroi and Sainflou, Minikin [11] also developed a simple expression to estimate the total force produced by breaking waves based on experiments performed by Bagnold [3]. Minikin's formula is:

$$F = \frac{101}{3} \frac{\rho g H^2 d}{L_D D} (d + D) \tag{11.6}$$

where d is the water depth at the toe of a vertical structure, L_D is the design wavelength, and D is the water depth at a distance of L_D away from the structure. However, Allsop et al. [1] demonstrated that Eq. (11.6) is fundamentally incorrect since it predicts a decrease in the wave force with increasing wavelength, thereby contradicting experimental observations.

Ultimately, it is important to realize that the equations of Hiroi, Sainflou, and Minikin can only address nonbreaking or breaking waves in isolation. It would not be until the

1970s when a general formulation would be developed that accounts for nonbreaking, breaking, and broken waves, while allowing for both overtopping and non-overtopping conditions, within a single framework.

11.3 The Goda Equation

In the 1970s, the Japanese harbor engineer Yoshimi Goda introduced a new formula that unified the treatment of nonbreaking, breaking, and broken waves for the determination of wave pressures on vertical breakwaters [6]. Similar to Fig. 11.6, the horizontal wave pressure takes a trapezoidal distribution with its maximum at the SWL. Unlike Eqs. (11.2)–(11.6), Goda's formula can account for waves approaching from any arbitrary direction relative to the structure. In particular, Goda utilized extensive physical testing to calibrate nondimensional parameters that address the transition from nonbreaking to breaking wave loads as a function of the bathymetry leading up to the structure. The Goda equation also explicitly accounts for the inclusion of an elevated rubble mound foundation as typical for caisson breakwaters (e.g., Fig. 11.4). Vertical forces acting along the base of the foundation are additionally incorporated to supplement horizontal wave loads.

11.3.1 Wave Pressure Distribution

Figure 11.7 illustrates an impermeable vertical structure B wide and h_w tall located on an elevated mound within water with a total depth h. A layer of armor blocks exists on the slope in front of the structure. The depth of water at the toe of the structure from the free surface to the top and bottom of the armor blocks are designated d and h', respectively, with $d = h'$ in the absence of armor blocks. In addition, the water depth to the top of the armor layer at a distance of five times the significant wave height ($D_m = 5H_s$) from the structure is denoted h_b. To mitigate wave overtopping, the top of the structure is located h_c above the free surface (referred to as the "freeboard"). Goda's equation for the distribution of lateral (p_1, p_2, and p_3) and uplift (p_u) wave pressures resulting from a design wave with height H and wavelength L, as shown in Fig. 11.7, is presented as follows:

$$p_1 = \frac{1}{2}(1 + \cos\beta)\left(\alpha_1 + \alpha_* \cos^2\beta\right)\rho g H \tag{11.7}$$

$$p_2 = \begin{cases} p_1\left(1 - \frac{h_c}{\eta^*}\right), & \eta^* > h_c \\ 0, & \eta^* \le h_c \end{cases} \tag{11.8}$$

$$p_3 = \alpha_3 p_1 \tag{11.9}$$

$$p_u = \frac{1}{2}(1 + \cos\beta)\alpha_1\alpha_3\rho g H \tag{11.10}$$

where:

$$\eta^* = 0.75(1 + \cos\beta)H \tag{11.11}$$

denotes the maximum elevation above the SWL to which wave pressures can be exerted due to wave runup. If waves are expected to overtop the structure, η^* will be greater than the height of the structure above the SWL (h_c). The angle of wave approach (β) is shown in Fig. 11.8, which may be taken as $\beta = 0$ for the most severe case of waves approaching perpendicular to the wall.

The dimensionless α parameters featured within Goda's equation are:

$$\alpha_* = \alpha_2 \tag{11.12}$$

$$\alpha_1 = 0.6 + \frac{1}{2}\left[\frac{\frac{4\pi h}{L}}{\sinh\left(\frac{4\pi h}{L}\right)}\right]^2 \tag{11.13}$$

$$\alpha_2 = \min\left[\frac{h_b - d}{3h_b}\left(\frac{H}{d}\right)^2, \frac{2d}{H}\right] \tag{11.14}$$

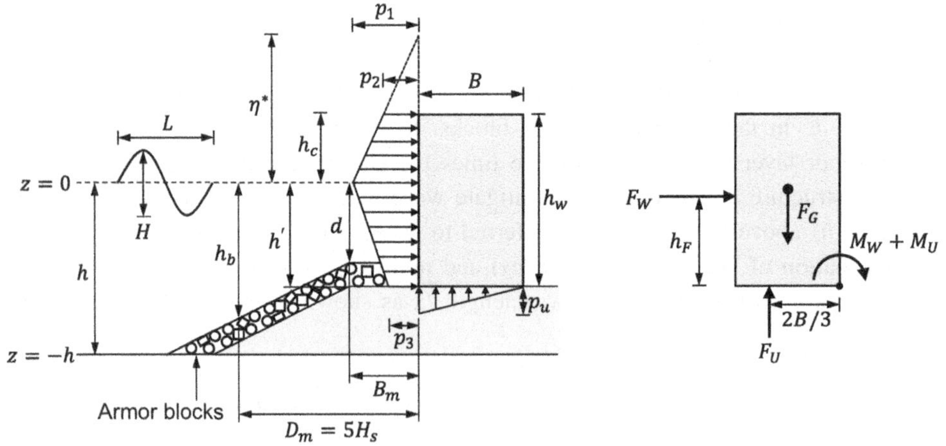

Fig. 11.7 Wave pressure distribution from Goda's equation and total resultant wave forces and moments exerted on a structure located on a sloped mound

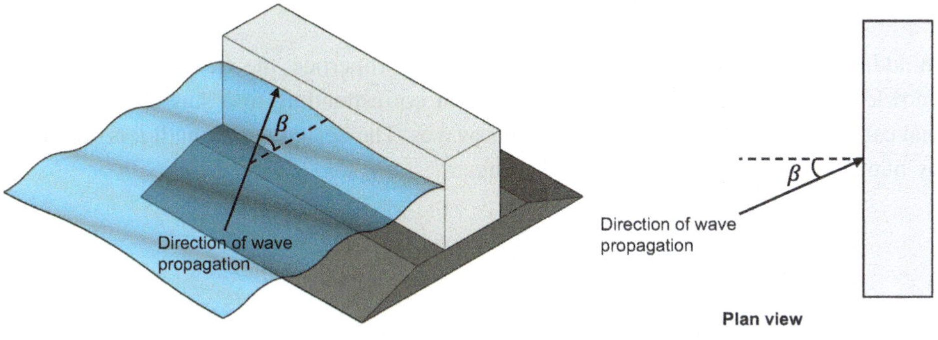

Fig. 11.8 3D and plan view of vertical structure showing angle of wave approach

$$\alpha_3 = 1 - \frac{h'}{h}\left[1 - \frac{1}{\cosh\left(\frac{2\pi h}{L}\right)}\right] \tag{11.15}$$

Here, α_1 accounts for the increase in wave pressure produced by waves with longer periods while α_2 accounts for the increase in wave pressure associated with higher mounds causing waves to potentially break on the structure. The parameter α_3 accounts for the linear variation in pressure from the SWL to the toe of the structure. Parameters α_1 and α_2 were obtained via experimental testing and address the contribution of the elevated mound to wave breaking. α_1 was calibrated for conditions where the seabed profile up to a distance D_m from the structure is flat (i.e., $h_b = d$). For this scenario, $\alpha_2 = 0$ and only nonbreaking or already broken waves are expected to impact the structure. α_2 represents the increased pressure at the SWL ($z = 0$) associated with shoaling and potential wave breaking which is assumed to increase parabolically with the H/d ratio. A limit of $2d/H$ is imposed in (11.14) to prevent the pressure from becoming artificially large if d is very small.

11.3.2 Design Wave Parameters

In addition to relevant structural and bathymetric properties, the Goda equation requires knowledge of the wavelength and wave height corresponding to a single "design wave" that can represent either regular or irregular waves. The design wavelength L is determined by numerically solving dispersion relation:

$$L = \frac{gT^2}{2\pi}\tanh\left(\frac{2\pi h}{L}\right) \tag{11.16}$$

or can be approximated using:

$$L \approx \frac{gT^2}{2\pi}\sqrt{\tanh\left(\frac{4\pi^2 h}{T^2 g}\right)} \tag{11.17}$$

Here, T is the "design wave period" which is simply taken as the significant wave period T_s for irregular waves. If T_s is not explicitly given, it can be approximated using:

$$T = T_s \approx 1.2T_m \approx 3.3H_s^{0.63} \tag{11.18}$$

Likewise, the H used in the Goda equation is referred to as the "design wave height". For irregular waves, H is generally taken as 1.8 times the significant wave height (H_s) if outside of the surf zone (i.e., the water is sufficiently deep such that waves are not expected to break due to shallow bathymetry). Note that H_s is taken at the vicinity of the structure, which, if needed, can be determined using the deep-water wave height per Eq. (9.9) to account for shoaling. Within the surf zone, H is selected to be the highest of the random breaking waves at a distance $D_m = 5H_s$ in front of the structure where the water depth is h_b (see Fig. 11.7). For simplicity, this could be taken as $H_b = 0.78h_b$ if the seabed slope is at or below 1:50 ($\theta \leq 1.15°$). Otherwise, H_b can be determined using Eq. (9.13). The design wave height H is therefore:

$$H = \min[1.8H_s, H_b] \tag{11.19}$$

11.3.3 Resultant Forces and Moments

The wave pressures determined using Goda's equation can next be used to obtain the total horizontal (F_W) and uplift force (F_U) per unit length of the structure associated with wave loading (see Fig. 11.7):

$$F_W = \frac{1}{2}(p_1 + p_2)h^* + \frac{1}{2}(p_1 + p_3)h' \tag{11.20}$$

$$F_U = \frac{1}{2} p_u B \tag{11.21}$$

where:

$$h^* = \min[h_c, \eta^*] \tag{11.22}$$

denotes the maximum elevation of wave pressure above the SWL. The point of application of the total resultant horizontal wave force above the base of the structure (h_F) is:

$$h_F = \frac{(p_3 + 2p_1)h'^2 + (p_1 + 2p_2)h^{*2} + 3h'(p_2 + p_1)h^*}{3[(p_2 + p_1)h^* + (p_1 + p_3)h']} \tag{11.23}$$

whereas the resultant uplift force (F_U) is applied at a distance $2B/3$ from the point of rotation at the heel due to its triangular pressure distribution. Subsequently, the wave moment at the heel per unit length (M_W) is simply the horizontal wave force F_W multiplied by its vertical lever arm to the base (h_F):

$$M_W = F_W h_F = \frac{1}{6}(2p_1 + p_3)h'^2 + \frac{1}{6}(p_1 + 2p_2)h^{*2} + \frac{1}{2}(p_1 + p_2)h^* h' \tag{11.24}$$

Likewise, the overturning moment at the heel produced by the uplift force is:

$$M_U = F_U\left(\frac{2B}{3}\right) = \frac{1}{3} p_u B^2 \tag{11.25}$$

If water exists only on one side of the structure (e.g., if acting as a flood barrier), an additional hydrostatic loading component as shown in Fig. 11.9 must be included when computing the net horizontal force. As discussed in Chap. 10, the total horizontal force per unit length due to hydrostatic pressure only (F_H) is:

$$F_H = \frac{1}{2}\rho g h'^2 \tag{11.26}$$

with the hydrostatic moment per unit length at the heel being:

$$M_H = \frac{1}{6}\rho g h'^3 \tag{11.27}$$

For water on one side only, the total horizontal force and heel moment arises from the sum of the wave and hydrostatic components $F_W + F_H$ and $M_W + M_U + M_H$. For water on both sides, only the wave components F_W, F_U, and $M_W + M_U$ need to be considered since the hydrostatic pressures on both sides of the wall will cancel each other out (assuming that the SWL is at the same elevation on both sides of the wall). The total forces acting on the structure are summarized as follows:

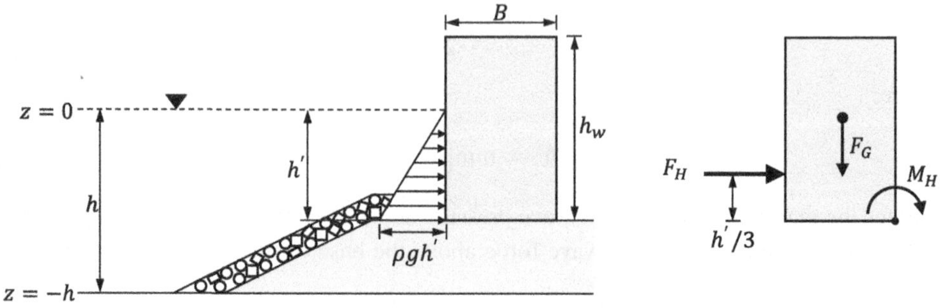

Fig. 11.9 Hydrostatic pressure distribution and resultant force and moment on a structure

$$F_T = \begin{cases} F_W + F_H, & \text{Water on one side only} \\ F_W, & \text{Water on both sides} \end{cases} \quad (11.28)$$

$$M_T = \begin{cases} M_W + M_U + M_H, & \text{Water on one side only} \\ M_W + M_U, & \text{Water on both sides} \end{cases} \quad (11.29)$$

where F_T and M_T denote the total horizontal force and overturning moment per unit length.

The forces from Eqs. (11.28) and (11.29) must be resisted by the structure's self-weight. If water exists on both sides of an impermeable structure (e.g., an offshore breakwater), Archimedes' principle will result in an upwards buoyant force that is equal to the weight of the displaced water (see Chap. 10). The total weight of a structure with width B, height h_w, and average density ρ_s per unit length will be:

$$F_G = \begin{cases} \rho_s g B h_w, & \text{Water on one side only} \\ \rho_s g B h_w - \rho g B h', & \text{Water on both sides} \end{cases} \quad (11.30)$$

where $\rho g B h'$ is the buoyant force counteracting the structure's self-weight. The resulting self-weight force thus provides a restoring moment (M_R) to prevent overturning of the structure:

$$M_R = F_G\left(\frac{B}{2}\right) \quad (11.31)$$

The factor of safety against overturning is defined as:

$$SF_{overturning} = \frac{M_R - M_H - M_U}{M_W} \quad (11.32)$$

where $M_H = 0$ if water exists on both sides. This factor of safety must be greater than 1 to prevent overturning of the structure. In addition, a structure must also resist sliding under wave impact. The force required to induce sliding of a caisson (F_S) is simply the total weight after accounting for buoyancy (F_G) multiplied by the coefficient of friction (μ) between the caisson and the surface it sits upon:

$$F_S = \mu F_G \tag{11.33}$$

The coefficient of friction is typically taken as $\mu = 0.6$ between concrete and rubble stones in the context of caisson breakwaters as recommended by the Technical Standards and Commentaries for Port and Harbor Facilities in Japan [7]. However, experiments by Hutchinson et al. [9] suggest that μ can be as high as 0.78 if the base of the caisson has been deliberately roughened. The factor of safety against sliding is:

$$SF_{sliding} = \frac{F_S}{F_T} \tag{11.34}$$

which must be greater than 1 to prevent sliding of the structure. For design applications, Goda [5] recommends that the factors of safety for both sliding and overturning should be greater than 1.2. A summary of forces and moments acting on a rectangular structure with width B, height h_w, and average density ρ_s exposed to wave attack is presented in Table 11.1.

11.3.4 Bearing Stress

In addition to sliding and overturning mechanisms of failure, the foundation must also be examined to avoid bearing capacity failure. The overturning moments induced at the heel combined with the structural self-weight results in a nonsymmetrical distribution of bearing stresses along the base of the caisson. This bearing stress varies linearly from a minimum at the toe to a maximum at the heel of the structure as shown in Fig. 11.10. To ensure vertical force equilibrium, the resultant bearing force (i.e., area of the bearing stress distribution) must equal the weight of the structure F_G. This force resultant acts at a distance of t_e from the heel and is determined by considering moment equilibrium about the heel:

$$t_e = \frac{M_R - M_T}{F_G} \tag{11.35}$$

where M_R and M_T denote the restoring and overturning moments, respectively, and F_G is the self-weight after accounting for buoyancy (see Table 11.1). The maximum (q_1) and minimum (q_2) bearing stresses as given by Takahashi [18] are:

$$q_1 = \begin{cases} \frac{2F_G}{B}\left(2 - \frac{3t_e}{B}\right), & t_e > B/3 \\ \frac{2F_G}{3t_e}, & t_e \leq B/3 \end{cases} \tag{11.36}$$

Table 11.1 Summary of total Goda wave forces and moments per unit length on a vertical structure

	Water on both sides	Water on one side only
Elevation view	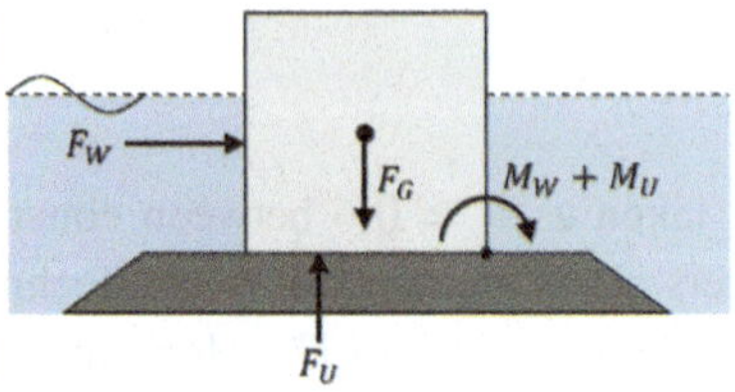	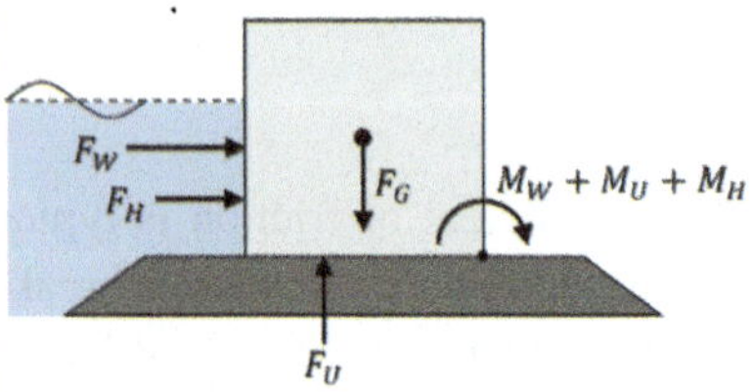
Total horizontal force	$F_T = F_W$ $F_W = \frac{1}{2}(p_1 + p_2)h^* + \frac{1}{2}(p_1 + p_3)h'$	$F_T = F_W + F_H$ $F_H = \frac{1}{2}\rho g h'^2$
Uplift force	$F_U = \frac{1}{2}p_u B$	
Total overturning moment	$M_T = M_W + M_U$ $M_W = \frac{1}{6}(2p_1 + p_3)h'^2$ $+ \frac{1}{6}(p_1 + 2p_2)h^{*2}$ $+ \frac{1}{2}(p_1 + p_2)h^* h'$ $M_U = \frac{1}{3}p_u B^2$	$M_T = M_W + M_U + M_H$ $M_H = \frac{1}{6}\rho g h'^3$
Self-weight	$F_G = \rho_s g B h_w - \rho g B h'$	$F_G = \rho_s g B h_w$
Restoring moment	$M_R = \frac{F_G B}{2}$	
Friction force	$F_S = \mu F_G$	

Note The structure's center of gravity is assumed to be at $B/2$ from each end

$$q_2 = \begin{cases} \frac{2F_G}{B}\left(\frac{3t_e}{B} - 1\right), & t_e > B/3 \\ 0, & t_e \leq B/3 \end{cases} \tag{11.37}$$

where t_e is from Eq. (11.35). To prevent bearing failure, the maximum bearing stress q_1 must be smaller than the bearing capacity of the foundation.

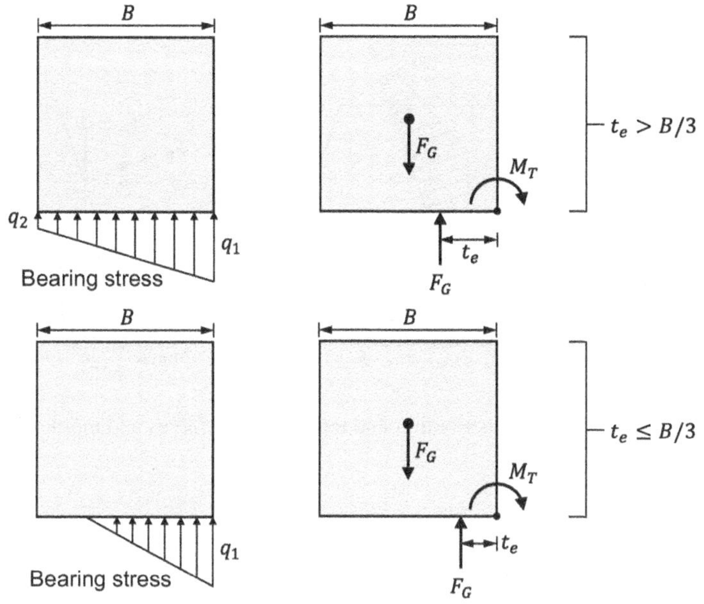

Fig. 11.10 Distribution of bearing stress and resultant forces on a caisson

11.4 The Goda Equation for Inclined Surfaces

Tanimoto and Kimura [20] demonstrated that the Goda wave pressure formulations given by Eqs. (11.7)–(11.10) can also be applied to inclined surfaces such as the trapezoidal structure shown in Fig. 11.11. Like with purely vertical walls, the maximum wave pressure p_1 will occur at the SWL. The resultant wave (F_θ) and uplift (F_U) forces per unit length acting on the structure becomes:

$$F_\theta = \frac{1}{\sin\theta}\left[\frac{1}{2}(p_1 + p_2)h^* + \frac{1}{2}(p_1 + p_3)h'\right] \tag{11.38}$$

$$F_U = \frac{1}{2}p_u B \tag{11.39}$$

where θ is the angle of inclination of the front wall exposed to wave attack as shown in Fig. 11.11. The point of application of the resultant horizontal wave force above the base of the structure (h_F) is determined via the same manner as for purely vertical walls:

$$h_F = \frac{(p_3 + 2p_1)h'^2 + (p_1 + 2p_2)h^{*2} + 3h'(p_2 + p_1)h^*}{3\left[(p_2 + p_1)h^* + (p_1 + p_3)h'\right]} \tag{11.40}$$

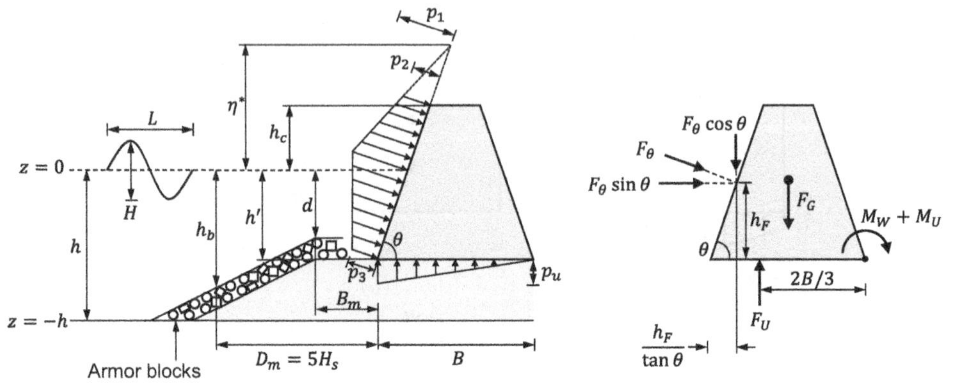

Fig. 11.11 Goda wave pressure distribution and total resultant forces and moments on an inclined wall

where:

$$h^* = \min\left[h_c, \eta^*\right] \tag{11.41}$$

The overturning wave moment at the heel per unit length is calculated from the sum of the moments generated by the horizontal and vertical components of the wave force (M_W) in combination with the moment induced by the uplift force (M_U):

$$M_W = h_F F_\theta \sin\theta - \left(B - \frac{h_F}{\tan\theta}\right) F_\theta \cos\theta \tag{11.42}$$

$$M_U = \left(\frac{2B}{3}\right) F_U \tag{11.43}$$

and is illustrated in Fig. 11.11. These dynamic forces from wave loading can then be superimposed with hydrostatic forces to obtain the total design loads.

11.5 The Goda Equation for Elevated Structures

Vertical piles or columns are often used to elevate properties beyond the reach of waves and floodwaters in coastal regions vulnerable to flood hazards [15]. However, excessively large waves combined with storm surge inundation from tropical cyclones can still cause significant damage if waves are able to reach the elevated structure. For example, wave loading from Hurricane Ike in 2008 caused extensive damage to elevated houses along the Texas coast [2], some of which are shown in Fig. 11.12. In regions susceptible to

Fig. 11.12 Wave damage to elevated houses in Bolivar Peninsula (left) and Galveston (right), Texas, after Hurricane Ike (2008). Photographs courtesy of the US National Archives

extreme storm hazards, the prediction of wave forces on elevated structures is naturally of great importance for engineering design.

Wiebe et al. [23] demonstrated that the Goda equation may also be used to model the distribution of nonbreaking, breaking, and broken wave forces on elevated structures. Figure 11.13 illustrates two examples pertaining to a partially elevated and fully elevated structure. For the partially elevated scenario, the still water level is located a distance a above the base of the structure ($a < 0$), whereas the base of the fully elevated structure is a above the still water level ($a > 0$). The distance a is termed the "air gap" and represents the distance from the SWL to the base of the structure, with $a < 0$ and $a > 0$ for partially and fully elevated structures, respectively. The pressures p_1 and p_2 are obtained from Eqs. (11.7) and (11.8) while the pressure p_4 at the base of the elevated structure is:

$$p_4 = \begin{cases} \left(1 - \frac{a}{\eta^*}\right)p_1, & a > 0 \\ \left(1 + \frac{a}{h'}\right)(p_1 - p_3) + p_3, & a \leq 0 \end{cases} \tag{11.44}$$

where p_3 is from Eq. (11.9). The total horizontal wave force per unit length of the elevated structure is therefore:

$$F_W = \begin{cases} \frac{1}{2}(p_2 + p_4)(h^* - a), & a > 0 \\ \frac{1}{2}(p_1 + p_2)h^* - \frac{1}{2}(p_1 + p_4)a, & a \leq 0 \end{cases} \tag{11.45}$$

where h^* is per Eq. (11.22). The point of application of the total horizontal force above the base of the elevated structure (h_F) is:

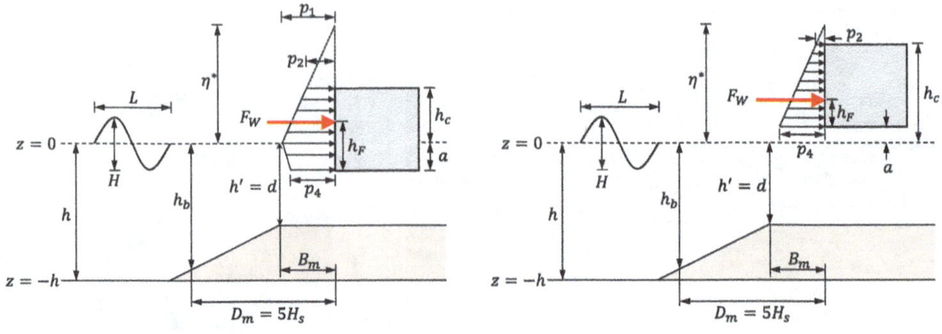

Fig. 11.13 Goda wave pressure distribution for partially (left) and fully (right) elevated structures

$$h_F = \begin{cases} \frac{(p_4+2p_2)(h^*-a)}{3(p_4+p_2)}, & a > 0 \\ \frac{(p_4+2p_1)a^2+(p_1+2p_2)h^{*2}-3a(p_2+p_1)h^*}{3[(p_2+p_1)h^*-(p_1+p_4)a]}, & a \leq 0 \end{cases} \quad (11.46)$$

and is shown in Fig. 11.13. Note that when $a = 0$, $p_4 = p_1$ and both equations in (11.45) and (11.46) will yield the same result.

However, it must be emphasized that the equations proposed by Wiebe et al. [23] can only be used to obtain the horizontal forces (F_W) acting on elevated structures. Wave flume experiments conducted by Park et al. [13] revealed that elevated structures may be exposed to significant vertical loads on its underside, especially when the air gap a is close to zero. Numerical simulations by Pawitan et al. [14] discovered that this phenomenon results from the opening of a small air gap below an elevated structure with $a \approx 0$ due to set-down of the incoming wave as shown in Fig. 11.14. This air gap then rapidly closes as the momentum of the wave passes below the structure, leading to a high intensity upwards pressure acting on its underside.

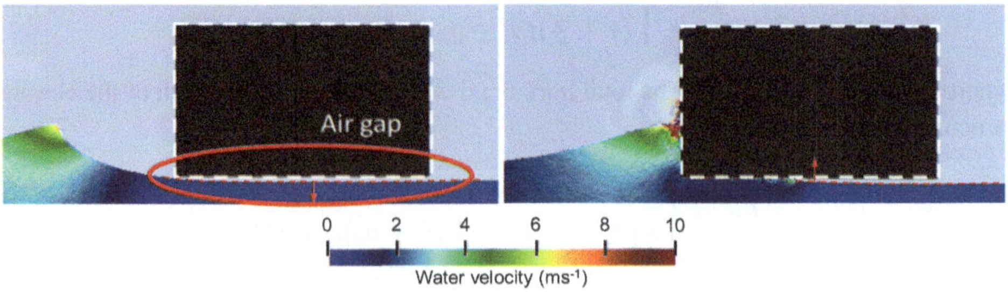

Fig. 11.14 Numerical simulation of wave impact on an elevated structure with zero air gap ($a = 0$)

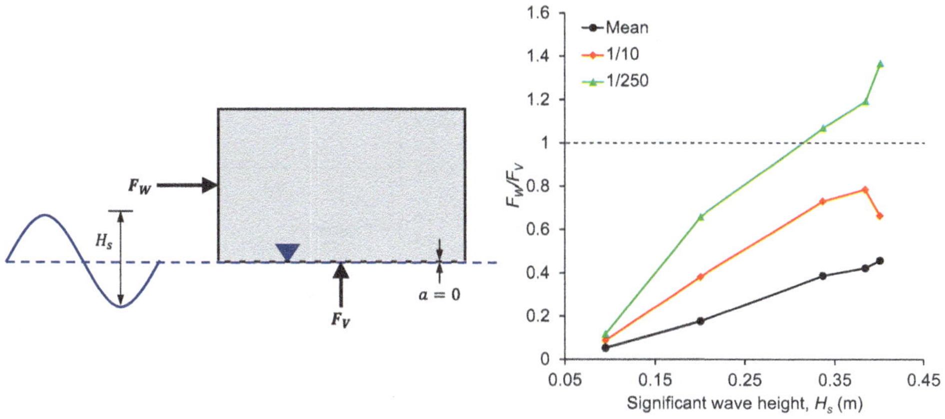

Fig. 11.15 Ratio between horizontal (F_W) and vertical (F_V) wave forces on an elevated structure with zero air gap ($a = 0$) as a function of significant wave height from experiments by Park et al. [13]

The relationship between horizontal (F_W) and vertical (F_V) wave forces on an elevated structure with zero air gap from the experiments of Park et al. [13] are summarized in Fig. 11.15. Here, the force ratios are given as the mean, average of the largest 1/10, and average of the largest 1/250 waves within an irregular wave train with significant wave height H_s. We see that the horizontal wave force becomes larger with respect to the vertical force with increasing significant wave height. However, smaller wave heights can see the vertical force exceed its horizontal component by 5–10 times.

11.6 Modified Goda Equation for Impulsive Loading

At this stage, it is important to recognize that Goda's equation is very much an approximation that only serves as a guide when determining wave pressures on walls. As seen in Goda's experimental results used to calibrate the parameters α_1 and α_2 (Fig. 11.16), correlations between the α parameters (dashed line) and experimental values (discrete data points) are rather coarse. Even Goda himself stated in his seminal 1974 paper that:

> Though the agreement between the formulae and the laboratory data is not excellent, the difference may be disregarded for the purpose of general formulation of wave pressure estimation.

In response, Takahashi et al. [19] sought to improve the capabilities of Goda's equation when predicting impulsive pressures from breaking waves. They aimed to address this by introducing another parameter α_I to supersede Goda's original α_2 parameter for the

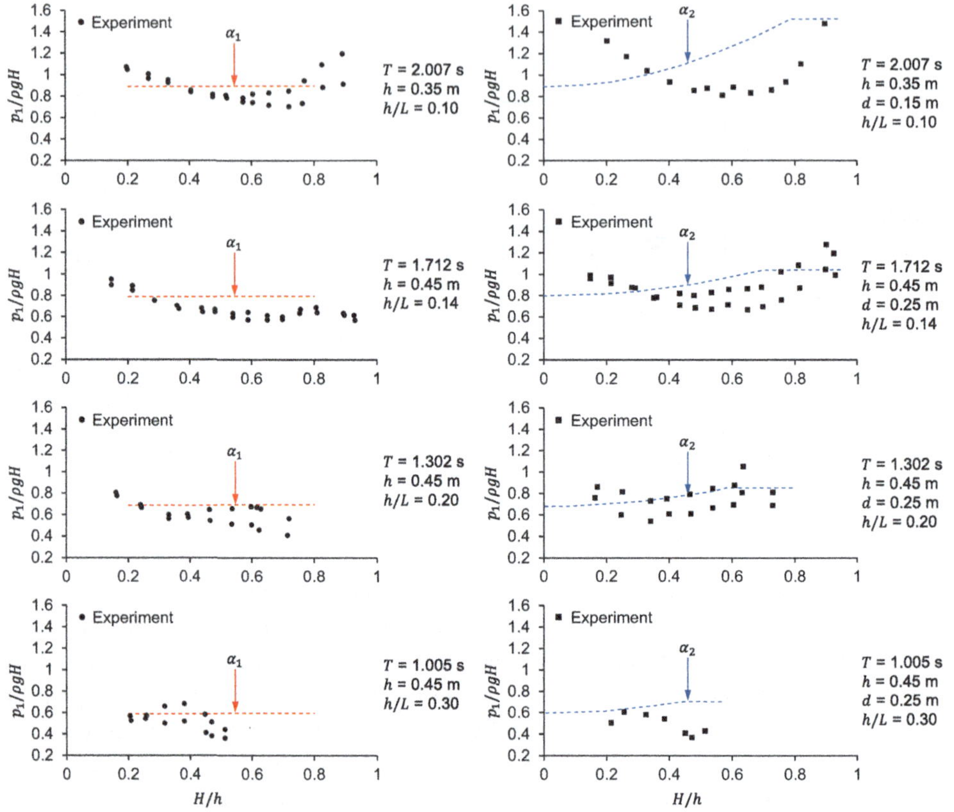

Fig. 11.16 An excerpt of Goda's experimental results used to obtain expressions for the α_1 and α_2 parameters (adapted from Goda [6])

treatment of wave breaking. Hence, the α_* parameter from Eq. (11.12) is re-expressed as:

$$\alpha_* = \max[\alpha_2, \alpha_I] \tag{11.47}$$

where α_2 is per (11.14). The α_I parameter was formulated based on extensive physical testing by Takahashi et al. [19] and is defined as:

$$\alpha_I = \alpha_{I0}\alpha_{I1} \tag{11.48}$$

where:

$$\alpha_{I0} = \begin{cases} \frac{H}{d}, & \frac{H}{d} \leq 2 \\ 2, & \frac{H}{d} > 2 \end{cases} \tag{11.49}$$

$$\alpha_{I1} = \begin{cases} \frac{\cos \delta_2}{\cosh \delta_1}, & \delta_2 \leq 0 \\ \frac{1}{\cosh \delta_1 \sqrt{\cosh \delta_2}}, & \delta_2 > 0 \end{cases} \tag{11.50}$$

and:

$$\delta_1 = \begin{cases} 20\delta_{11}, \ \delta_{11} \leq 0 \\ 15\delta_{11}, \ \delta_{11} > 0 \end{cases} \tag{11.51}$$

$$\delta_{11} = 0.93\left(\frac{B_m}{L} - 0.12\right) + 0.36\left(\frac{h-d}{h} - 0.6\right) \tag{11.52}$$

$$\delta_2 = \begin{cases} 4.9\delta_{22}, \ \delta_{22} \leq 0 \\ 3\delta_{22}, \quad \delta_{22} > 0 \end{cases} \tag{11.53}$$

$$\delta_{22} = -0.36\left(\frac{B_m}{L} - 0.12\right) + 0.93\left(\frac{h-d}{h} - 0.6\right) \tag{11.54}$$

constitute the parameters accounting for impulsive loading.

11.7 Structural Modification Factors for the Goda Equation

The Goda equation presented thus far can only be used to compute wave pressures acting on flat surfaces. However, the Technical Standards and Commentaries for Port and Harbor Facilities in Japan [7] presents a method to modify the Goda pressures in order to accommodate different structural geometries (e.g., curved surfaces). Here, the horizontal pressure at the SWL (p_1) is expressed as:

$$p_1 = \frac{1}{2}(1 + \cos\beta)\left(\lambda_1\alpha_1 + \lambda_2\alpha_* \cos^2\beta\right)\rho g H \tag{11.55}$$

while the uplift pressure (p_u) is:

$$p_u = \frac{1}{2}(1 + \cos\beta)\lambda_3\alpha_1\alpha_3\rho g H \tag{11.56}$$

where λ_1, λ_2, and $\lambda_3$are modification factors applied to the baseline Goda equation to account for different surface topologies. The remaining pressures p_2 and p_3 are given by Eqs. (11.8) and (11.9), respectively, while the α constants remain unchanged from (11.12) to (11.15). However, the maximum wave runup (η^*) is now:

$$\eta^* = 0.75(1 + \cos\beta)\lambda_1 H \tag{11.57}$$

which has been modified through the introduction of the λ_1 parameter. For flat surfaces, $\lambda_1 = \lambda_2 = \lambda_3 = 1$, thus returning the unmodified Goda equation. For other topologies subject to wave forcing, the values of the three λ parameters must be calibrated for the

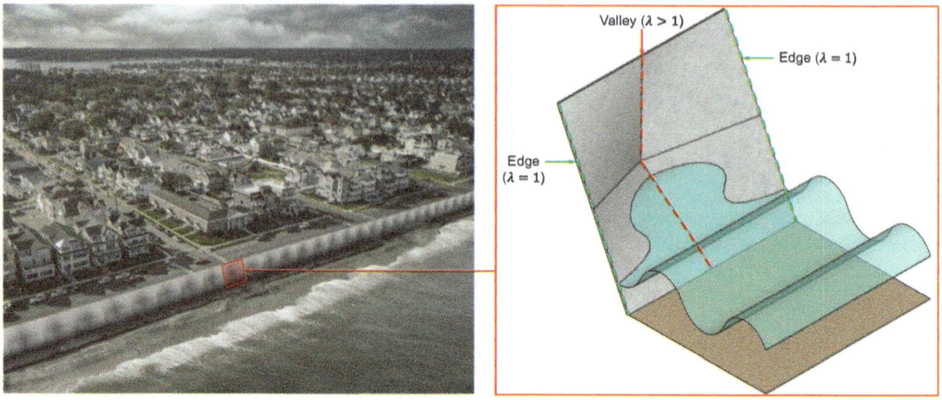

Fig. 11.17 Wave impact on a hypothetical seawall composed of hyperbolic paraboloid modules

particular geometry of interest. As an example, numerical simulations by Wu et al. [24] discovered that wave pressures acting along the valley of seawalls adopting hyperbolic paraboloid (hypar) geometries were much greater than that predicted by Goda's equation, thus necessitating $\lambda > 1$ along the line shown in Fig. 11.17. However, wave pressures on the edges yielded magnitudes comparable to the Goda equation with $\lambda = 1$. The λ parameters for such a geometry must therefore be expressed as a function of position relative to the valley in order to fully capture the distribution of pressures over the entire surface. Nevertheless, it is recommended that experimental or numerical methods of analysis be employed for the hydrodynamic design and analysis of unconventional topologies such as hypars or other curved surfaces [21, 22].

11.8 Example: Wave Forces on an Offshore Breakwater

An example is provided to illustrate the use of Goda's equation for the determination of wave forces on an offshore breakwater. The breakwater we will examine is based on the sand-filled concrete caissons present at Ibaraki Port near the city of Hitachinaka, Japan (Fig. 11.18). The dimensions of this breakwater were provided by Suh et al. [17].

The breakwater has a total width of $B = 22\,\text{m}$, height of $h_w = 28\,\text{m}$, and average density $\rho_s = 1900\ \text{kg m}^{-3}$. The structure sits in water $h = 24\,\text{m}$ deep and is elevated 5.5 m above the seabed by a rubble mound covered with concrete armor blocks 1.5 m thick. Figure 11.19 presents a cross-section of the Hitachinaka breakwater which has been simplified for the purpose of this example. The breakwater will be evaluated against waves with significant wave height $H_s = 8.3\,\text{m}$ and significant wave period $T_s = 14\,\text{s}$. The direction of wave propagation is perpendicular to the breakwater such that $\beta = 0°$ (this

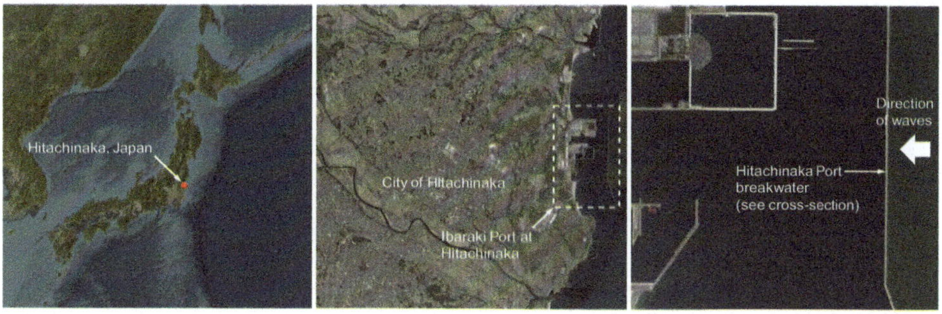

Fig. 11.18 Location of breakwater at Ibaraki Port, Hitachinaka, Japan. Satellite images courtesy of USGS

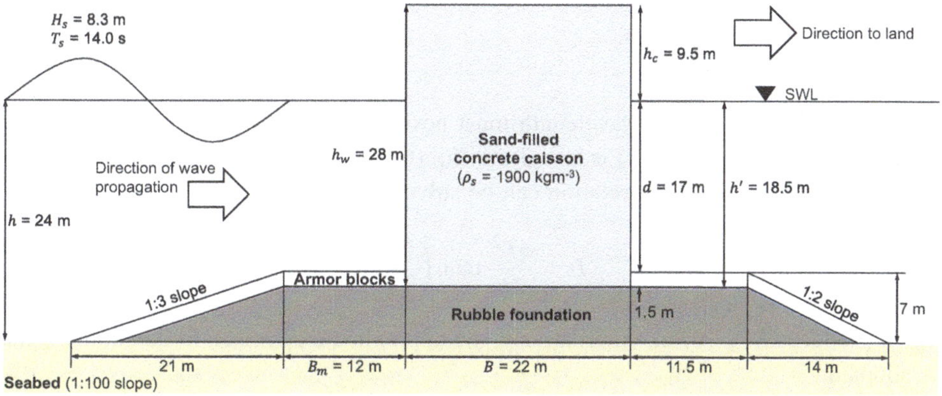

Fig. 11.19 Idealized cross-section of Hitachinaka breakwater (dimensions from Suh et al. [17])

will give the largest wave force). Table 11.2 summarizes the properties of the breakwater and waves used in this example.

11.8.1 Design Wave Parameters

The design wave height H for the given significant wave height of 8.3 m must first be determined. According to Eq. (11.19):

$$H = \min[1.8H_s, H_b] = \min[1.8 \times 8.3, 0.78 \times 24] = 14.9\,\mathrm{m}$$

Note that in this case, the water depth a distance $5H_s$ from the breakwater (h_b) is simply equal to the total water depth h since $5 \times 8.3 = 41.5\,\mathrm{m}$. This distance is beyond the maximum extent of the rubble mound ($21 + 12 = 33\,\mathrm{m}$) hence its slope will not have

Table 11.2 Breakwater and wave properties adopted for the example

Component	Property	Variable	Value
Breakwater	Breakwater height	h_w	28 m
	Breakwater width	B	22 m
	Breakwater density	ρ_s	1900 kg m^{-3}
	Height from SWL to armor layer	d	17 m
	Height of breakwater below SWL	h'	18.5 m
	Height of breakwater above SWL	h_c	9.5 m
	Distance from breakwater to slope of rubble foundation	B_m	12 m
Waves	Water depth	h	24 m
	Significant wave height	H_s	8.3 m
	Significant wave period	T_s	14 s
	Wave incident angle	β	0°

an effect on h_b. The design wavelength must next be calculated based on the significant wave period of $T_s = 14\,\text{s}$. If T_s is not given, Eq. (6.13) may be used as an approximation. For $T_s = 14\,\text{s}$, the dispersion relation can be solved iteratively to yield L:

$$L = \frac{gT_s^2}{2\pi}\tanh\left(\frac{2\pi h}{L}\right)$$

which produces $L = 197\,\text{m}$. Alternatively, the approximate approach to determine L may be adopted:

$$L \approx \frac{gT_s^2}{2\pi}\sqrt{\tanh\left(\frac{4\pi^2 h}{T^2 g}\right)} = \frac{9.81 \times 14^2}{2\pi}\sqrt{\tanh\left(\frac{4\pi^2 \times 24}{14^2 \times 9.81}\right)} = 207\,\text{m}$$

which is 5% off from the exact solution. We will use the exact solution of $L = 197\,\text{m}$ for this example.

11.8.2 Wave Pressures

We wish to determine the distribution of wave pressure on the front wall and the uplift pressure on the base of the caisson as shown in Fig. 11.7. This involves calculating p_1, p_2, p_3, and p_u per Eqs. (11.7)–(11.10), respectively. Hence, we must first solve for the dimensionless α parameters α_1, α_2, and α_3 via Eqs. (11.13)–(11.15). To account for potential impulsive loads from wave breaking, we must also determine α_* via (11.47). First determining α_I:

$$
\begin{aligned}
\delta_{22} &= -0.36\left(\frac{B_m}{L} - 0.12\right) + 0.93\left(\frac{h-d}{h} - 0.6\right) \\
&= -0.36\left(\frac{12}{197} - 0.12\right) + 0.93\left(\frac{24-17}{24} - 0.6\right) \\
&= -0.265
\end{aligned}
$$

Since $\delta_{22} \leq 0$:

$$\delta_2 = 4.9\delta_{22} = 4.9 \times -0.265 = -1.3$$

Solving for δ_1:

$$
\begin{aligned}
\delta_{11} &= 0.93\left(\frac{B_m}{L} - 0.12\right) + 0.36\left(\frac{h-d}{h} - 0.6\right) \\
&= 0.93\left(\frac{12}{197} - 0.12\right) + 0.36\left(\frac{24-17}{24} - 0.6\right) \\
&= -0.166
\end{aligned}
$$

Since $\delta_{11} \leq 0$:

$$\delta_1 = 20\delta_{11} = 20 \times -0.166 = -3.32$$

Since $\delta_2 \leq 0$:

$$\alpha_{I1} = \frac{\cos\delta_2}{\cosh\delta_1} = \frac{\cos(-1.3)}{\cosh(-3.32)} = 0.0193$$

Since $H/d = 14.9/17 = 0.88 \leq 2$:

$$\alpha_{I0} = \frac{H}{d} = \frac{14.9}{17} = 0.88$$

Therefore:

$$\alpha_I = \alpha_{I0}\alpha_{I1} = 0.88 \times 0.0193 = 0.017$$

Next determining α_2:

$$\alpha_2 = \min\left[\frac{h_b - d}{3h_b}\left(\frac{H}{d}\right)^2, \frac{2d}{H}\right] = \min\left[\frac{24-17}{3 \times 24}\left(\frac{14.9}{17}\right)^2, \frac{2 \times 17}{14.9}\right] = 0.075$$

The parameter α_* is therefore:

$$\alpha_* = \max[\alpha_2, \alpha_I] = \max[0.075, 0.017] = 0.075$$

which implies that impulsive effects predicted by Takahashi et al. [19] will not apply for this example. Computing the remaining α parameters:

$$\alpha_1 = 0.6 + \frac{1}{2}\left[\frac{\frac{4\pi h}{L}}{\sinh\left(\frac{4\pi h}{L}\right)}\right]^2 = 0.6 + \frac{1}{2}\left[\frac{\frac{4\pi\times 24}{197}}{\sinh\left(\frac{4\pi\times 24}{197}\right)}\right]^2 = 0.84$$

$$\alpha_3 = 1 - \frac{h'}{h}\left[1 - \frac{1}{\cosh\left(\frac{2\pi h}{L}\right)}\right] = 1 - \frac{18.5}{24}\left[1 - \frac{1}{\cosh\left(\frac{2\pi\times 24}{197}\right)}\right] = 0.82$$

The wave pressures on the caisson can now be determined. The pressure at the still water level (p_1) is:

$$\begin{aligned} p_1 &= \frac{1}{2}(1 + \cos\beta)\left(\alpha_1 + \alpha_* \cos^2\beta\right)\rho g H \\ &= \frac{1}{2}(1 + \cos 0)\left(0.84 + 0.075\cos^2 0\right) \times 1025 \times 9.81 \times 14.9 \\ &= 137 \times 10^3\ \text{Pa} \end{aligned}$$

To obtain the pressure at the top of the breakwater (p_2), we must first determine the wave runup height (η^*):

$$\eta^* = 0.75(1 + \cos\beta)H = 0.75(1 + \cos 0) \times 14.9 = 22.35\ \text{m}$$

Since $\eta^* > h_c$, waves are expected to overtop the breakwater and:

$$p_2 = p_1\left(1 - \frac{h_c}{\eta^*}\right) = 137 \times 10^3 \times \left(1 - \frac{9.5}{22.35}\right) = 79 \times 10^3\ \text{Pa}$$

The pressure at the base of the breakwater (p_3) is:

$$p_3 = \alpha_3 p_1 = 0.82 \times 137 \times 10^3 = 112 \times 10^3\ \text{Pa}$$

whereas the maximum uplift pressure (p_u) is:

$$\begin{aligned} p_u &= \frac{1}{2}(1 + \cos\beta)\alpha_1\alpha_3\rho g H \\ &= \frac{1}{2}(1 + \cos 0) \times 0.84 \times 0.82 \times 1025 \times 9.81 \times 14.9 = 103 \times 10^3\ \text{Pa} \end{aligned}$$

11.8.3 Resultant Forces and Overturning Moments

The distribution of horizontal and uplift wave pressures on the caisson is visualized in Fig. 11.20. Using the computed pressures, the total horizontal (F_W) and uplift (F_U) wave force per unit length of the breakwater can also be determined via:

$$F_W = \frac{1}{2}(p_1 + p_2)h^* + \frac{1}{2}(p_1 + p_3)h'$$
$$= \frac{1}{2}(137 + 79) \times 9.5 + \frac{1}{2}(137 + 112) \times 18.5 = 3330\,\text{kN/m}$$

$$F_U = \frac{1}{2}p_u B = \frac{1}{2} \times 103 \times 22 = 1130\,\text{kN/m}$$

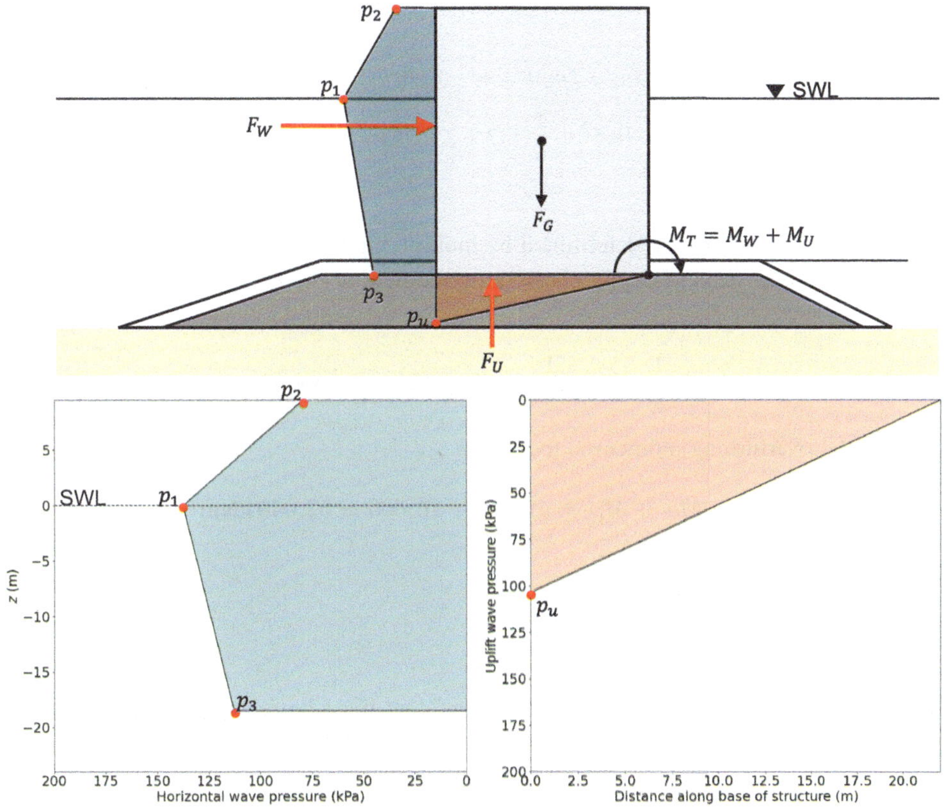

Fig. 11.20 Distribution of horizontal and uplift wave pressures on the caisson

where:

$$h^* = \min\left[h_c, \eta^*\right] = \min[9.5, 22.35] = 9.5\,\text{m}$$

is the height of wave runup on the wall. Since waves are expected to overtop the wall, h^* will simply equal the freeboard h_c. The point of application of the horizontal force F_W above the base is:

$$\begin{aligned} h_F &= \frac{(p_3 + 2p_1)h'^2 + (p_1 + 2p_2)h^{*2} + 3h'(p_2 + p_1)h^*}{3\left[(p_2 + p_1)h^* + (p_1 + p_3)h'\right]} \\ &= \frac{(112 + 2 \times 137) \times 18.5^2 + (137 + 2 \times 79) \times 9.5^2 + 3 \times 18.5 \times (79 + 137) \times 9.5}{3[(79 + 137) \times 9.5 + (137 + 112) \times 18.5]} \\ &= 13.6\,\text{m} \end{aligned}$$

which is approximately located at half the total height of the breakwater. The overturning moment associated with the horizontal wave force F_W is:

$$\begin{aligned} M_W &= \frac{1}{6}(2p_1 + p_3)h'^2 + \frac{1}{6}(p_1 + 2p_2)h^{*2} + \frac{1}{2}(p_1 + p_2)h^*h' \\ &= \frac{1}{6}(2 \times 137 + 112) \times 18.5^2 + \frac{1}{6}(137 + 2 \times 79) \times 9.5^2 + \frac{1}{2}(137 + 79) \times 9.5 \times 18.5 \\ &= 45{,}440\,\text{kNm/m} \end{aligned}$$

This moment can also be determined by multiplying the horizontal force with its lever arm $M_W = F_W h_F$. Likewise, the overturning moment at the heel due to the uplift force F_U is:

$$M_U = \frac{1}{3}p_u B^2 = \frac{1}{3} \times 103 \times 22^2 = 16{,}600\,\text{kNm/m}$$

The total overturning moment is therefore:

$$M_T = M_W + M_U = 45{,}440 + 16{,}600 = 62{,}040\,\text{kNm/m}$$

11.8.4 Factors of Safety

The horizontal wave force and overturning moment must be resisted by the self-weight (F_G) of caisson. The total self-weight after accounting for buoyancy (due to water being present on both sides) per unit length of the breakwater is:

$$\begin{aligned} F_G &= \rho_s g B h_w - \rho g B h' \\ &= 1900 \times 9.81 \times 22 \times 28 - 1025 \times 9.81 \times 22 \times 18.5 \\ &= 7390 \times 10^3\,\mathrm{N/m} \end{aligned}$$

which yields a total restoring moment (M_R) per unit length of:

$$M_R = F_G\left(\frac{B}{2}\right) = 7390 \times \left(\frac{22}{2}\right) = 81{,}290\,\mathrm{kNm/m}$$

The factor of safety against overturning of the Hitachinaka breakwater is:

$$SF_{overturning} = \frac{M_R - M_U}{M_W} = \frac{81{,}290 - 16{,}600}{45{,}440} = 1.42 > 1$$

Hence, the caisson will not overturn under the design wave. The force required to initiate sliding of the caisson is:

$$F_S = \mu F_G = 0.6 \times 7390 = 4434\,\mathrm{kN/m}$$

when assuming $\mu = 0.6$. The factor of safety against sliding of the Hitachinaka breakwater is therefore:

$$SF_{sliding} = \frac{F_S}{F_W} = \frac{4434}{3330} = 1.33 > 1$$

Which implies that the caisson will not slide under the design wave.

11.8.5 Bearing Stress

We can next determine the maximum bearing stress. To do so, we first calculate the horizontal distance from the heel to the bearing force resultant:

$$t_e = \frac{M_R - M_T}{F_G} = \frac{81{,}290 - (16{,}600 + 45{,}440)}{7390} = 2.60\,\mathrm{m}$$

Since $t_e < B/3$ (7.33 m), the maximum bearing pressure will be:

$$q_1 = \frac{2F_G}{3t_e} = \frac{2 \times 7390}{3 \times 2.6} = 1895\,\mathrm{kPa}$$

The bearing capacity of the foundation must therefore exceed this stress to avoid bearing failure.

Table 11.3 Calculation of Goda wave pressures on a wall

Step 1	Determine design wave height H via (11.19)	$H = \min\left[1.8H_s, H_b\right]$ where H_b is from (9.13) at a depth h_b
Step 2	Determine design wave period T via (11.18)	$T = T_s \approx 1.2T_m \approx 3.3H_s^{0.63}$
Step 3	Determine design wavelength L via (11.16) or (11.17)	$L = \frac{gT^2}{2\pi}\tanh\left(\frac{2\pi h}{L}\right) \approx \frac{gT^2}{2\pi}\sqrt{\tanh\left(\frac{4\pi^2 h}{T^2 g}\right)}$
Step 4	Determine wave runup via (11.11)	$\eta^* = 0.75(1+\cos\beta)H$
Step 5	Determine Goda wave pressures on wall via (11.7)–(11.10) after solving for α coefficients	$p_1 = \frac{1}{2}(1+\cos\beta)\left(\alpha_1 + \alpha_* \cos^2\beta\right)\rho g H, \quad p_3 = \alpha_3 p_1$ $p_2 = \begin{cases} p_1\left(1 - \frac{h_c}{\eta^*}\right), & \eta^* > h_c \\ 0, & \eta^* \le h_c \end{cases}$ $p_u = \frac{1}{2}(1+\cos\beta)\alpha_1\alpha_3\rho g H$

11.9 Summary: Goda Wave Forces on Vertical Walls

The procedure used to determine Goda wave pressures from waves with a design wave height H and period T is presented in Table 11.3. Table 11.4 then summarizes the calculation procedure to determine the stability of a vertical caisson placed on an elevated mound using the wave pressures obtained from Table 11.3. Note that the following parameters are required before carrying out the calculations (see Fig. 11.21):

Wave parameters	Structural and bathymetric parameters
• Significant wave height (H_s) • Significant wave period (T_s) • Angle of wave approach (β) • Water density (ρ)	• Height of caisson (h_w) • Width of caisson (B) • Average caisson density (ρ_s) • Water depth at toe of mound (h) • Water depth at top of armor layer (d) • Water depth at bottom of armor layer (h') • Water depth at $D_m = 5H_s$ from caisson (h_b) • Distance from toe to slope of mound (B_m)

Table 11.4 Stability assessment of a vertical caisson

Step 1	Determine p_1, p_2, p_3, and p_u per Table 11.3	
Step 2	Determine resultant forces due to wave and hydrostatic effects via (11.20), (11.21), and (11.26)	$F_W = \frac{1}{2}(p_1 + p_2)h^* + \frac{1}{2}(p_1 + p_3)h', \quad h^* = \min[h_c, \eta^*]$ $F_U = \frac{1}{2}p_u B, \quad F_H = \frac{\rho g h'^2}{2}$
Step 3	Determine point of application of resultant force on wall via (11.23)	$h_F = \frac{(p_3+2p_1)h'^2+(p_1+2p_2)h^{*2}+3h'(p_2+p_1)h^*}{3[(p_2+p_1)h^*+(p_1+p_3)h']}$
Step 4	Determine wave and hydrostatic overturning moments via (11.24), (11.25), and (11.27)	$M_W = h_F F_W, \quad M_U = \frac{p_u B^2}{3}$ $M_H = \frac{1}{6}\rho g h'^3$
Step 5	Determine total horizontal force and overturning moment on structure via (11.28) and (11.29)	$F_T = \begin{cases} F_W + F_H, & \text{Water on one side only} \\ F_W, & \text{Water on both sides} \end{cases}$ $M_T =$ $\begin{cases} M_W + M_U + M_H, & \text{Water on one side only} \\ M_W + M_U, & \text{Water on both sides} \end{cases}$
Step 6	Determine weight of structure after accounting for buoyancy via (11.30)	$F_G =$ $\begin{cases} \rho_s g B h_w, & \text{Water on one side only} \\ \rho_s g B h_w - \rho g B h', & \text{Water on both sides} \end{cases}$
Step 7	Determine restoring moment and sliding force via (11.31) and (11.33)	$M_R = F_G\left(\frac{B}{2}\right), \quad F_S = \mu F_G$
Step 8	Determine safety factors for overturing and sliding via (11.32) and (11.34)	$SF_{overturning} = \frac{M_R - M_H - M_U}{M_W}, \quad SF_{sliding} = \frac{F_S}{F_T}$
Step 9	Determine maximum bearing pressure via (11.36) and compare with bearing capacity of foundation	$q_1 = \begin{cases} \frac{2F_G}{B}\left(2 - \frac{3t_e}{B}\right), & t_e > B/3 \\ \frac{2F_G}{3t_e}, & t_e \leq B/3 \end{cases}, \quad t_e = \frac{M_R - M_T}{F_G}$

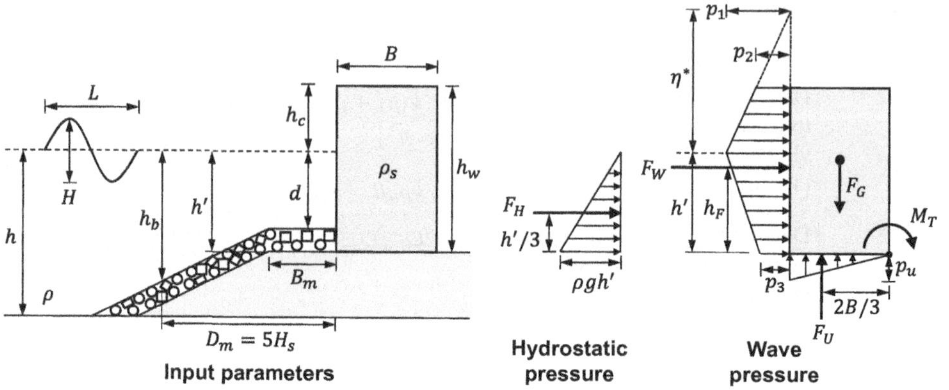

Fig. 11.21 Summary of input parameters along with hydrostatic and wave pressures

References

1. Allsop, N.W.H., Vicinanza, D., McKenna, J.E., 1996. Wave forces on vertical and composite breakwaters (Strategic Research Report SR 443). HR Wallingford, Wallingford, UK.
2. Amini, M., Memari, A.M., 2020. Review of literature on performance of coastal residential buildings under hurricane conditions and lessons learned. Journal of Performance of Constructed Facilities 34.
3. Bagnold, R.A., 1939. Interim report on wave-pressure research. Journal of the Institution of Civil Engineers 12, 202–226.
4. Gerstner, F.J., 1809. Theorie der Wellen. Annalen der Physik 32, 412–445.
5. Goda, Y., 2000. Random seas and design of maritime structures. World Scientific Publishing Co., Singapore.
6. Goda, Y., 1974. New wave pressure formulae for composite breakwaters, in: Proceedings of the 14th International Coastal Engineering Conference. pp. 1702–1720.
7. Goda, Y., Takahashi, S., Yagyu, T., Yamamoto, S., 2009. The Technical Standards and Commentaries for Port and Harbour Facilities in Japan.
8. Hiroi, I., 1920. The force and power of waves. The Engineer 184–187.
9. Hutchinson, S., Young, M., Macleod, A., 2011. Caisson breakwater design for sliding. Coastal Engineering Proceedings 1.
10. Kortenhaus, A., Oumeraci, H., 1998. Classification of wave loading on monolithic coastal structures, in: Coastal Engineering Proceedings. ASCE, pp. 867–880.
11. Minikin, R.R., 1963. Winds, waves and maritime structures. Griffin, London.
12. Oumeraci, H., 1994. Review and analysis of vertical breakwater failures - lessons learned. Coastal Engineering 22, 3–29.
13. Park, H., Tomiczek, T., Cox, D.T., van de Lindt, J.W., Lomonaco, P., 2017. Experimental modeling of horizontal and vertical wave forces on an elevated coastal structure. Coastal Engineering 128, 58–74.
14. Pawitan, K.A., Garlock, M., Wang, S., 2024. Multiphase SPH Analysis of a Breaking Wave Impact on Elevated Structures with Vertical and Inclined Walls. Applied Ocean Research 142.

15. Robertson, I.N., Riggs, H.R., Yim, S.C.S., Young, Y.L., 2007. Lessons from Hurricane Katrina storm surge on Bridges and Buildings. Journal of Waterway, Port, Coastal, and Ocean Engineering 133, 463–483.
16. Sainflou, G., 1928. Essai sur les diques maritimes verticales. Annales des Ponts et Chaussees 98, 5–48.
17. Suh, K.-D., Kim, S.-W., Mori, N., Mase, H., 2012. Effect of climate change on performance-based design of caisson breakwaters. Journal of Waterway, Port, Coastal, and Ocean Engineering 138, 215–225.
18. Takahashi, S., 2002. Design of vertical breakwaters. Port and Airport Research Institute, Japan.
19. Takahashi, S., Tanimoto, K., Shimosako, K., 1994. A proposal of impulsive pressure coefficient for the design of composite breakwaters, in: Proc. Int. Conf. Hydro-Technical Engrg. for Port and Harbor Constr. Yokosuka, Japan, pp. 489–504.
20. Tanimoto, K., Kimura, K., 1985. A hydraulic experimental study on trapezoidal caisson breakwaters. Technical note of the Port and Harbour Research Institute, Ministry of Transport, Japan.
21. Wang, S., Garlock, M., Deike, L., Glisic, B., 2022. Feasibility of Kinetic Umbrellas as deployable flood barriers during landfalling hurricanes. Journal of Structural Engineering 148.
22. Wang, S., Notario, V., Garlock, M., Glisic, B., 2021. Parameterization of hydrostatic behavior of deployable hypar umbrellas as flood barriers. Thin-Walled Structures 163.
23. Wiebe, D.M., Park, H., Cox, D.T., 2014. Application of the Goda Pressure Formulae for Horizontal Wave Loads on Elevated Structures. KSCE Journal of Civil Engineering 18, 1573–1579.
24. Wu, G., Garlock, M., Wang, S., 2022. A decoupled SPH-FEM analysis of hydrodynamic wave pressure on hyperbolic-paraboloid thin-shell coastal armor and corresponding structural response. Engineering Structures 268.

Extra Topics on Breaking Wave Forces on Walls

12

The high-intensity impulsive load produced by breaking waves (F_I in Fig. 11.2) will often need to be considered during the design of many different types of coastal and offshore structures such as seawalls or breakwaters (e.g., Fig. 12.1). To demonstrate the intense forces produced by breaking waves, Fig. 12.2 presents a numerical simulation of such a wave impinging on a vertical surface. The force time history graph clearly reflects a rapid increase in the total force as the wave crest impacts the wall, which then reaches a magnitude several times greater than the following quasi-static component due to wave runup. Due to its extreme nature, many attempts have been made since the introduction of Goda's equation to improve the prediction of maximum forces on vertical walls that are susceptible to wave breaking. Beyond coastal engineering, the quantification of impact forces is also important in the aeronautics, shipbuilding, offshore oil and gas, and marine renewable energy industries [7]. In this chapter, we will briefly introduce a few alternative procedures that can be used to calculate forces from breaking or broken waves on walls.

12.1 Breaking Wave Forces from PROVERBS

In the 1990s, an extensive campaign involving small and large-scale wave flume experiments were undertaken in Germany and the United Kingdom under the "Probabilistic design tools for vertical breakwaters" (PROVERBS) project. The results of this project yielded a plethora of design guidelines for offshore breakwaters from a hydraulic, structural, and geotechnical perspective [12]. Of particular interest was the development of predictive equations for the impulsive peak force F_I. One such equation as presented by Allsop and Vicinanza [3] is:

S. Wang, *Applied Ocean Mechanics for Structural Engineering and Climate Adaptation*, Synthesis Lectures on Ocean Systems Engineering,
https://doi.org/10.1007/978-3-032-18199-2_12

Fig. 12.1 Breaking wave impacting against a seawall in Scheveningen, South Holland, circa 1955 (courtesy of the National Archives of the Netherlands)

$$F_I = 15\rho g d^2 \left(\frac{H_s}{d}\right)^{3.134} \tag{12.1}$$

where d is the water depth at the toe of the structure and H_s is the significant wave height. Note that Eq. (12.1) represents the force produced by a 1/250 wave associated with a given H_s. This means that for a series of 500 random waves, F_I would correspond to the average of the largest two waves. In addition to the deterministic method of Eq. (12.1), a probabilistic approach for the estimation of impulsive forces was given by Oumeraci et al. [12]:

$$F_I = F^* \rho g H_b^2 \tag{12.2}$$

Here, H_b is the breaking wave height shown in Fig. 12.3, which can be obtained from Eq. (9.13) or Fig. 9.5. The term F^* denotes the relative wave force which was assumed to follow a Generalized Extreme Value (GEV) distribution given as:

$$F^* = \frac{\alpha}{\gamma}\left\{1 - \left[-\ln P^*\right]^{\gamma}\right\} + \beta \tag{12.3}$$

where α, γ, and β are respectively the scale, shape, and location parameters of the GEV distribution. These parameters are expressed as a function of the seabed slope immediately leading up to the structure and are summarized in Table 12.1. In addition, P^* takes a

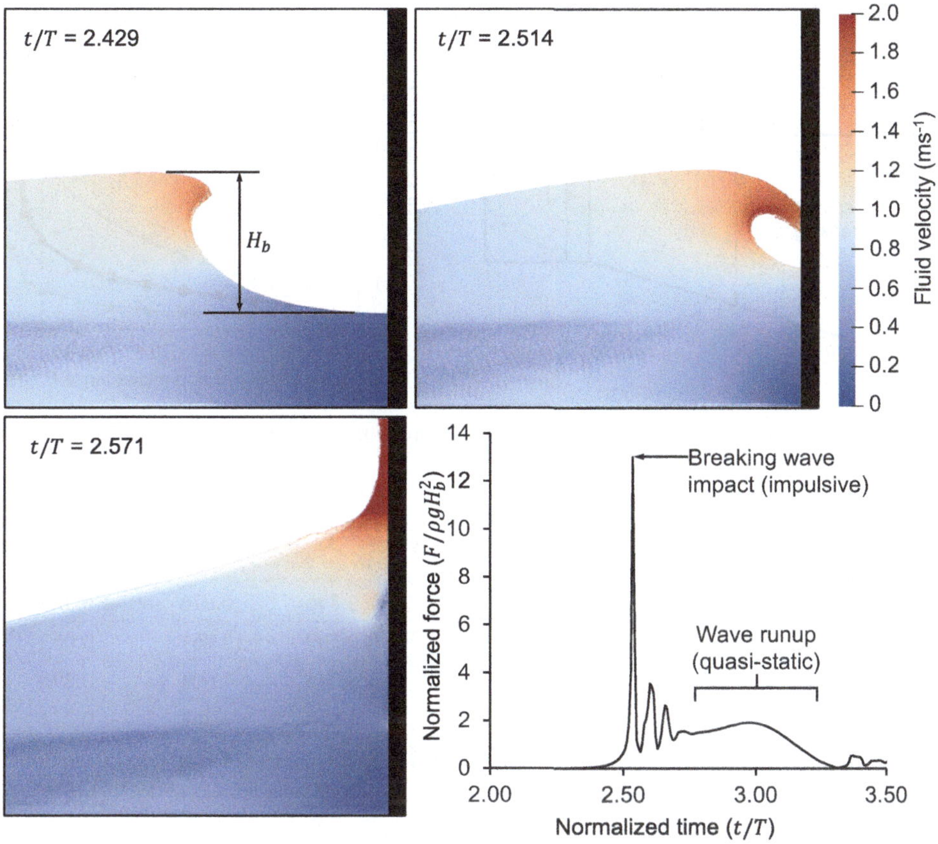

Fig. 12.2 Numerical simulation of a breaking wave impacting a vertical wall and its force time history

value between 0 and 1 which represents the probability of non-exceedance of the impact force F_I, to which $P^* = 0.9$ (i.e., there is a 90% chance that the computed F_I will not be exceeded) is recommended for design. The relationships between F^* and P^* across different slope angles are shown in Fig. 12.3.

12.1.1 Horizontal Impulsive Force

The time taken to reach the maximum impulsive force F_I is referred to as the "rise time" t_r. For breaking wave impact on vertical walls, Oumeraci et al. [12] gives the rise time as:

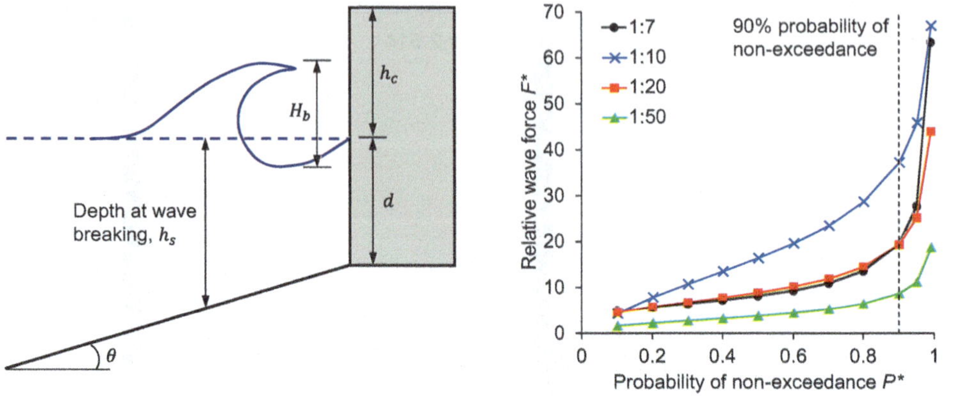

Fig. 12.3 Breaking wave impact on a vertical structure with relationship between F^* and P^* for different slope angles

Table 12.1 GEV parameters as a function of bed slope for use with Eq. (12.3)

Slope	α	β	γ
1:7 ($\theta = 8.13°$)[a]	2.896	6.976	– 0.526
1:10 ($\theta = 5.71°$)[a]	10.209	12.761	– 0.063
1:20 ($\theta = 2.86°$)	3.745	7.604	– 0.295
1:50 ($\theta = 1.15°$)	1.910	3.268	– 0.232

[a] Care should be exercised for slopes of 1:7 or 1:10 due to limited experimental data

$$t_r = \frac{k'}{F^*}\sqrt{\frac{d}{g}} \tag{12.4}$$

where the parameter k' can be described using a log-normal distribution with a mean of 0.086 and a standard deviation of 0.084. The total loading duration of the impact t_d is:

$$t_d = -\frac{c}{\ln t_r} \tag{12.5}$$

where c is normally distributed with a mean of 2.17 and a standard deviation of 1.08. Equations (12.4) and (12.5) effectively idealize the force-time relationship of the impulsive load as a triangle with the peak corresponding to F_I as illustrated in Fig. 12.4. The time leading up to the peak is t_r, while the duration corresponding to the base of the triangle is t_d. The force at any given time t during the impact is then:

$$F(t) = \begin{cases} \frac{F_I}{t_r}t, & 0 \le t \le t_r \\ F_I\left(\frac{1-t/t_d}{1-t_r/t_d}\right), & t_r \le t \le t_d \end{cases} \tag{12.6}$$

where $t = 0$ denotes the initiation of impact. As expected, $F(t) = F_I$ when $t = t_r$.

The distribution of breaking wave pressures giving rise to the total force $F(t)$ at any time $0 \le t \le t_d$ follows a trapezoidal shape (similar to the Goda equation) and is shown in Fig. 12.5. The maximum pressure at the SWL is:

$$p_1(t) = \frac{F(t)}{0.4H_b + 0.7(d + d_c)} \tag{12.7}$$

where $d_c = h' - d$ is the height of the armor blocks. The pressure at the base of the wall is:

$$p_3(t) = 0.45 p_1(t) \tag{12.8}$$

At the crest of the structure, the pressure p_2 depends on whether or not the breaking wave will overtop the wall. This pressure is given as:

$$p_2(t) = \begin{cases} 0, & \eta^* < h_c \\ \frac{\eta^* - h_c}{\eta^*} p_1(t), & \eta^* \ge h_c \end{cases} \tag{12.9}$$

where $\eta^* = 0.8H_b$ is the breaking wave runup. However, it is important to understand that the prediction of maximum breaking wave forces F_I via PROVERBS were largely based on non-overtopping waves. If overtopping is expected to occur (i.e., $\eta^* \ge h_c$), the maximum force from Eqs. (12.1) and (12.2) must be reduced by the additional force that would have been present if no overtopping were to occur. The maximum total force in the presence of overtopping ($F_{I,ov}$) is then:

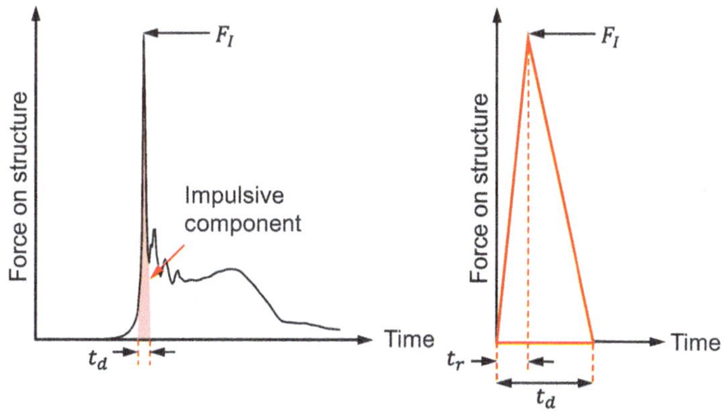

Fig. 12.4 Left: Typical force time history of breaking wave impact. Right: Idealization of the impulsive component

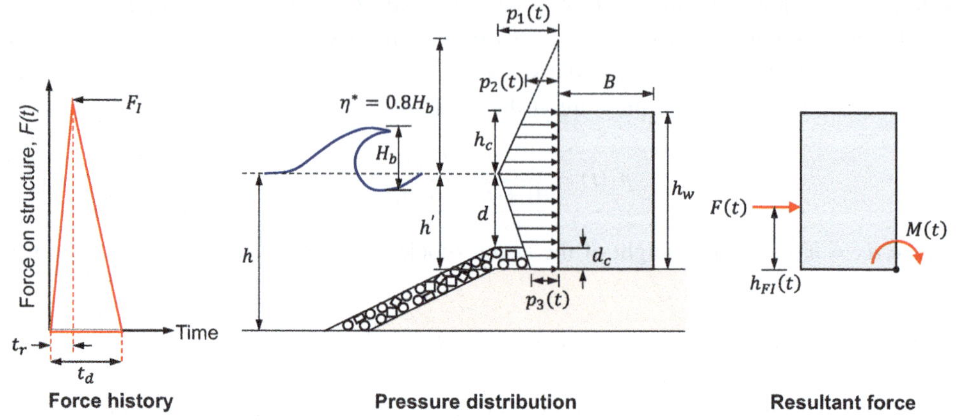

Fig. 12.5 Distribution of breaking wave pressures and force resultant on a vertical wall

$$F_{I,ov} = F_I - \frac{1}{2}(\eta^* - h_c)p_2(t_r), \quad \eta^* \geq h_c \tag{12.10}$$

where p_2 is taken when $t = t_r$. In addition, the statistical parameters presented in Table 12.1 may no longer be valid in the case of overtopping waves. The point of application of the force resultant $F(t)$ above the base of the wall is finally given as:

$$h_{FI}(t) = \frac{(p_3 + 2p_1)h'^2 + (p_1 + 2p_2)h^{*2} + 3h'(p_2 + p_1)h^*}{3\left[(p_2 + p_1)h^* + (p_1 + p_3)h'\right]} \tag{12.11}$$

where

$$h^* = \min\left[h_c, \eta^*\right] \tag{12.12}$$

The overturning moment due to the impact force is thus:

$$M(t) = F(t)h_{FI}(t) \tag{12.13}$$

with F and h_{FI} given by Eqs. (12.6) and (12.11), respectively, and are illustrated in Fig. 12.5.

12.1.2 Uplift Impulsive Force

In addition to the horizontal impact force, the maximum uplift force below the structure due to breaking wave impact can also be determined using Eqs. (12.2) and (12.3), albeit with different statistical parameters of $\alpha = 2.17$, $\beta = 4.384$, and $\gamma = -0.11$. Using these parameters, Eq. (12.3) becomes:

$$F_U^* = -19.727\left\{1 - \left[-\ln P^*\right]^{-0.11}\right\} + 4.384 \tag{12.14}$$

with P^* being the probability of non-exceedance. The total force is then:

$$F_{UI} = F_U^* \rho g H_b^2 \tag{12.15}$$

Note that Eq. (12.15) represents an upper limit of the uplift force. Oumeraci et al. [12] recommends that this limit be compared with an alternative upper bound where the maximum force is determined from an assumed pressure distribution over the base of the structure. Here, the bottom pressures decrease linearly from a value of p_3 on the seaward side to $0.2p_3$ on the leeward side of the structure. This results in a force of:

$$F_{UI} = \frac{B(p_3 + 0.2p_3)}{2} = 0.6p_3 B \tag{12.16}$$

The maximum uplift force is then taken as the minimum of the two approaches, being:

$$F_{UI} = \min\begin{cases} F_U^* \rho g H_b^2 \\ 0.6 p_3 B \end{cases} \tag{12.17}$$

where p_3 is from Eq. (12.8) when $t = t_r$ (i.e., the maximum pressure at the base) and B is the width of the structure. The rise time of the uplift force can next be obtained using Eq. (12.4), albeit with k' taking a mean value of 0.16 and standard deviation 0.17. The mean rise time for F_{UI} is then:

$$t_{Ur} = \frac{0.16}{F_U^*}\sqrt{\frac{d}{g}} \tag{12.18}$$

Likewise, the total loading duration is from Eq. (12.5) where the parameter c exhibits a mean of 1.88 and standard deviation 0.99. The mean loading duration of F_{UI} thus becomes:

$$t_{Ud} = -\frac{1.88}{\ln t_{Ur}} \tag{12.19}$$

Similar to Eq. (12.6), the uplift force at any given time t during the impact is then:

$$F_U(t) = \begin{cases} \frac{F_{UI}}{t_{Ur}} t, & 0 \le t \le t_{Ur} \\ F_{UI}\left(\frac{1 - t/t_{Ud}}{1 - t_{Ur}/t_{Ud}}\right), & t_{Ur} \le t \le t_{Ud} \end{cases} \tag{12.20}$$

where $t = 0$ denotes the start of impact and $F_U(t) = F_{UI}$ when $t = t_{Ur}$.

After obtaining the maximum uplift force F_{UI}, the uplift pressure distribution below the structure as illustrated in Fig. 12.6 can next be found. At the leeward side, the uplift pressure is:

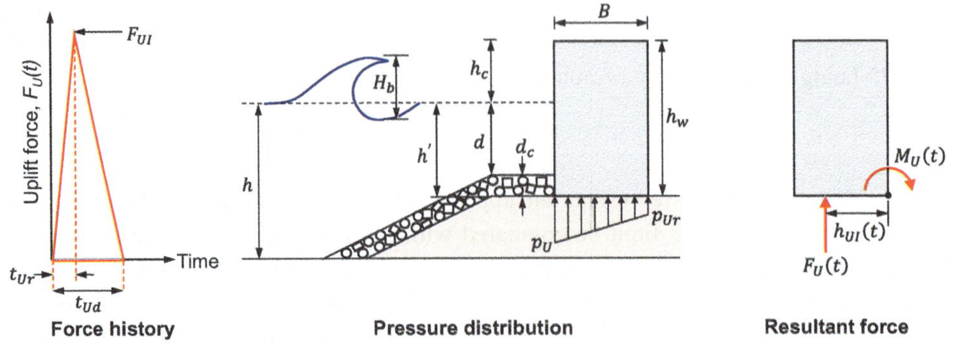

Fig. 12.6. Distribution of breaking wave uplift pressures and force resultant below a structure

$$p_{Ur} = \rho g H_b \left(\frac{H_b}{h} - 0.1 \right) \tag{12.21}$$

Using p_{Ur}, the uplift pressure at the seaward side can then be calculated:

$$p_U(t) = \frac{2F_U(t)}{B} - p_{Ur} \tag{12.22}$$

where B denotes the width of the structure. Finally, the horizontal lever arm between the resultant uplift force F_U and the leeward side of the structure is determined via:

$$h_{UI}(t) = \frac{B^2(p_{Ur} + 2p_U)}{6F_U(t)} \tag{12.23}$$

The overturning moment due to the uplift force is then:

$$M_U(t) = F_U(t) h_{UI}(t) \tag{12.24}$$

with F_U and h_{UI} given by Eqs. (12.20) and (12.23), respectively, and are shown in Fig. 12.6.

12.2 Breaking Wave Forces from VOWS

Cuomo et al. [6] leveraged additional wave flume experiments under the "Violent Overtopping by Waves at Seawalls" (VOWS) project to formulate expressions that predict both the maximum impulsive force (F_I) and the succeeding quasi-static component (F_{QS}) shown in Fig. 11.2. By measuring the 1/250 wave from a train of 1000 irregular waves (i.e., average of the four largest forces), the impulsive force peak is given as:

$$F_I = C_r^{1.65} \rho g H_s L_s \left(1 - \frac{|h_b - d|}{d}\right) \tag{12.25}$$

where L_s is the wavelength at the toe of the structure and h_b is the water depth at breaking when evaluated using the Miche [11] breaking criteria:

$$h_b = \frac{1}{k}\text{arctanh}\left(\frac{H_s}{0.14L_s}\right) \tag{12.26}$$

with $k = 2\pi/L_s$ being the wavenumber corresponding to L_s. The parameter C_r is known as the reflection coefficient which is dependent on the freeboard-to-significant wave height (h_c/H_s) ratio. Allsop et al. [2] suggests:

$$C_r = \begin{cases} 1.0, & \frac{h_c}{H_s} > 3.0 \\ 0.9, & 1.0 < \frac{h_c}{H_s} < 3.0 \\ 0.8 + \frac{0.1h_c}{H_s}, & 0.4 < \frac{h_c}{H_s} < 1.0 \end{cases} \tag{12.27}$$

where h_c is the elevation of the structure above the SWL (see Fig. 12.3). The quasi-static force maximum which follows the initial impact force is given as:

$$F_{QS} = 4.8 \rho g H_s^2 \tag{12.28}$$

Note however, that Eqs. (12.25) and (12.28) for F_I and F_{QS}, respectively, were developed using relatively small-scale experiments under the following wave conditions:

- Significant wave height: $0.2 < H_s < 0.7$ m
- Water depth at toe of structure: $0.5 < d < 1.3$ m
- Mean wave period: $2.0 < T_m < 3.7$ s.

Care should be exercised when extrapolating wave forces using parameters that lie significantly beyond these limits. Figure 12.7 demonstrates the impulsive and quasi-static forces computed using (12.25) and (12.28) when assuming a significant wave height of $H_s = 0.5$ m and wavelength $L_s = 8.0$ m. The depth at the toe of the structure d was varied from 0.5 to 1.3 m and C_r was maintained at 1.0.

12.3 Broken Wave Forces

Some overland structures might be placed beyond the reach of breaking waves such that only broken waves can approach them. Camfield [5] introduced a procedure to calculate the maximum force from a broken wave on vertical structures:

$$F = 0.18 \rho g H_b^2 \left(1 - \frac{X_1 \tan\theta}{R_A}\right)^2 \tag{12.29}$$

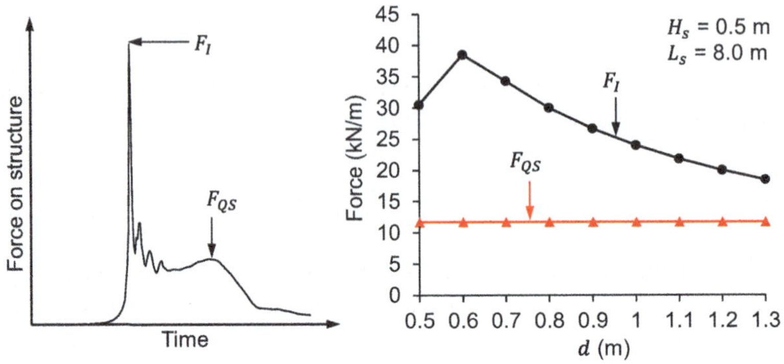

Fig. 12.7. Impulsive and quasi-static force predicted using Eqs. (12.25) and (12.28) for $H_s = 0.5$ m and $L_s = 8.0$ m

where X_1 is the horizontal distance from the shoreline to the structure as shown in Fig. 12.8 and H_b is the breaking wave height which can be calculated using Eq. (9.13) or Fig. 9.5. R_A is the maximum wave runup on a slope with angle θ in the absence of a structure. Ahren and Heimbaugh [1] give the runup for rough slopes as:

$$R_A = \frac{a\xi}{1 + b\xi} H_s \tag{12.30}$$

where $a = 1.022$ and $b = 0.247$, and:

$$\xi = \frac{\tan\theta}{\sqrt{\frac{2\pi H_s}{gT_p^2}}} \tag{12.31}$$

with H_s and T_p denoting the significant wave height and peak period, respectively.

As an example, suppose that we wish to determine the force from broken waves impacting a structure placed $X_1 = 5$ m from the shoreline on a 1:10 slope. The waves have a significant wave height of $H_s = 4.5$ m and $T_p = 12.8$ s, thus matching the example from Sect. 9.3. We must first determine the maximum runup R_A via:

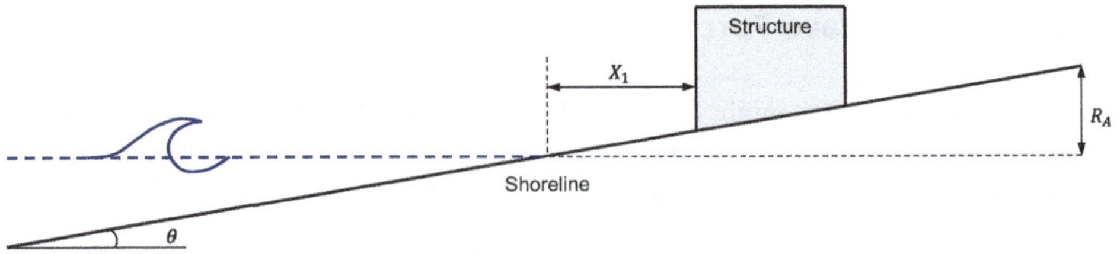

Fig. 12.8 Broken wave impact on a coastal structure

$$\xi = \frac{\tan\theta}{\sqrt{\frac{2\pi H_s}{gT_p^2}}} = \frac{\tan 5.71^\circ}{\sqrt{\frac{2\pi\times 4.5}{9.81\times 12.8^2}}} = 0.754$$

giving:

$$R_A = \frac{a\xi}{1+b\xi}H_s = \frac{1.022\times 0.754}{1+0.247\times 0.754}\times 4.5 = 2.92\ \text{m}$$

From Sect. 9.3, we have previously calculated the breaking wave height to be $H_b = 6.75$ m. The broken wave force on the structure is therefore:

$$\begin{aligned} F &= 0.18\rho g H_b^2\left(1-\frac{X_1\tan\theta}{R_A}\right)^2 \\ &= 0.18\times 1025\times 9.81\times 6.75^2\left(1-\frac{5\tan 5.71^\circ}{2.92}\right) = 68\times 10^3\ \text{N/m} \end{aligned}$$

We can compare this result with the maximum breaking wave force predicted using Eq. (12.1). When assuming the structure is located in water 5.1 m deep such that it is now susceptible to wave breaking, the impulsive force will be:

$$F_I = 15\rho g d^2\left(\frac{H_s}{d}\right)^{3.134} = 15\times 1025\times 9.81\times 5.1^2\left(\frac{4.5}{5.1}\right)^{3.134} = 2650\times 10^3\ \text{N/m}$$

which is significantly greater than that produced by broken waves. Note, however, that this impulsive force occurs over a much shorter duration and should therefore be treated as a dynamic load.

12.4 Example: Breaking Wave Forces on a Vertical Wall

In this example, we will demonstrate how the PROVERBS approach described in Sect. 12.1 can be leveraged to predict the impact forces of a breaking wave impinging a vertical structure. For simplicity, we will assume that the structure is situated on a nearly flat seabed ($\theta \approx 0$) with a constant water depth of 5 m. The impacted caisson is to be 10 m tall and 8 m wide and is shown in Fig. 12.9.

12.4.1 Horizontal Wave Force and Moment

To use the PROVERBS formulas, we must first determine the breaking wave height H_b. Since our bathymetry is essentially flat, Eq. (9.13) gives:

$$H_b = 0.78h = 0.78\times 5 = 3.9\ \text{m}$$

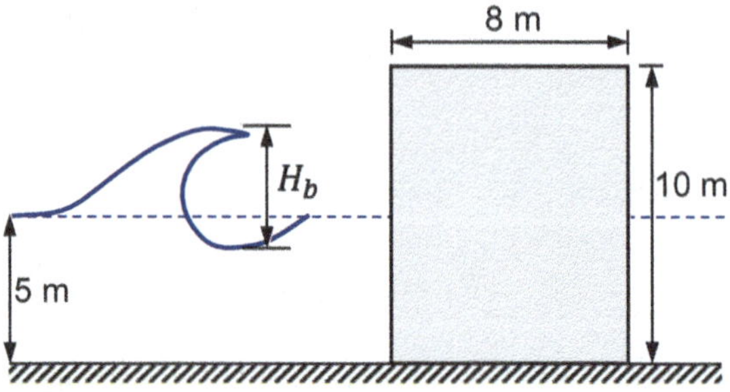

Fig. 12.9 Breaking wave impact on a vertical caisson

as the breaking wave height. Assuming that we are interested a 90% probability of non-exceedance, the relative wave force on the wall is:

$$F^* = \frac{\alpha}{\gamma}\left\{1 - \left[-\ln P^*\right]^{\gamma}\right\} + \beta = \frac{1.910}{-0.232}\left\{1 - \left[-\ln 0.9\right]^{-0.232}\right\} + 3.268 = 8.912$$

where we have used the statistical parameters for a 1:50 slope. This gives a maximum horizontal wave force of:

$$F_I = F^* \rho g H_b^2 = 8.912 \times 1025 \times 9.81 \times 3.9^2 = 1363 \times 10^3 \mathrm{N/m}$$

The average rise time of this force is then:

$$t_r = \frac{k'}{F^*}\sqrt{\frac{d}{g}} = \frac{0.086}{8.912}\sqrt{\frac{5}{9.81}} = 0.00689 \text{ s}$$

and the total loading duration is:

$$t_d = -\frac{c}{\ln t_r} = -\frac{2.17}{\ln 0.00689} = 0.436 \text{ s}$$

We can now determine the pressure distribution acting on the wall. Assuming that we are only interested in the pressure at the maximum instant of loading, i.e., when $F(t_r) = F_I$, the pressure at the SWL is:

$$p_1 = \frac{F_I}{0.4H_b + 0.7(d + d_c)} = \frac{1363 \times 10^3}{0.4 \times 3.9 + 0.7(5 + 0)} = 269 \times 10^3 \text{ Pa}$$

while at the base of the wall, the pressure is:

$$p_3 = 0.45 p_1 = 0.45 \times 269 \times 10^3 = 121 \times 10^3 \text{ Pa}$$

To calculate the pressure at the crest of the structure, we need to know the maximum runup:

$$\eta^* = 0.8H_b = 0.8 \times 3.9 = 3.12 \text{ m}$$

since η^* is smaller than the freeboard ($h_c = 5$ m), no overtopping is expected and:

$$p_2 = 0$$

The point of application of F_I above the base is then:

$$\begin{aligned} h_{FI} &= \frac{(p_3 + 2p_1)h'^2 + (p_1 + 2p_2)h^{*2} + 3h'(p_2 + p_1)h^*}{3\left[(p_2 + p_1)h^* + (p_1 + p_3)h'\right]} \\ &= \frac{(121 + 2 \times 269) \times 5^2 + (269 + 2 \times 0) \times 3.12^2 + 3 \times 5 \times (0 + 269) \times 3.12}{3[(0 + 269) \times 3.12 + (269 + 121) \times 5]} \\ &= 3.79 \text{ m} \end{aligned}$$

where:

$$h^* = \min\left[h_c, \eta^*\right] = 3.12 \text{ m}$$

The overturning moment at the heel of the structure due to the horizontal force of impact is therefore:

$$M = F_I h_{FI} = 1363 \times 3.79 = 5166 \text{ kNm/m}$$

Note that this represents a dynamic moment occurring over an extremely short duration.

12.4.2 Uplift Force and Moment

We can next calculate the maximum uplift force acting on the base of the caisson. When adopting a 90% probability of non-exceedance ($P^* = 0.9$), the normalized peak uplift force is:

$$F_U^* = -19.727\left\{1 - \left[-\ln P^*\right]^{-0.11}\right\} + 4.384 = 9.925$$

The maximum uplift force is then calculated as:

$$\begin{aligned} F_{UI} &= \min\begin{cases} F_U^* \rho g H_b^2 \\ 0.6 p_3 B \end{cases} = \min\begin{cases} 9.925 \times 1025 \times 9.81 \times 3.9^2 = 1518 \times 10^3 \\ 0.6 \times 121 \times 10^3 \times 8 = 581 \times 10^3 \end{cases} \\ &= 581 \times 10^3 \text{ N/m} \end{aligned}$$

This uplift force occurs following a mean rise time of:

$$t_{Ur} = \frac{0.16}{F_U^*}\sqrt{\frac{d}{g}} = \frac{0.16}{9.925}\sqrt{\frac{5}{9.81}} = 0.0115 \text{ s}$$

In addition, the total mean duration of the uplift force is:

$$t_{Ud} = -\frac{1.88}{\ln t_{Ur}} = -\frac{1.88}{\ln 0.0115} = 0.421 \text{ s}$$

We notice that while the mean total duration of the uplift force (t_{Ud}) is similar to that of the horizontal breaking wave force (t_d), the mean rise time of the uplift force (t_{Ur}) is approximately 1.7 times longer than that of the horizontal wave force (t_r). To convert F_{UI} into a pressure distribution, we first find the pressure on the leeward side of the structure:

$$p_{Ur} = \rho g H_b\left(\frac{H_b}{h} - 0.1\right) = 1025 \times 9.81 \times 3.9 \times \left(\frac{3.9}{5} - 0.1\right) = 26670 \text{ Pa}$$

which gives an uplift pressure on the seaward side of:

$$p_U = \frac{2F_{UI}}{B} - p_{Ur} = \frac{2 \times 581 \times 10^3}{8} - 26670 = 118580 \text{ Pa}$$

The horizontal lever arm corresponding to the peak uplift force can now be determined via:

$$h_{UI} = \frac{B^2(p_{Ur} + 2p_U)}{6F_{UI}} = \frac{8^2 \times (26670 + 2 \times 118580)}{6 \times 581 \times 10^3} = 4.84 \text{ m}$$

which gives an overturning moment of:

$$M_U = F_{UI} h_{UI} = 581 \times 4.84 = 2812 \text{ kNm/m}$$

12.4.3 Distribution of Impact Forces on Structure

It is important to recognize that due to the different rise times between the horizontal and uplift forces, F_{UI} and M_U will not occur concurrently with F_I and M produced by the horizontal component of wave impact. Figure 12.10 illustrates the distribution of horizontal and uplift wave pressures on the caisson as well as the time history of the resultant impact forces.

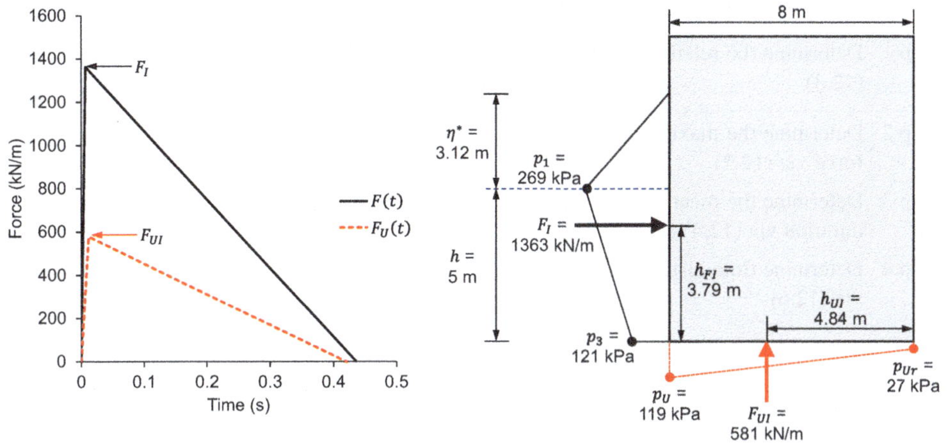

Fig. 12.10 Left: Time history of horizontal and uplift forces. Right: Pressure distribution at maximum force

12.5 Summary: Breaking Wave Forces from PROVERBS Procedure

The procedure used to determine the horizontal and uplift impact forces and pressures using the methodology developed by the PROVERBS project is summarized in this section. In addition to the dimensions of the impacted structure and bathymetry, only the breaking wave height H_b, determined using Eq. (9.13), and the probability of non-exceedance P^* are required as input. Tables 12.2 and 12.3 summarize the procedure used to determine the horizontal and uplift components of loading, respectively. Figure 12.11 illustrates the distribution of these forces and pressures on a vertical caisson. Once the time-varying forces and moments are determined, they can then be used for dynamic analysis to assess the performance of structures subjected to impulsive breaking waves.

12.6 ASCE 7 Guidelines for Nonbreaking and Breaking Waves

The Goda equation or the alternative formulations presented in this chapter may be used for all onshore and offshore structures interacting with waves. However, for overland structures subject to flood loads where the water is not excessively deep, the pressure distribution and total force acting on walls may be estimated based on more simplified methods of analysis. For this purpose, the American Society of Civil Engineers (ASCE) stipulates procedures to determine loads that must be accounted for during the design of vertical walls impacted by floods via ASCE 7-22 [4].

Table 12.2 Calculation of horizontal breaking wave forces and pressures via PROVERBS

Step 1	Determine the relative wave force via (12.3)	$F^* = \frac{\alpha}{\gamma}\left\{1 - \left[-\ln P^*\right]^{\gamma}\right\} + \beta$ where α, β, and γ are given in Table 12.1
Step 2	Determine the maximum horizontal wave force via (12.2)	$F_I = F^* \rho g H_b^2$
Step 3	Determine the mean rise time and impact duration via (12.4) and (12.5)	$t_r = \frac{0.086}{F^*}\sqrt{\frac{d}{g}},\ t_d = -\frac{2.17}{\ln t_r}$
Step 4	Determine time history of impact force via (12.6)	$F(t) = \begin{cases} \frac{F_I}{t_r} t, & 0 \le t \le t_r \\ F_I\left(\frac{1-t/t_d}{1-t_r/t_d}\right), & t_r \le t \le t_d \end{cases}$
Step 5	Determine the maximum pressure at the SWL and base via (12.7) and (12.8)	$p_1(t) = \frac{F(t)}{0.4H_b+0.7(d+d_c)},\ p_3(t) = 0.45 p_1(t)$
Step 6	Determine the maximum pressure at the crest via (12.9)	$p_2(t) = \begin{cases} 0, & \eta^* < h_c \\ \frac{\eta^*-h_c}{\eta^*} p_1(t), & \eta^* \ge h_c \end{cases},\ \eta^* = 0.8H_b$
Step 7	Adjust F_I from Step 2 if overtopping occurs via (12.10)	$F_{I,ov} = F_I - \frac{1}{2}(\eta^* - h_c) p_2(t_r) \quad , \quad \eta^* \ge h_c$
Step 8	Determine height of the force resultant via (12.11)	$h_{FI}(t) = \frac{(p_3+2p_1)h'^2 + (p_1+2p_2)h^{*2} + 3h'(p_2+p_1)h^*}{3\left[(p_2+p_1)h^* + (p_1+p_3)h'\right]}$ where $h^* = \min\left[h_c, \eta^*\right]$
Step 9	Determine the dynamic moment via (12.13)	$M(t) = F(t) h_{FI}(t)$

12.6.1 Nonbreaking Waves

For nonbreaking waves, all dynamic effects associated with waves and moving water are handled entirely by a hydrostatic approach, where hydrostatic pressures are increased due to an additional surcharge depth (d_h) added to the local still water depth (d_s). In addition, the free surface d_s is to be further elevated by 0.3 m to account for wave effects. The equivalent hydrostatic pressure distribution to be sustained by walls exposed to nonbreaking or broken waves is shown via Fig. 12.12, with the maximum pressure at the base being:

$$p_{max} = \rho g(d_s + 0.3 + d_h) \tag{12.32}$$

The surcharge height d_h is a function of the average water velocity V and drag coefficient a associated with the geometry of the structure:

Table 12.3 Calculation of uplift breaking wave forces and pressures via PROVERBS

Step 1	Determine the relative wave force via (12.14)	$F_U^* = -19.727\left\{1 - \left[-\ln P^*\right]^{-0.11}\right\} + 4.384$
Step 2	Determine the maximum uplift force via (12.17)	$F_{UI} = \min\begin{cases} F_U^* \rho g H_b^2 \\ 0.6 p_3 B \end{cases}$
Step 3	Determine the mean rise time and impact duration via (12.18) and (12.19)	$t_{Ur} = \frac{0.16}{F_U^*}\sqrt{\frac{d}{g}}, \quad t_{Ud} = -\frac{1.88}{\ln t_{Ur}}$
Step 4	Determine time history of uplift force via (12.20)	$F_U(t) = \begin{cases} \frac{F_{UI}}{t_{Ur}} t, & 0 \le t \le t_{Ur} \\ F_{UI}\left(\frac{1-t/t_{Ud}}{1-t_{Ur}/t_{Ud}}\right), & t_{Ur} \le t \le t_{Ud} \end{cases}$
Step 5	Determine the uplift pressure on the leeward side via (12.21)	$p_{Ur} = \rho g H_b\left(\frac{H_b}{h} - 0.1\right)$
Step 6	Determine the uplift pressure on the seaward side via (12.22)	$p_U(t) = \frac{2F_U(t)}{B} - p_{Ur}$
Step 7	Determine the horizontal lever arm of the uplift force via (12.23)	$h_{UI}(t) = \frac{B^2(p_{Ur} + 2p_U)}{6F_U(t)}$
Step 8	Determine the dynamic uplift moment via (12.24)	$M_U(t) = F_U(t) h_{UI}(t)$

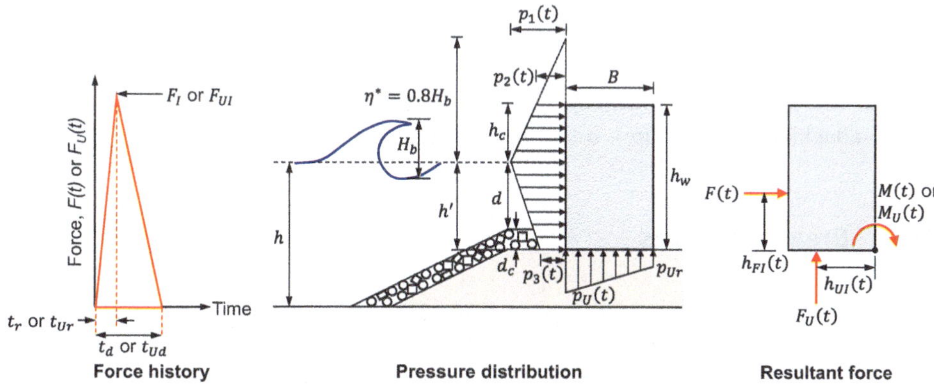

Fig. 12.11 Distribution of horizontal and uplift forces and pressures per PROVERBS

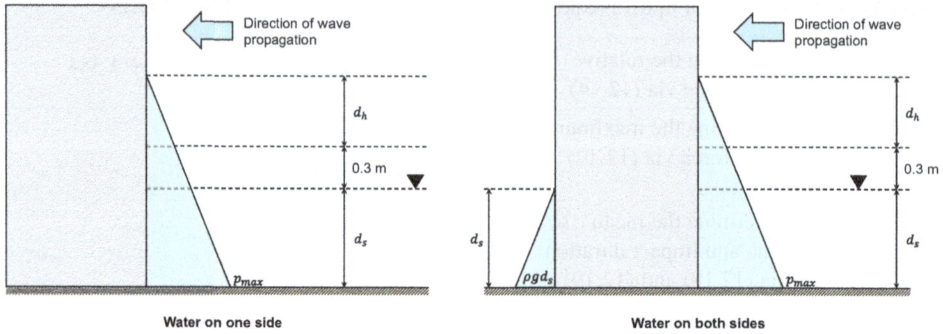

Fig. 12.12 Distribution of nonbreaking wave pressures per ASCE 7-22

$$d_h = \frac{aV^2}{2g} \tag{12.33}$$

The average water velocity V can be estimated based on the local depth d_s following the procedure stipulated by Federal Emergency Management Agency (FEMA) Coastal Construction Manual [8]. Here, the range of values for V falls between:

$$d_s \leq V \leq \sqrt{gd_s} \tag{12.34}$$

with an additional caveat that V cannot exceed 3.05 ms^{-1} for the equivalent hydrostatic method shown in Fig. 12.12 to be adopted for design purposes. If the upper boundary of $\sqrt{gd_s}$ is selected, Eq. (12.32) is only valid up to a maximum inundation of 0.95 m, whereas for the lower limit of d_s, (12.32) may be used for flood levels up to 3.05 m. The drag coefficient a depends on the shape of the impacted surface and should not be less than 1.25 per ASCE guidelines. For a wall perpendicular to the flow (i.e., waves are directly attacking the structure), $a = 2$ may be adopted.

12.6.2 Breaking Waves

The ASCE guidelines for wave pressures associated with breaking waves were based on a technical report prepared by the USACE [13], which in turn derives its procedure from a study by Hom-ma and Horikawa [10]. This approach assumes that waves approaching a wall will be reflected such that its crest is located $1.2d_s$ above the still water level. Hence, a hydrostatic pressure distribution will exist at the wall which varies linearly from 0 at $2.2d_s$ above ground to $2.2\rho g d_s$ at ground level. An additional dynamic pressure distribution resulting from breaking waves will be added to this hydrostatic pressure. The impulsive dynamic pressure varies linearly from 0 at $2.2d_s$ to a maximum of $C_p \rho g d_s$ at the still water level. This pressure then reduces back down to 0 at the ground level in a linear manner as shown in Fig. 12.13. The maximum pressure due to wave breaking at the still water level due to both hydrostatic and hydrodynamic components is therefore:

$$p_{max} = 1.2\rho g d_s + C_p \rho g d_s \tag{12.35}$$

where the first and second terms in Eq. (12.35) refers to the static and dynamic components, respectively. The total breaking wave force per unit length of the structure is therefore:

$$F_t = 1.1 C_p \rho g d_s^2 + 2.4 \rho g d_s^2 \tag{12.36}$$

The dynamic pressure coefficient C_p varies between 1.6 and 3.5 to which its selection is based on the probability of exceedance from experiments performed by Fuhrboter [9]. ASCE 7-22 then correlates this probability of exceedance to the building risk category as summarized in Table 12.4.

Note that care should be exercised when adopting the ASCE 7-22 procedure for determining breaking wave forces on nearshore structures. In fact, even the USACE report

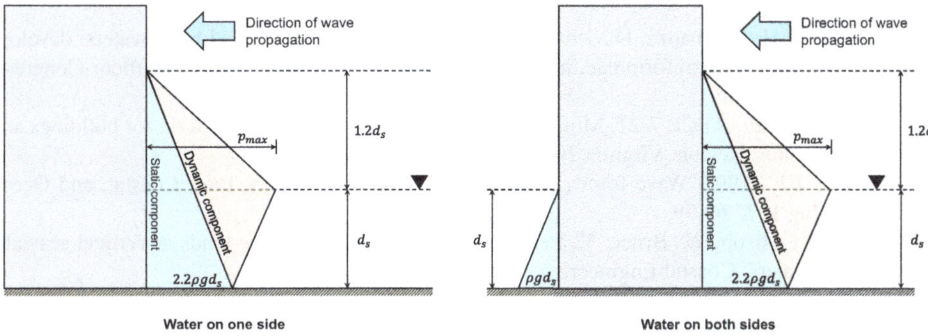

Fig. 12.13 Distribution of breaking wave pressures per ASCE 7-22

Table 12.4 Dynamic pressure coefficients for ASCE 7-22 risk categories

Risk category[a]	Dynamic pressure coefficient (C_p)	Approximate probability of exceedance (%)
I	1.6	50
II	2.8	1
III	3.2	0.25
IV	3.5	0.1

[a] Refer to ASCE 7-22 for detailed definitions of risk categories

[13] from which Eqs. (12.35) and (12.36) were ultimately recommended by ASCE 7-22 states:

> The limitations of these equations are not known; therefore, none of the equations can be used with any level of confidence outside the limited conditions for which they were formulated... It should be recognized that forces calculated via the Hom-ma and Horikawa method may be low compared with many measured shock forces, but reasonable guidance on determination of such shock forces is still not within state-of-the-art coastal engineering predictive ability.

References

1. Ahrens, J.P., Heimbaugh, M.S., 1988. Approximate upper limit of irregular wave runup on riprap (Technical Report CERC-88-5). US Army Engineer Waterways Experiment Station, Vicksburg, MS.
2. Allsop, N.W.H., McBride, M.W., Colombo, D., 1994. The reflection performance of vertical walls and 'low reflection' alternatives: results of wave flume tests, in: Proceedings of the 3rd Project Workshop, MAST II, MCS-Project: Monolithic (Vertical) Coastal Structures. The Netherlands.
3. Allsop, N.W.H., Vicinanza, D., 1996. Wave impact loadings on vertical breakwaters: development of new prediction formulae, in: Proceedings of the 11th International Harbour Congress. Antewerp, Belgium.
4. ASCE/SEI, 2022. ASCE 7-22. Minimum design loads and associated criteria for buildings and other structures. Reston, Virginia 20191.
5. Camfield, F.E., 1991. Wave forces on a wall. Journal of Waterway, Port, Coastal, and Ocean Engineering 117, 76–79.
6. Cuomo, G., Allsop, W., Bruce, T., Pearson, J., 2010. Breaking wave loads at vertical seawalls and breakwaters. Coastal Engineering 57, 424–439.
7. Dias, F., Ghidaglia, J.-M., 2018. Slamming: Recent Progress in the Evaluation of Impact Pressures. Annual Review of Fluid Mechanics 50, 243–273.
8. FEMA, 2000. Coastal Construction Manual (No. P-55). Federal Emergency Management Agency, Washington D.C.
9. Fuhrboter, A., 1986. Model and prototype tests for wave impact and runup on a uniform 1:4 slope. Coastal Engineering 10, 49–84.
10. Hom-ma, M., Horikawa, K., 1964. Wave forces against sea wall, in: Proceedings of the 9th Coastal Engineering Conference. American Society of Civil Engineers, pp. 490–503.
11. Miche, R., 1944. Mouvements ondulatoires des mers en profondeur constante on decroissante. Les Annales des Ponts et Chaussees.
12. Oumeraci, H., Kortenhaus, A., Allsop, N.W.H., De Groot, M.B., Crouch, R.S., Vrijling, J.K., Voortman, H.G., 2001. Probabilistic design tools for vertical breakwaters. Balkema, Rotterdam.
13. Walton, T.L., Ahrens, J.P., Truitt, C.L., Dean, R.G., 1989. Criteria for evaluating coastal flood-protection structures (No. Tech. Report CERC 89-15). U.S. Army Corps of Engineers, Vicksburg, MS.

13 Wave Forces on Piles

Many nearshore and offshore structures are supported by vertical piles to elevate them beyond the reach of water and waves (Fig. 13.1). In addition to carrying the axial load from the weight of the structure, piles must also resist horizontal forces induced by waves which lead to bending and shear stresses within the pile and its foundation. This chapter will introduce methods for the estimation of forces on vertical piles imparted by linear and nonlinear waves, across both nonbreaking and breaking regimes.

13.1 Forcing Components

The total force produced by waves interacting with a pile arises from the summation of the following two components:

- **Drag force (F_D)**: produced by flow separation around the pile.
- **Inertia force (F_I)**: from the acceleration of the flow.

Both the drag and inertia components of the total force vary with time t and depth z down the pile. The distribution of the total pile force $F = F_D + F_I$ (per unit length) along the with total force acting on the entire pile (F_T) and base moment (M_T) are illustrated in Fig. 13.2.

From Chap. 3, we know that waves can be classified as either linear, nonlinear, or breaking per Le Méhauté's abacus (Fig. 3.9). The determination of wave forces resulting from such waves will be discussed in the following sections. Similar to the treatment of wave forces on walls, this chapter will use H and T to denote the design wave height and

S. Wang, *Applied Ocean Mechanics for Structural Engineering and Climate Adaptation*, Synthesis Lectures on Ocean Systems Engineering,
https://doi.org/10.1007/978-3-032-18199-2_13

Fig. 13.1 Left: Free-standing piles in front of a pile-elevated pier in San Francisco. Right: Pile-supported footbridge in Napier, New Zealand

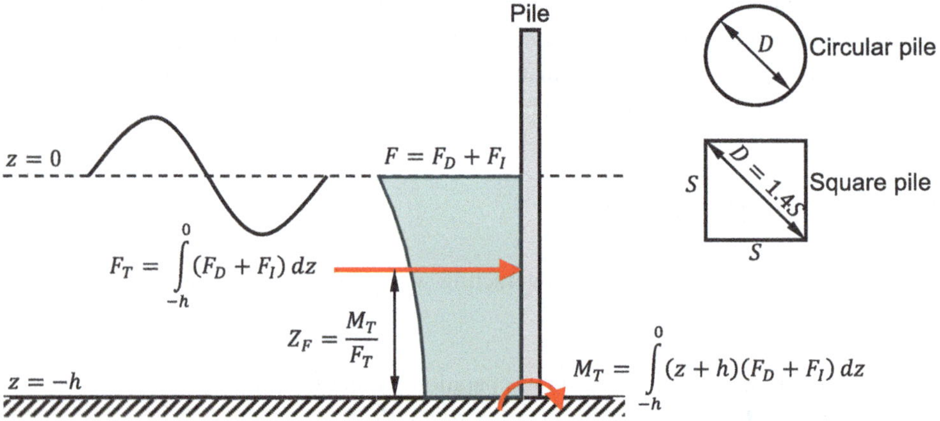

Fig. 13.2 Wave force distribution on a pile and its representation as a concentrated force

period, respectively. For regular waves, this is simply the wave height and period of the wave. However, for irregular waves with significant wave height H_s and significant wave period T_s, the design wave height may be taken as:

$$H = \min[1.8H_s,\ H_b] \tag{13.1}$$

where H_b can be determined from Eq. (9.13) and the design wave period is simply the significant wave period, $T = T_s$. This practice is consistent with the treatment of irregular wave forces on walls (see Chap. 11). In addition, the equations presented in this chapter are valid only for slender piles where the wavelength is greater than 20 times the pile diameter, i.e., $L \geq 20D$.

13.2 Linear Wave Forces on Piles

13.2.1 Drag and Inertia Forces

The Morison equation [8] is a semi-empirical approach that allows for the determination of forces exerted by linear waves on vertical piles as a function of depth and time. For a pile with diameter D, the drag force is:

$$F_D = \frac{1}{2} C_D \rho D u|u| \tag{13.2}$$

Note that D refers to the diameter of a circular pile. For square piles, D is taken as length along the diagonal which equates to 1.4 times the length of each side (see Fig. 13.2). C_D is the drag coefficient, the density of seawater is $\rho = 1025\ \text{kgm}^{-3}$, and u is the horizontal particle velocity (recall from Chap. 3 on linear wave theory) given as:

$$u = \frac{\partial \phi}{\partial x} = A\omega \left(\frac{\cosh[k(h+z)]}{\sinh(kh)} \right) \cos(kx - \omega t) \tag{13.3}$$

where $|u|$ is the absolute value of u. The term $u|u|$ is used in (13.3) instead of u^2 in order to ensure that the correct sign is produced (since u^2 can only yield positive values). Substituting (13.3) into (13.2), the drag force can therefore be expressed as:

$$F_D(z,t) = \frac{C_D \rho D \omega^2 A^2}{2} \left(\frac{\cosh[k(h+z)]}{\sinh(kh)} \right)^2 \cos(kx - \omega t)|\cos(kx - \omega t)| \tag{13.4}$$

The inertia force is:

$$F_I = C_M \rho \left(\frac{\pi D^2}{4} \right) a_x \tag{13.5}$$

where C_M is the inertia coefficient and a_x is the horizontal particle acceleration from Eq. (3.48):

$$a_x = \frac{\partial u}{\partial t} = A\omega^2 \left(\frac{\cosh[k(h+z)]}{\sinh(kh)} \right) \sin(kx - \omega t) \tag{13.6}$$

Substituting (13.6) into (13.5), the inertia force can therefore be expressed as:

$$F_I(z,t) = C_M \frac{\rho \pi D^2}{4} A\omega^2 \left(\frac{\cosh[k(h+z)]}{\sinh(kh)} \right) \sin(kx - \omega t) \tag{13.7}$$

The drag and inertia coefficients are determined according to the Reynolds number (R_e) of the flow. The Reynolds number gives the ratio of inertial to viscous forces and is calculated via:

$$R_e = \frac{u_{max} D}{\nu} \tag{13.8}$$

where $\nu = 1 \times 10^{-6}$ m^{2}s^{-1} is the kinematic viscosity of water and u_{max} is the maximum particle velocity:

$$u_{max} = A\omega \left(\frac{\cosh[k(h+z)]}{\sinh(kh)} \right) \tag{13.9}$$

Note that while u_{max} is depth dependent, one approach to obtain a constant R_e would be to compute u_{max} at $z = -h/2$ (i.e., at a location halfway down the submerged pile). The Reynolds number can then be used to estimate C_D and C_M per recommendations from Reeve et al. [9]:

$$C_D = \begin{cases} 1.2, R_e < 4 \times 10^5 \\ 0.7, R_e > 4 \times 10^5 \end{cases} \tag{13.10}$$

$$C_M = \begin{cases} 2, & R_e < 2.5 \times 10^5 \\ 2.5 - \frac{R_e}{5 \times 10^5}, & 2.5 \times 10^5 < R_e < 5 \times 10^5 \\ 1.5, & R_e > 5 \times 10^5 \end{cases} \tag{13.11}$$

Since pile forces are typically calculated at $x = 0$, The complete Morison equation for the pile force (F) per unit length of the pile in N/m as a function of depth (z) and time (t) is given by:

$$\begin{aligned} F(z,t) &= F_D + F_I \\ &= \frac{C_D \rho D \omega^2 A^2}{2} \left(\frac{\cosh[k(h+z)]}{\sinh(kh)} \right)^2 \cos(-\omega t)|\cos(-\omega t)| \\ &+ \frac{C_M \rho \pi D^2 A \omega^2}{4} \left(\frac{\cosh[k(h+z)]}{\sinh(kh)} \right) \sin(-\omega t) \end{aligned} \tag{13.12}$$

Examining (13.12), it is seen that the maximum drag force occurs at $t = 0$ which corresponds to the wave crest. However, the maximum inertia force occurs at $t = -T/4$ which is one quarter of a wave period before the arrival of the wave crest.

13.2.2 Total Pile Force and Base Moment

The time-varying total force (F_T) with units of N acting on the entire pile and bending moment at the base (M_T) with units of Nm assuming the pile acts as a vertical cantilever can be determined through the integration of (13.12) from $z = -h$ to $z = 0$:

$$F_T(t) = \int_{-h}^{0} (F_I + F_D)dz \tag{13.13}$$

$$M_T(t) = \int_{-h}^{0} (z + h)(F_I + F_D)dz \tag{13.14}$$

which becomes:

$$F_T(t) = \frac{C_M A\omega^2 \rho\pi D^2}{4\sinh(kh)} \sin(-\omega t) \int_{-h}^{0} \cosh[k(h+z)]dz + \frac{C_D \rho D\omega^2 A^2}{2\sinh^2(kh)} \cos(\omega t)|\cos(\omega t)| \int_{-h}^{0} \cosh^2[k(h+z)]dz \tag{13.15}$$

giving:

$$F_T(t) = \frac{C_M A\omega^2 \rho\pi D^2}{4\sinh(kh)} \sin(-\omega t)K_1 + \frac{C_D \rho D\omega^2 A^2}{2\sinh^2(kh)} \cos(\omega t)|\cos(\omega t)|K_2 \tag{13.16}$$

for the total horizontal force. The base moment is:

$$M_T(t) = \frac{C_M A\omega^2 \rho\pi D^2}{4\sinh(kh)} \sin(-\omega t) \int_{-h}^{0} (z+h)\cosh[k(h+z)]dz + \frac{C_D \rho D\omega^2 A^2}{2\sinh^2(kh)} \cos(\omega t)|\cos(\omega t)| \int_{-h}^{0} (z+h)\cosh^2[k(h+z)]dz \tag{13.17}$$

which yields:

$$M_T(t) = \frac{C_M A\omega^2 \rho\pi D^2}{4\sinh(kh)} \sin(-\omega t)K_3 + \frac{C_D \rho D\omega^2 A^2}{2\sinh^2(kh)} \cos(\omega t)|\cos(\omega t)|K_4 \tag{13.18}$$

The K_1, K_2, K_3, and K_4 parameters resolve to the following:

$$K_1 = \int_{-h}^{0} \cosh[k(h+z)]dz = \frac{\sinh(kh)}{k} \tag{13.19}$$

$$K_2 = \int_{-h}^{0} \cosh^2[k(h+z)]dz = \frac{2kh + \sinh(2kh)}{4k} \tag{13.20}$$

$$K_3 = \int_{-h}^{0} (z+h)\cosh[k(h+z)]dz = \frac{kh\sinh(kh) - \cosh(kh) + 1}{k^2} \tag{13.21}$$

$$K_4 = \int_{-h}^{0} (z+h)\cosh^2[k(h+z)]dz = \frac{2(kh)^2 + 2kh\sinh(2kh) - \cosh(2kh) + 1}{8k^2} \tag{13.22}$$

Since both F_T and M_T are time-varying, the maximum total force and moment on the pile to be used for design purposes are:

$$F_{TM} = \max_t F_T(t) \tag{13.23}$$

$$M_{TM} = \max_t M_T(t) \tag{13.24}$$

which are determined by plotting Eqs. (13.16) and (13.18) over a single wave period (from $t = 0$ to $t = T$) and then identifying the maximum value. The point of application of the maximum pile force F_{TM} above the seabed (Z_F) is then:

$$Z_F = \frac{M_{TM}}{F_{TM}} \tag{13.25}$$

Note that the value calculated for the base moment M_{TM} will only be valid for cantilevered piles. However, the total maximum pile force F_{TM} and its point of application Z_T determined using (13.25) will hold for all boundary conditions. A safety factor ranging from 1.5 to 2 is also commonly applied to forces and moments determined via the Morison equation when used for engineering applications.

13.3 Nonlinear Wave Forces on Piles

The Morison equation was derived under the assumption of linear wave theory where the wave height H was assumed small relative to the water depth h and wavelength L (see Chap. 3). Hence, they are most useful when waves fall within the linear region of Le Méhauté's abacus shown via Fig. 13.3. In reality however, this condition may not always be satisfied, and the use of higher order wave theories will produce better results. While detailed discussions into higher order wave theories are beyond the scope of this text, the maximum total inertia (F_{IM}) and drag (F_{DM}) forces acting on the pile from nonlinear waves may be determined as follows:

$$F_{IM} = C_M \rho g \frac{\pi D^2}{4} H K_{Im} \tag{13.26}$$

$$F_{DM} = \frac{1}{2} C_D \rho g D H^2 K_{Dm} \tag{13.27}$$

where K_{Im} and K_{Dm} are given by Fig. 13.4. The maximum total inertia (M_{IM}) and drag (M_{DM}) moments at the seabed ($z = -h$) are:

$$M_{IM} = C_M \rho g \frac{\pi D^2}{4} H K_{Im} h S_{Im} = F_{IM} h S_{Im} \tag{13.28}$$

$$M_{DM} = \frac{1}{2} C_D \rho g D H^2 K_{Dm} h S_{Dm} = F_{DM} h S_{Dm} \tag{13.29}$$

with S_{Im} and S_{Dm} determined from Fig. 13.5 and h is the water depth. Each curve on these charts were produced using stream-function theory by Dean [4] and presents the wave height H as a fraction of the breaking wave height H_b for a given normalized depth h/gT^2. The breaking wave height is given by the line demarcating the boundary between nonlinear and breaking waves in Le Méhauté's abacus (Fig. 13.3). To obtain H_b, the normalized depth is first calculated and correlated with the value denoting the normalized wave height H/gT^2 on the vertical axis which can be used to determine the unknown H. This will be the breaking wave height H_b used in Figs. 13.4 and 13.5 to identify the appropriate curve. For wave heights in between two curves, linear interpolation may be used.

Furthermore, it is important to recognize that Eqs. (13.26) to (13.29) give the maximum wave force or moment associated with the inertia or drag components of wave loading. Since both components vary with time, they may not occur simultaneously at the same instant. Hence, if using:

$$F_{TM} = F_{DM} + F_{IM} \tag{13.30}$$

$$M_{TM} = M_{DM} + M_{IM} \tag{13.31}$$

to determine the total wave force or moment, the result may be greater than the actual force produced. Nevertheless, they serve as a useful approximation if a conservative design solution is desired. The USACE Coastal Engineering Manual (Part VI Chapter 5) [10] contains additional design charts for the determination of true maximum wave forces and moments associated with nonlinear waves on piles. A safety factor ranging from 1.5 to 2 is also commonly applied to nonlinear wave forces and moments for engineering applications.

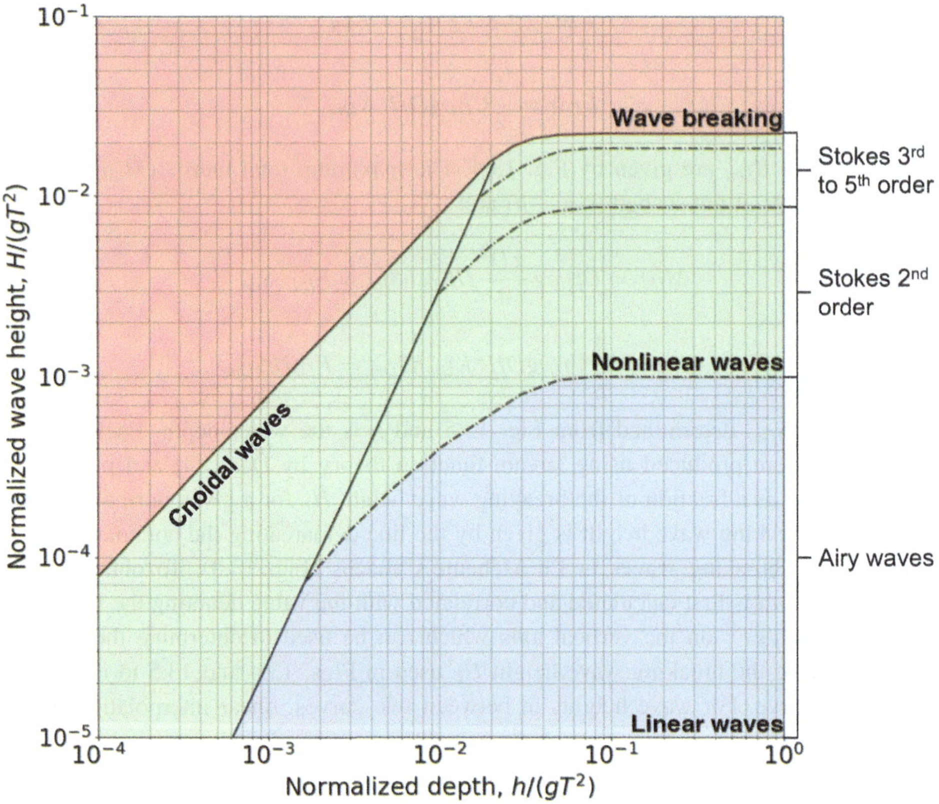

Fig. 13.3 Le Méhauté's abacus showing regions denoting linear, nonlinear, and breaking waves

13.4 Breaking Wave Forces on Piles

Given the extreme nonlinearity of wave breaking, Morison's equation cannot be used to predict impact forces on vertical piles. The treatment of breaking wave forces can follow either a fully dynamic approach that varies with time $F_{BI}(t)$ or approximated as an equivalent static load F_B.

13.4.1 Dynamic Approach

Wienke and Oumeraci [12] performed an extensive experimental campaign to capture the impact force induced by breaking waves on a circular pile via a series of pressure transducers embedded into the frontal face of the pile (Fig. 13.6). The results were used to develop an analytical model that idealizes the total force history as two step functions

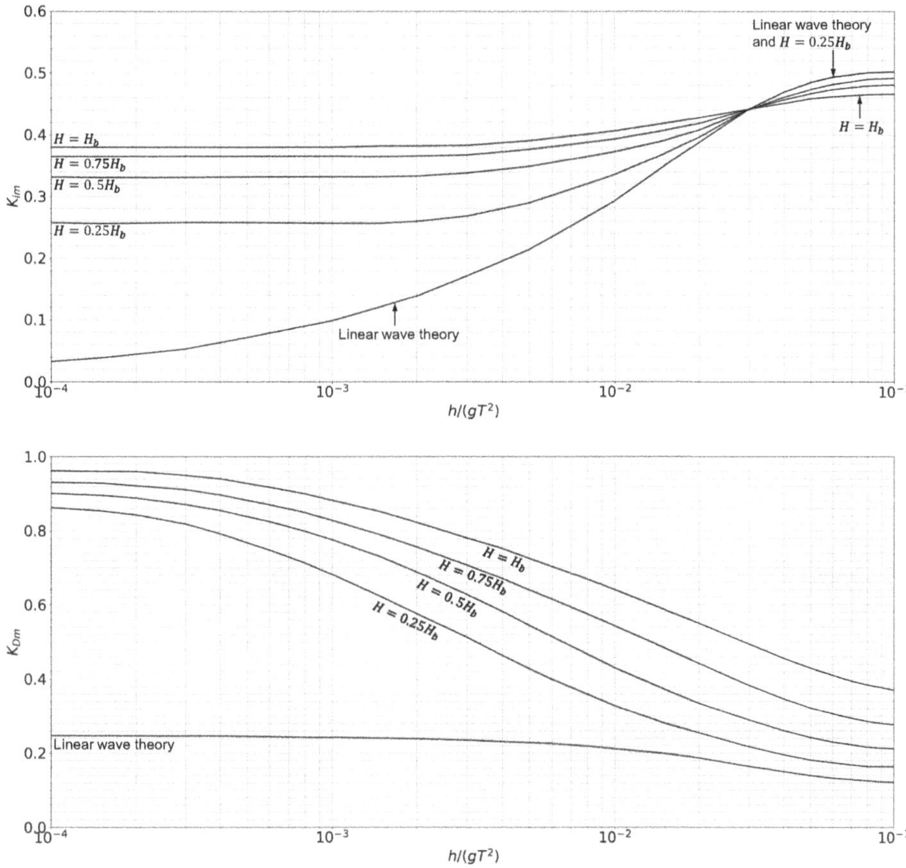

Fig. 13.4 K_{Im} (top) and K_{Dm} (bottom) values for linear and nonlinear waves

that decay parabolically with time. A comparison between the model and experimental measurements is shown in the right figure of Fig. 13.6. The total force is assumed to act over a height of $\lambda_b \eta_b$ on the pile, where η_b is the height of the breaking wave crest above the SWL, as shown in the left figure of Fig. 13.6. The parameter λ_b is referred to as the "curling factor" which describes the proportion of η_b constituting the impacted surface. The horizontal impact force F_{BI} is assumed to be uniformly distributed over this height with its resultant acting at the midpoint.

If breaking waves are expected to occur, the breaking wave height H_s can be determined using Eq. (9.13). For plunging breakers, Wienke and Oumeraci [12] observed that approximately 70–80% of H_b will be above the SWL while Goda et al. [5] suggests a conservative value of:

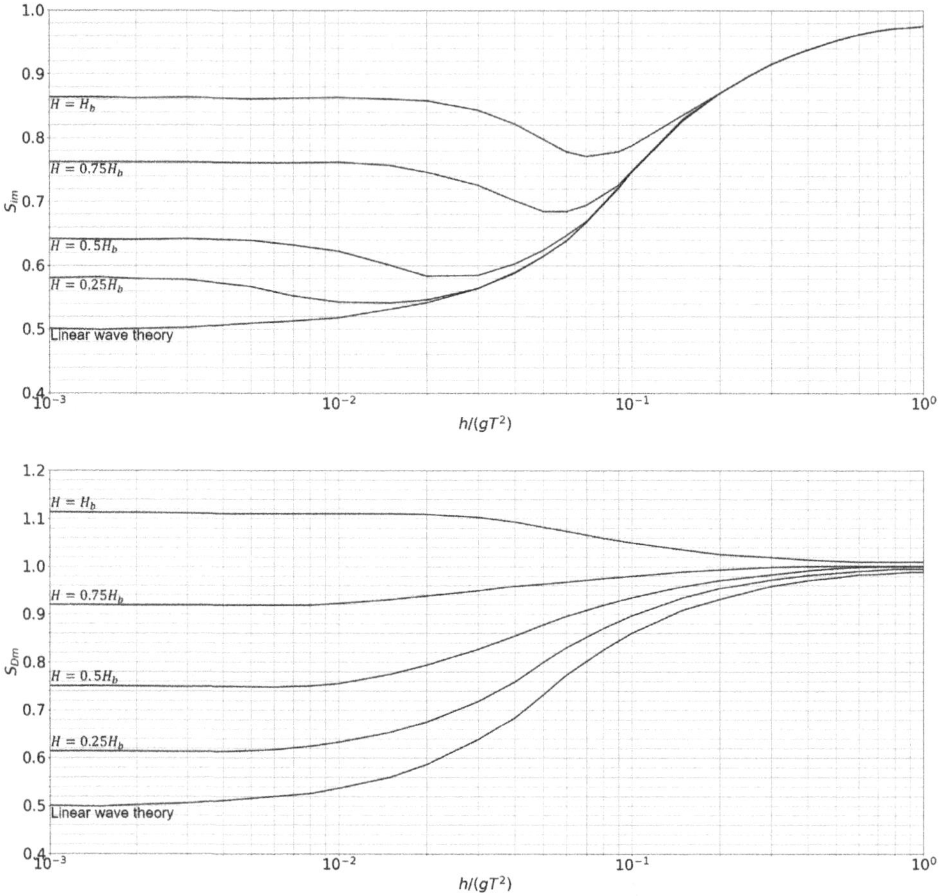

Fig. 13.5 S_{Im} (top) and S_{Dm} (bottom) values for linear and nonlinear waves

$$\eta_b \approx 0.8H_b \tag{13.32}$$

In addition, a curling factor of $\lambda_b \approx 0.45$ is recommended for plunging breakers. As previously mentioned, Goda et al. [12] idealizes the total impact force history into two step functions over the entire impact duration t_{max}. This total duration is given as:

$$t_{max} = \frac{13R}{32C_b} \tag{13.33}$$

where $R = D/2$ is the pile's radius and C_b is the velocity of the breaking wave crest. Goda et al. [5] recommends the following for shallow and deep-water waves:

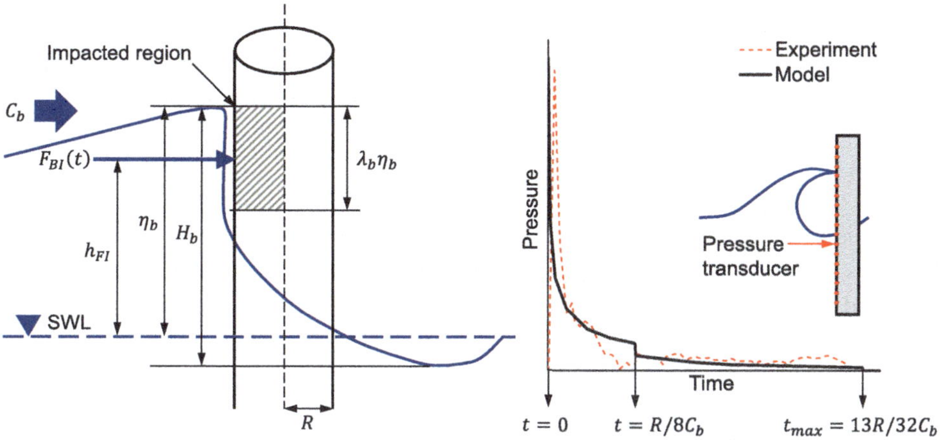

Fig. 13.6 Dynamic breaking wave impact on a vertical pile

$$C_b = \begin{cases} \sqrt{g(h+\eta_b)}, & \text{Shallow water} \\ \sqrt{1.34gH_b}, & \text{Deep water} \end{cases} \tag{13.34}$$

The step function describing the impact force as a function of time t on a vertical pile is then given as:

$$F_{BI}(t) = F_{BI}^*\left[2\pi - 2\sqrt{\frac{C_bt}{R}}\left(\text{arctanh}\sqrt{1-\frac{C_bt}{4R}}\right)\right], \quad 0 \leq t \leq \frac{R}{8C_b} \tag{13.35}$$

$$F_{BI}(t) = F_{BI}^*\left[\pi\sqrt{\frac{R}{6C_bt'}} - \left(\frac{8C_bt'}{3R}\right)^{\frac{1}{4}}\text{arctanh}\sqrt{1-\frac{C_bt'}{R}\sqrt{\frac{6C_bt'}{R}}}\right],$$
$$\frac{R}{8C_b} \leq t \leq \frac{13R}{32C_b} \tag{13.36}$$

$$F_{BI}(t) = 0, \quad \frac{13R}{32C_b} < t \tag{13.37}$$

where:

$$t' = t - \frac{R}{32C_b} \tag{13.38}$$

and:

$$F_{BI}^* = \lambda_b\eta_b\rho RC_b^2 \tag{13.39}$$

Note that when $t = 0$, the impact force is:

$$F_{BI} = 2\pi\lambda_b\eta_b\rho RC_b^2, \quad t = 0 \tag{13.40}$$

which is consistent with the theoretical solution of Wagner [11] for the maximum breaking wave force on a vertical pile. The point of application of F_I above the SWL is:

$$h_{FI} = \frac{\eta_b(2 - \lambda_b)}{2} \tag{13.41}$$

Evidently, Eqs. (13.35) to (13.37) give the dynamic force which can be used to compute the pile's dynamic structural response as informed by its natural period of vibration. Interested readers are encouraged to consult texts pertaining to structural dynamics such as Chopra [3].

13.4.2 Equivalent Static Approach

ASCE 7-22 [1] provides an alternative procedure for the treatment of breaking wave forces as an equivalent static load. This force (F_B) is applied as a static horizontal load at the still water level per Fig. 13.7 with the following magnitude:

$$F_B = \frac{1}{2}\rho g C_{DB} D H_b^2 \tag{13.42}$$

where D is the diameter of a circular pile or 1.4 times the length of each side for square piles and H_b is the breaking wave height. C_{DB} is the drag coefficient for breaking waves which depends on the geometry of the pile:

$$C_{DB} = \begin{cases} 1.75, & \text{Circular piles} \\ 2.25, & \text{Square piles} \end{cases} \tag{13.43}$$

The base moment from breaking waves is then:

$$M_B = F_B h \tag{13.44}$$

where h is the water depth.

13.4.3 Comparison Between Dynamic and Static Forces

To compare the forces from Eqs. (13.35)–(13.37) and (13.42), consider a circular pile with $D = 0.5$ m placed in water with a depth of $h = 5$ m. Assuming shallow water conditions with flat bathymetry, the breaking wave height will be:

$$H_b = 0.78h = 0.78 \times 5 = 3.9 \text{ m}$$

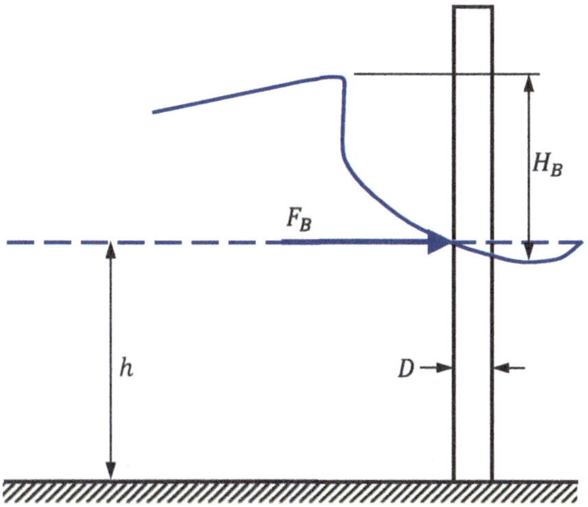

Fig. 13.7 Breaking wave force applied as a static load

Using the ASCE 7-22 approach, the equivalent static load is:

$$F_B = \frac{1}{2}\rho g C_{DB} D H_b^2 = \frac{1}{2} \times 1025 \times 9.81 \times 1.75 \times 0.5 \times 3.9^2 = 66900 \text{ N}$$

When using the dynamic procedure proposed by Wienke and Oumeraci [12], the height of the wave crest above the SWL can be estimated as:

$$\eta_b \approx 0.8 H_b = 0.8 \times 3.9 = 3.12 \text{ m}$$

which results in a crest velocity of:

$$C_b = \sqrt{g(h + \eta_b)} = \sqrt{9.81(5 + 3.12)} = 8.93 \text{ ms}^{-1}$$

The total impact duration can now be determined from Eq. (13.33):

$$t_{max} = \frac{13R}{32C_b} = \frac{13 \times 0.25}{32 \times 8.93} = 0.0114 \text{ s}$$

At the beginning of impact ($t = 0$), the maximum impact force is:

$$\begin{aligned} F_{BI} &= 2\pi \lambda_b \eta_b \rho R C_b^2 \\ &= 2\pi \times 0.45 \times 3.12 \times 1025 \times 0.25 \times 8.93^2 = 180266 \text{ N} \end{aligned}$$

Figure 13.8 compares the dynamic impact force history determined using Eqs. (13.35) to (13.37) with the ASCE 7-22 equivalent static force from Eq. (13.42). While the maximum force is significantly larger for the dynamic approach (180 kN vs. 67 kN), this impact occurs over an extremely short duration.

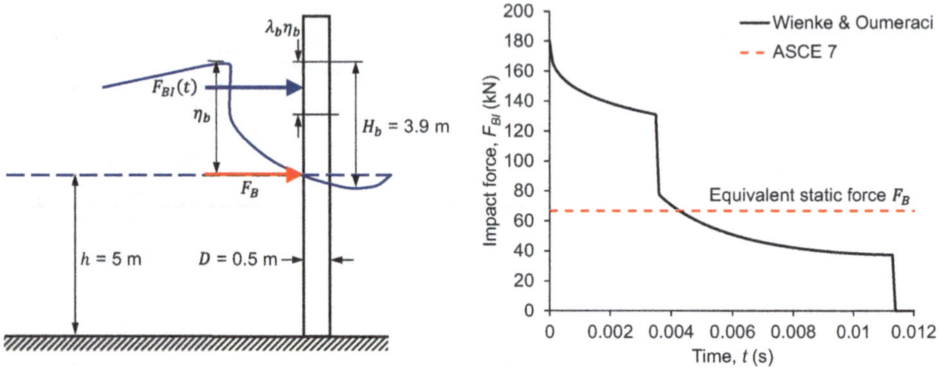

Fig. 13.8 Comparison between Wienke and Oumeraci [12] and ASCE 7-22 forces for $H_b = 3.9$ m on a vertical pile with $D = 0.5$ m

13.5 Example: Wave Forces on an Isolated Pile

An example is provided to illustrate the application of the linear and nonlinear equations used to determine nonbreaking wave forces on a single pile. Consider a pile 0.5 m in diameter placed in water of depth 10 m. Waves with a design wave height of $H = 0.3$ m and period $T = 10$ s are interacting with the pile (Fig. 13.9). The normalized water depth and wave height are:

$$\frac{h}{gT^2} = \frac{10}{9.81 \times 10^2} = 0.01, \quad \frac{H}{gT^2} = \frac{0.3}{9.81 \times 10^2} = 0.0003$$

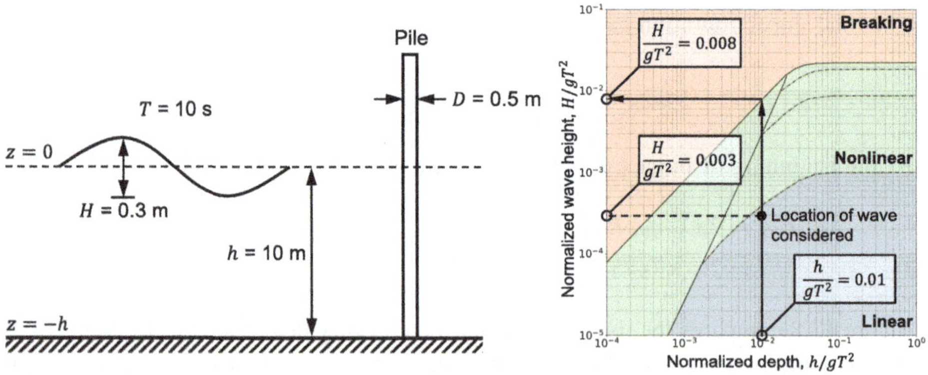

Fig. 13.9 Example problem demonstrating the Morison equation

which places it within the linear region as shown in Fig. 13.9. Hence, the Morison equation can be reliably applied to determine the wave force and associated base moment. At this depth of $h = 10$ m, it is also observed that the normalized wave height required to initiate wave breaking is 0.008 which corresponds to:

$$H_b = 0.008gT^2 = 0.008 \times 9.81 \times 10^2 = 7.8 \text{ m}$$

The wavenumber k must first be determined using the wave period T via the dispersion relation:

$$\omega = \frac{2\pi}{T} = \sqrt{gk \tanh(kh)}$$

which can be rewritten as:

$$\left(\frac{2\pi}{T}\right)^2 \frac{1}{gk} = \tanh(kh)$$

This equation must be solved numerically. We can try different guesses for k such that the LHS of the above equation is close enough to the RHS as shown in Table 13.1.

Alternatively, the wavelength L and wavenumber k may be estimated using the approximate solution:

$$L = \frac{2\pi}{k} \approx \frac{gT^2}{2\pi}\sqrt{\tanh\left(\frac{4\pi^2 h}{T^2 g}\right)} = \frac{9.81 \times 10^2}{2\pi}\sqrt{\tanh\left(\frac{4\pi^2 \times 10}{10^2 \times 9.81}\right)} = 96.5 \text{ m}$$

We will adopt the exact solution that is $k = 0.068 \text{ m}^{-1}$ and $L = 2\pi/k = 92.4$ m. Since the free surface elevation follows that of a cosine function $\eta = A\cos(kx - \omega t)$, the times ($t$) corresponding to a wave crest are integer multiples of the wave period T. Taking $x = 0$, the free surface of the wave as a function of time is described via:

$$\eta(t) = \frac{0.3}{2}\cos(-0.628t)$$

Table 13.1 Iteratively solving for k

Guess for k	LHS: $\left(\frac{2\pi}{T}\right)^2 \frac{1}{gk}$	RHS: $\tanh(kh)$	Comment
0.05	0.805	0.462	LHS > RHS, increase k
0.06	0.671	0.537	LHS > RHS, increase k
0.07	0.575	0.604	LHS < RHS, decrease k
0.065	0.619	0.572	LHS > RHS, increase k
0.068	0.592	0.592	OK

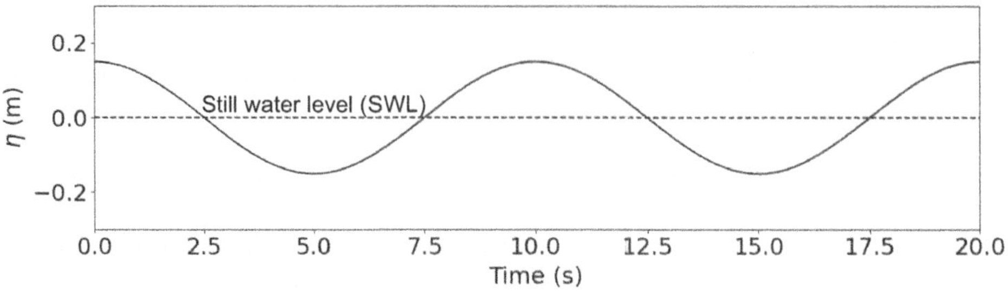

Fig. 13.10 Wave profile with time

where $\omega = 2\pi/T = 0.628$ rads^{-1} is the angular frequency. The wave is shown in Fig. 13.10 over a total time of $2T = 20$s.

Since the wave has $T = 10$ s, the wave crest occurs at $t = 0, 10, 20, 30, 40$ s, etc. Next, we need to determine the Reynolds number to estimate the drag and inertia coefficients. Computing u_{max} at $z = -h/2 = -5$ m, the maximum particle velocity is:

$$u_{max} = A\omega\left(\frac{\cosh[k(h+z)]}{\sinh(kh)}\right) = \frac{0.3}{2} \times \left(\frac{2\pi}{10}\right) \times \left(\frac{\cosh[0.068(10-5)]}{\sinh(0.068 \times 10)}\right) = 0.14 \text{ ms}^{-1}$$

where $A = H/2$ and $\omega = 2\pi/T$. The Reynolds number is then:

$$R_e = \frac{0.14 \times 0.5}{1 \times 10^{-6}} = 0.7 \times 10^5$$

which gives $C_D = 1.2$ and $C_M = 2.0$ per Eqs. (13.10) and (13.11), respectively. When plotting the drag (13.4), inertia (13.7), and total (13.12) linear force at the still water level ($z = 0$) and seabed ($z = -h$) as a function of time across $2T$, Fig. 13.11 is produced.

We observe that the drag and inertia forces are out of phase by $T/4$, equal to 2.5 s for this example, such that F_D is at a maximum and F_I is zero under a wave crest. This is also true under a wave trough, but the force acts opposite to the direction of wave propagation. We are only interested in the time t that produces the maximum total force and moment on the pile. Plotting Eqs. (13.16) and (13.18) over a single wave period from $t = 0$ to $t = 10$ s gives Fig. 13.12.

Figure 13.12 indicates that the maximum force and moment occurs at approximately $t = 7.5$ s. This translates to being halfway between a crest and a trough. The values for the maximum force and moment (from Fig. 13.12) are:

$$F_{TM} = 350 \text{ N}$$
$$M_{TM} = 1820 \text{ Nm}$$

When treating F_M as a single horizontal point load, its point of application above the seabed is:

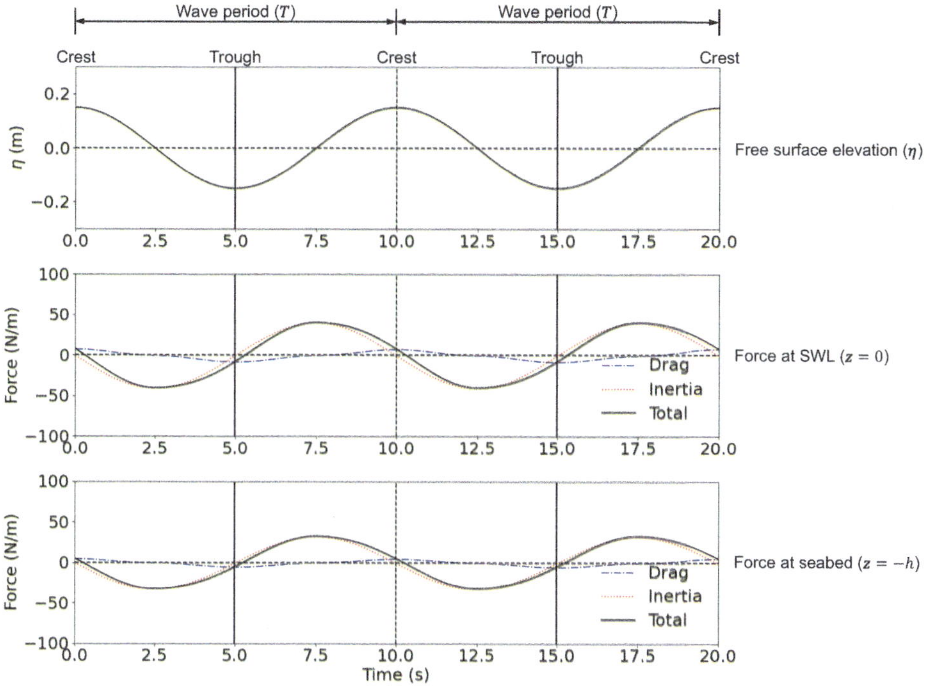

Fig. 13.11 Drag and inertia forces on pile

$$Z_F = \frac{M_{TM}}{F_{TM}} = \frac{1820}{350} = 5.2 \text{ m}$$

Plotting the force distribution with depth at $t = 7.5$ s produces Fig. 13.13:

The total maximum wave force and base moment can also be estimated using Figs. 13.4 and 13.5 via the curves for linear waves. From Fig. 13.4, it is seen that a normalized depth of $h/gT^2 = 0.01$ gives $K_{Im} = 0.29$ and $K_{Dm} = 0.2$. Using Eqs. (13.26) and (13.27), we obtain the following maximum inertia and drag forces:

$$\begin{aligned} F_{IM} &= C_M \rho g \frac{\pi D^2}{4} H K_{Im} \\ &= 2 \times 1025 \times 9.81 \times \frac{\pi \times 0.5^2}{4} \times 0.3 \times 0.29 = 344 \text{ N} \end{aligned}$$

$$\begin{aligned} F_{DM} &= \frac{1}{2} C_D \rho g D H^2 K_{Dm} \\ &= \frac{1}{2} \times 1.2 \times 1025 \times 9.81 \times 0.5 \times 0.3^2 \times 0.2 = 54 \text{ N} \end{aligned}$$

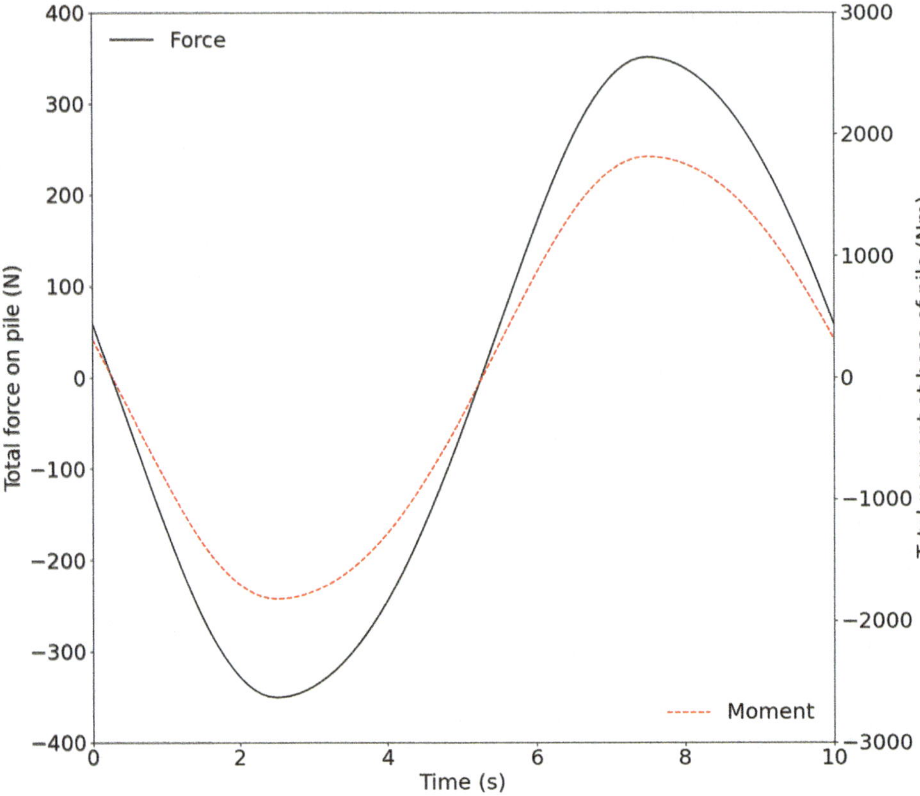

Fig. 13.12 Total force (F_T) and base moment (M_T) on a pile

A direct summation of the above gives a total pile force of:

$$F_{TM} = F_{IM} + F_{DM} = 344 + 54 = 398\text{N}$$

which is 14% larger than the 350 N produced via the Morison equation. This overprediction is due to the maximum inertia and drag forces not occurring at the same time as illustrated by Fig. 13.11. Likewise, Fig. 13.5 yields $S_{Im} = 0.52$ and $S_{Dm} = 0.54$ for a normalized depth of $h/gT^2 = 0.01$. Equations (13.28) and (13.29) then give the following maximum inertial and drag moments at the base of the pile:

$$M_{IM} = F_{IM} h S_{Im} = 344 \times 10 \times 0.52 = 1789 \text{ Nm}$$
$$M_{DM} = F_{DM} h S_{Dm} = 54 \times 10 \times 0.54 = 292 \text{ Nm}$$

A direct summation of the above gives a total pile moment of:

$$M_{TM} = M_{IM} + M_{DM} = 1789 + 292 = 2081\text{Nm}$$

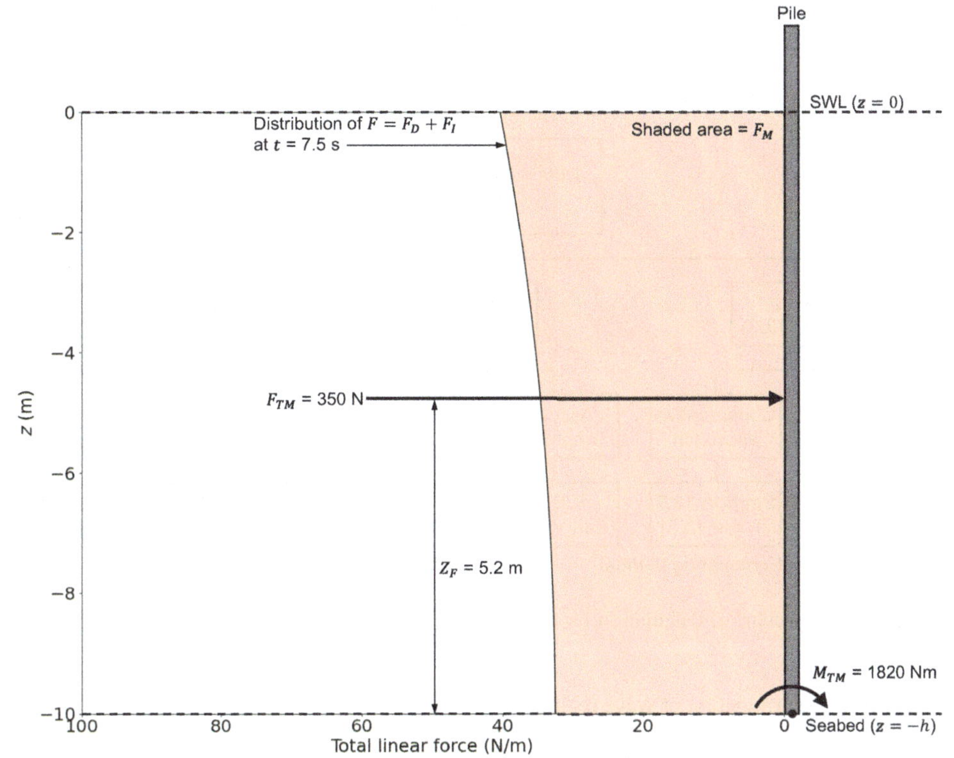

Fig. 13.13 Distribution of Morison wave force down the pile

which, just like the pile force, is 14% larger than the 1820 Nm produced via the Morison equation.

13.6 Summary: Wave Forces on Vertical Piles

A summary of the procedure used to determine forces resulting from waves propagating with a design wave height H and design wave period T in water of depth h on piles with diameter D is presented in Fig. 13.14. A summary of this process is illustrated in Tables 13.2, 13.3, 13.4, 13.5, 13.6 and 13.7:

- Table 13.2: Determination of wave regime
- Table 13.3: Calculation of linear or nonlinear wave forces on piles
- Table 13.4: Calculation of linear wave forces via the Morison equation
- Table 13.5: Calculation of linear or nonlinear wave forces via the conservative approach

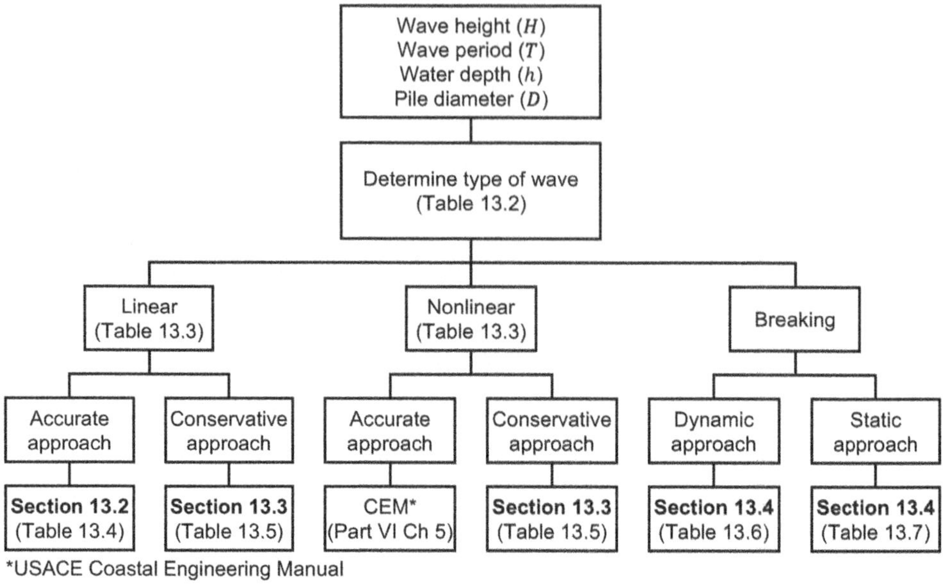

Fig. 13.14 Flowchart of calculation procedure for pile forces

Table 13.2 Determination of wave regime

Step 1	Determine the normalized water depth and wave height from H, h, and T: h/gT^2, H/gT^2
Step 2	Use Le Méhauté's abacus (Fig. 13.3) to determine whether waves are linear, nonlinear, or breaking based on the normalized water depth and wave height
Step 3	If waves are **linear** or **nonlinear**, go to Table 13.3. If waves are **breaking**, go to Table 13.6 for the dynamic approach or Table 13.7 for the static approach

- Table 13.6: Calculation of dynamic breaking wave forces
- Table 13.7: Calculation of equivalent static breaking wave forces.

13.7 Wave Forces on Pile Groups

The equations presented in this chapter are used to determine wave forces on single isolated piles. For piles acting as part of a group (e.g., Fig. 13.15), the Morison or nonlinear equations can also be applied provided that the clear spacing S_G between each individual pile is greater than three times the pile diameter (i.e., $S_G/D \geq 3$). If this condition is not satisfied, the computed nonbreaking pile force must be multiplied by an additional group effect factor (K_G) such that:

Table 13.3 Calculation of linear or nonlinear wave forces on piles

Step 1		Determine the wavelength L using one of the following methods: $L = \frac{gT^2}{2\pi}\tanh\left(\frac{2\pi h}{L}\right)$ or $L \approx \frac{gT^2}{2\pi}\sqrt{\tanh\left(\frac{4\pi^2 h}{T^2 g}\right)}$	
Step 2		Determine the wavenumber via $k = 2\pi/L$ and then find u_{max} and R_e: $u_{max} = A\omega\left(\frac{\cosh[k(h+z)]}{\sinh(kh)}\right), \quad R_e = \frac{u_{max}D}{\nu}$ where $A = H/2$, $\omega = 2\pi/T$, $z = -h/2$, and $\nu = 1 \times 10^{-6}$ $m^2 s^{-1}$	
Step 3		Use R_e to find the drag (C_D) and inertia (C_M) coefficients: $C_D = \begin{cases} 1.2, & R_e < 4 \times 10^5 \\ 0.7, & R_e > 4 \times 10^5 \end{cases}$ $C_M = \begin{cases} 2, & R_e < 2.5 \times 10^5 \\ 2.5 - \frac{R_e}{5 \times 10^5}, & 2.5 \times 10^5 < R_e < 5 \times 10^5 \\ 1.5, & R_e > 5 \times 10^5 \end{cases}$	
For linear waves		**For nonlinear waves**	
Accurate approach	Conservative approach	Accurate approach	Conservative approach
Go to Table 13.4	Go to Table 13.5	Refer to USACE Coastal Engineering Manual (Part VI Chapter 5) [10]	Go to Table 13.5

Table 13.4 Calculation of linear wave forces on piles via the Morison equation

Step 1	Plot the total wave force from $t = 0$ to $t = T$: $F_T(t) = \frac{C_M A\omega^2 \rho\pi D^2}{4\sinh(kh)}\sin(-\omega t)K_1 + \frac{C_D\rho D\omega^2 A^2}{2\sinh^2(kh)}\cos(\omega t)\lvert\cos(\omega t)\rvert K_2$ where $\rho = 1025$ kgm^{-3} for seawater and: $K_1 = \frac{\sinh(kh)}{k}$, $K_2 = \frac{2kh+\sinh(2kh)}{4k}$
Step 2	Plot the total base moment from $t = 0$ to $t = T$: $M_T(t) = \frac{C_M A\omega^2 \rho\pi D^2}{4\sinh(kh)}\sin(-\omega t)K_3 + \frac{C_D\rho D\omega^2 A^2}{2\sinh^2(kh)}\cos(\omega t)\lvert\cos(\omega t)\rvert K_4$ where $\rho = 1025$ kgm^{-3} for seawater and: $K_3 = \frac{kh\sinh(kh) - \cosh(kh) + 1}{k^2}$, $K_4 = \frac{2(kh)^2 + 2kh\sinh(2kh) - \cosh(2kh) + 1}{8k^2}$
Step 3	Determine the maximum wave force and moment from the plots: $F_{TM} = \max_t F_T(t)$, $M_{TM} = \max_t M_T(t)$
Step 4	Use F_{TM} and M_{TM} to determine the point of application of the wave force on the pile: $Z_F = \frac{M_{TM}}{F_{TM}}$
Step 5	Determine bending and shear forces in the pile due to F_{TM} applied at a distance Z_F above the base of the pile for design after applying a safety factor of 1.5 to 2 to the maximum force

$$F_{group} = K_G F_{single} \tag{13.45}$$

where F_{group} is the force acting on a pile within a group and F_{single} is the force on an equivalent isolated pile as determined via Eq. (13.16) for linear waves or Eq. (13.30) for nonlinear waves. The group effect factor K_G is a function of the pile arrangement pattern, the relative spacing (S_G/D), and the Keulegan-Carpenter number (K_C) for which appropriate equations are presented by Bonakdar et al. [2]. This nondimensional parameter describes the significance of drag forces relative to inertial forces when considering fluid–structure interaction [6]. For large K_C, drag forces resulting from turbulence dominates. The Keulegan-Carpenter number is calculated as follows:

$$K_C = \frac{u_{max}T}{D} \tag{13.46}$$

where u_{max} is the maximum horizontal fluid velocity per Eq. (13.9), T is the wave period, and D is the pile diameter. The K_G factor need only to be considered if:

$$\frac{S_G}{D} < 3 \tag{13.47}$$

Table 13.5 Calculation of linear or nonlinear wave forces via the conservative approach

Step 1	Determine F_{IM} and F_{DM}: $F_{IM} = C_M \rho g \frac{\pi D^2}{4} H K_{Im}$, $F_{DM} = \frac{1}{2} C_D \rho g D H^2 K_{Dm}$ where K_{Im} and K_{Dm} are from Fig. 13.4
Step 2	Determine the maximum total pile force: $F_{TM} = F_{IM} + F_{DM}$
Step 3	Determine M_{IM} and M_{DM}: $M_{IM} = C_M \rho g \frac{\pi D^2}{4} H K_{Im} h S_{Im} = F_{IM} h S_{Im}$, $M_{DM} = \frac{1}{2} C_D \rho g D H^2 K_{Dm} h S_{Dm} = F_{DM} h S_{Dm}$ where S_{Im} and S_{Dm} are from Fig. 13.5
Step 4	Determine the maximum pile base moment: $M_{TM} = M_{IM} + M_{DM}$
Step 5	Use F_{TM} and M_{TM} to determine the point of application of the wave force on the pile: $Z_F = \frac{M_{TM}}{F_{TM}}$
Step 6	Determine bending and shear forces in the pile due to F_{TM} applied at a distance Z_F above the base of the pile for design after applying a safety factor of 1.5 to 2 to the maximum force

Hence, for piles spaced more than three diameters apart, $K_G = 1$ can be adopted.

In addition to the K_C number, Bonakdar et al. [2] developed several equations for the K_G parameter based on the following four pile grouping configurations (see Fig. 13.15):

- Side-by-side
- Tandem
- Staggered
- 2×2.

Figure 13.16 presents the experimental data and predictive relationships for K_G as a function of K_C and the S_G/D ratio for the "side-by-side" and "tandem" configurations. Note that the predictive formulae for each S_G/D were developed by Bonakdar et al. [2] using a machine learning model that leverages the genetic programming (GP) method [7]. We notice that for piles placed side-by-side, a decrease in pile spacing (i.e., smaller S_G/D) results in an increase of the wave force acting on the instrumented pile. However, piles placed in tandem parallel to the direction of wave propagation results in a decrease in pile force with reductions in pile spacing and increasing K_C. Figure 13.17 illustrates the relationship between K_C, S_G/D, and K_G for piles placed in the "staggered" and "2×2" configurations. For staggered piles, we observe no general trend with all data points falling

Table 13.6 Calculation of dynamic breaking wave forces via Wienke and Oumeraci [12]

Step 1	Determine the elevation of the wave crest from the SWL and its velocity: $\eta_b \approx 0.8H_b, \; C_b = \begin{cases} \sqrt{g(h+\eta_b)}, & \text{Shallow water} \\ \sqrt{1.34gH_b}, & \text{Deep water} \end{cases}$
Step 2	Determine F_{BI}^*: $F_{BI}^* = \lambda_b \eta_b \rho R C_b^2$ where: $\lambda_b \approx 0.45, \; R = D/2$
Step 3	Determine the impact force as a function of time: $F_{BI}(t) = F_I^*\left[2\pi - 2\sqrt{\frac{C_b t}{R}}\left(\text{arctanh}\sqrt{1 - \frac{C_b t}{4R}}\right)\right], \; 0 \le t \le \frac{R}{8C_b}$ $F_{BI}(t) = F_I^*\left[\pi\sqrt{\frac{R}{6C_b t'}} - \left(\frac{8C_b t'}{3R}\right)^{\frac{1}{4}} \text{arctanh}\sqrt{1 - \frac{C_b t'}{R}\sqrt{\frac{6C_b t'}{R}}}\right], \; \frac{R}{8C_b} \le t \le \frac{13R}{32C_b}$ $F_{BI}(t) = 0, \; \frac{13R}{32C_b} < t$ where: $t' = t - \frac{R}{32C_b}$
Step 4	Determine the point of application of the impact force above the SWL: $h_{FI} = \frac{\eta_b(2-\lambda_b)}{2}$
Step 5	Determine the dynamic response of the pile via its dynamic properties (e.g., natural period of vibration) from principles of structural dynamics

Table 13.7 Calculation of equivalent static breaking wave forces via ASCE 7-22

Step 1	Determine the drag coefficient for breaking waves based on pile geometry: $C_{DB} = \begin{cases} 1.75, & \text{Circular piles} \\ 2.25, & \text{Square piles} \end{cases}$
Step 2	Determine the total breaking wave force: $F_B = \frac{1}{2}\rho g C_{DB} D H_b^2$
Step 3	Determine bending and shear forces in the pile due to F_B applied at the still water level ($z = 0$) for design

largely between $K_G = 0.9$ and 1.1. As such, Bonakdar et al. [2] recommends adopting $K_G = 1$ for all staggered configurations irrespective of spacing or K_C. Similar to the tandem case, piles placed in the 2×2 configuration generally exhibit a decrease in wave force with decreasing S_G/D and increasing K_C. However, discrepancies are observed when $S_G/D = 2$. Figure 13.18 and Table 13.8 summarize the predictive equations developed by

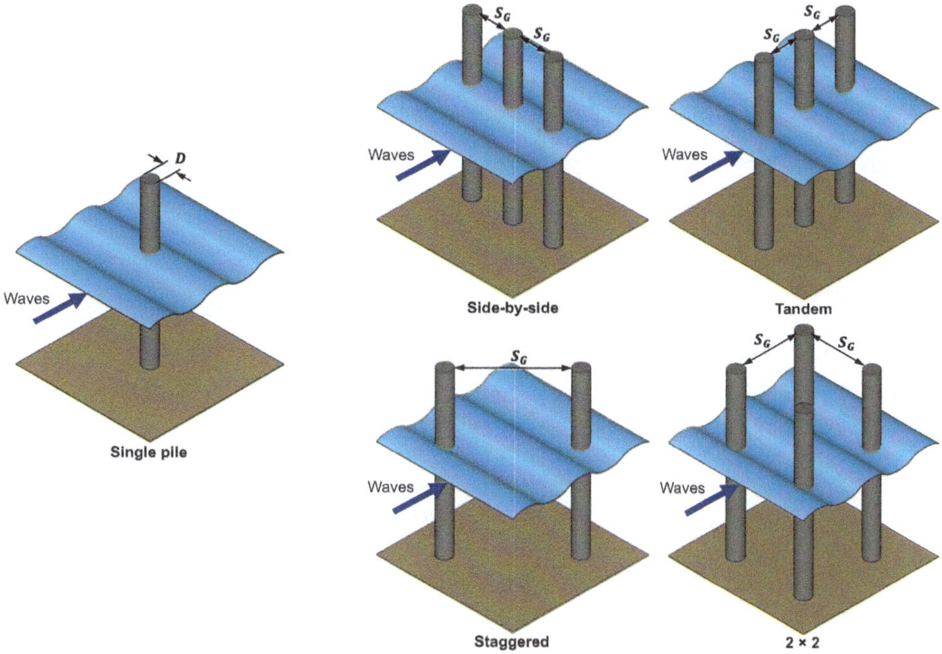

Fig. 13.15 Single pile versus pile groups. Spacing effects need only be considered if $S_G/D < 3$

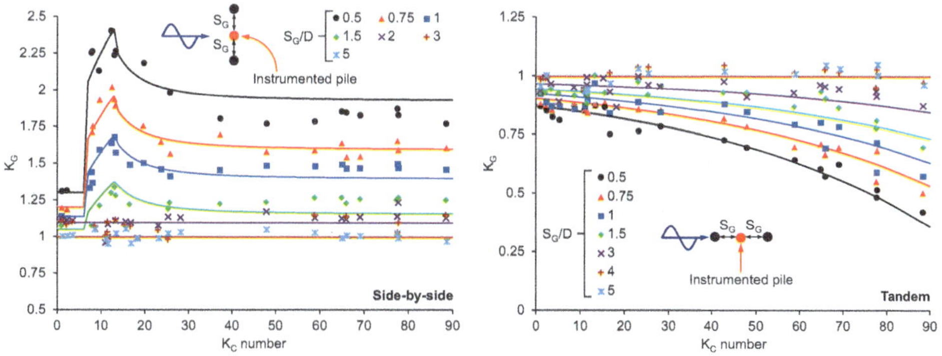

Fig. 13.16 Experimental data (markers) and predictive formulae (lines) for side-by-side (left) and tandem (right) pile groups (adapted from Bonakdar et al. [2])

Bonakdar et al. [2] for all grouping configurations based on the results from Figs. 13.16 and 13.17.

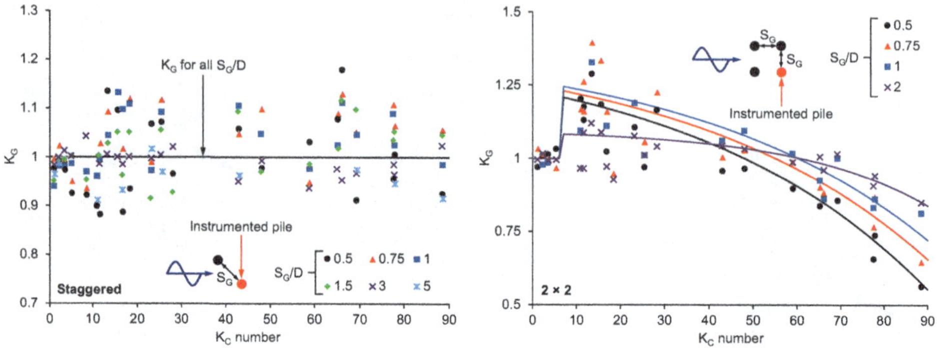

Fig. 13.17 Experimental data (markers) and predictive formulae (lines) for staggered (left) and 2 × 2 (right) pile groups (adapted from Bonakdar et al. [2])

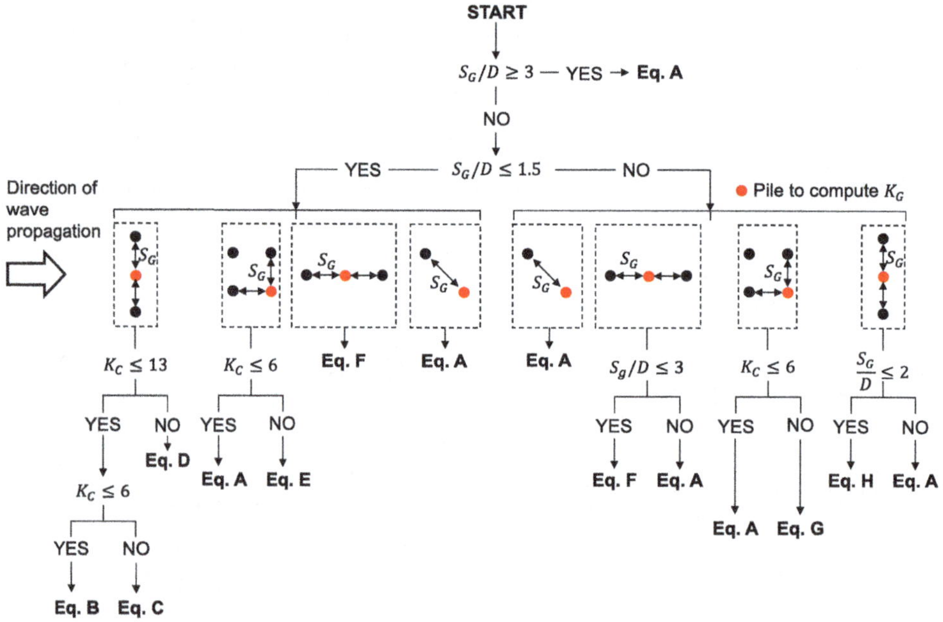

Fig. 13.18 Equations to determine K_G factors for pile groups per Bonakdar et al. [2]

Table 13.8 Equations to determine K_G factors following the procedure in Fig. 13.18

Eq. A	$K_G = 1$	Eq. E	$K_G = 1.4 - 0.136\left(\frac{S_G}{D}\right)^{-0.32}\exp\left(\frac{K_C}{56}\right)$
Eq. B	$K_G = 1.14\left(\frac{S_G}{D}\right)^{-0.19}$	Eq. F	$K_G = 1 - 0.074\left(\frac{S_G}{D}\right)^{-0.8}\exp\left(\frac{K_C}{56}\right)$
Eq. C	$K_G = 0.87\left(\frac{S_G}{D}\right)^{-0.51}K_C^{0.26}$	Eq. G	$K_G = 1.1 - 0.013\exp\left(\frac{K_C}{30}\right)$
Eq. D	$K_G = 1.4\left(\frac{S_G}{D}\right)^{-0.46}\exp\left(52.7K_C^{-2.22}\right)$	Eq. H	$K_G = 1.1$

References

1. ASCE/SEI, 2022. ASCE 7-22. Minimum design loads and associated criteria for buildings and other structures. Reston, Virginia 20191.
2. Bonakdar, L., Oumeraci, H., Etemad-Shahidi, A., 2015. Wave load formulae for prediction of wave-induced forces on a slender pile within pile groups. Coastal Engineering 102, 49–68.
3. Chopra, A.K., 2020. Dynamics of structures: Theory and applications to earthquake engineering, 5th Edition. ed. Pearson Education Limited, United Kingdom.
4. Dean, R.G., 1974. Evaluation and development of water wave theories for engineering application (No. Special Report 1, Vol I and II). Army Coastal Engineering Research Center, Washington D.C.
5. Goda, Y., Haranaka, S., Kitahata, M., 1966. Study of impulsive breaking wave forces on piles. Report of the Port and Harbour Research Institute 5.
6. Keulegan, G.H., Carpenter, L.H., 1958. Forces on cylinders and plates in an oscillating fluid. Journal of Research of the National Bureau of Standards 60, 423–440.
7. Koza, J.R., 1992. Genetic Programming: on the Programming of Computers by Means of Natural Selection. MIT Press, Cambridge, MA.
8. Morison, J.R., Johnson, J.W., Schaaf, S.A., 1950. The force exerted by surface waves on piles. Journal of Petroleum Technology 2, 149–154.
9. Reeve, D., Chadwick, A., Fleming, C., 2018. Coastal engineering: processes, theory & design practice. Taylor and Francis Group, Boca Raton, FL.
10. USACE, 2015. Coastal Engineering Manual. U.S. Army Corps of Engineers, Washington D.C.
11. Wagner, H., 1932. Über Stoß-und Gleitvorgänge an der Oberfläche von Flüssigkeiten. Zeitschrift fur angewandte Mathematik und Mechanik 12, 193–215.
12. Wienke, J., Oumeraci, H., 2005. Breaking wave impact force on a vertical and inclined slender pile - theoretical and large-scale model investigations. Coastal Engineering 52, 435–462.

14 Tsunami Forces on Overland Structures

Thus far, we have only discussed surface waves generated by wind blowing over a body of water (i.e., wind waves). This chapter will address the forces exerted upon overland structures subject to tsunami inundation based on Energy Grade Line Analysis (EGLA) first introduced in ASCE 7-16 [3]. Tsunamis are surface gravity waves produced by the sudden displacement of a large volume of water. Common triggering sources induce undersea earthquakes (e.g., 2004 Indian Ocean tsunami—Fig. 14.1) and volcanic eruptions (e.g., 2021–22 Hunga Tonga–Hunga Ha'apai tsunami—Fig. 14.1). Landslides that deposit large quantities of rock and soil into the water can also generate tsunami waves propagating outwards from the source of impact (e.g., 1958 Lituya Bay tsunami—Fig. 14.2). When considering geologic timescales, asteroid impacts were responsible for producing some of the largest tsunamis ever to occur on earth (e.g., the Chicxulub impactor 66 million years ago—Fig. 14.2). Unlike wind waves with periods on the order of seconds, tsunamis can have wavelengths exceeding hundreds of kilometers and periods of over an hour.

14.1 Tsunami Runup

To determine the forces exerted by a tsunami on overland structures, we must first determine the water depth at the structure's location (referred to as the inundation depth, h) and the flow velocity (u) associated with this depth. These properties depend upon the maximum distance tsunami floodwaters propagate inland from the shore (inundation distance, x_R) which is governed by the maximum runup elevation (R) as shown in Fig. 14.3. Note that all elevations are taken relative to the mean high water (MHW) datum at the shoreline in question.

S. Wang, *Applied Ocean Mechanics for Structural Engineering and Climate Adaptation*, Synthesis Lectures on Ocean Systems Engineering,
https://doi.org/10.1007/978-3-032-18199-2_14

Fig. 14.1 Left: Aftermath of the 2004 Indian Ocean tsunami in Aceh, Indonesia (courtesy of AusAID). Right: 2021–22 Hunga Tonga–Hunga Ha'apai underwater volcano eruption captured by the NOAA GOES-17 satellite (courtesy of NOAA)

Fig. 14.2 Left: Aftermath of a rockslide contributing to the 1958 Lituya Bay tsunami in Alaska. Right: Artist impression of the Chicxulub asteroid impact event. Images courtesy of NASA

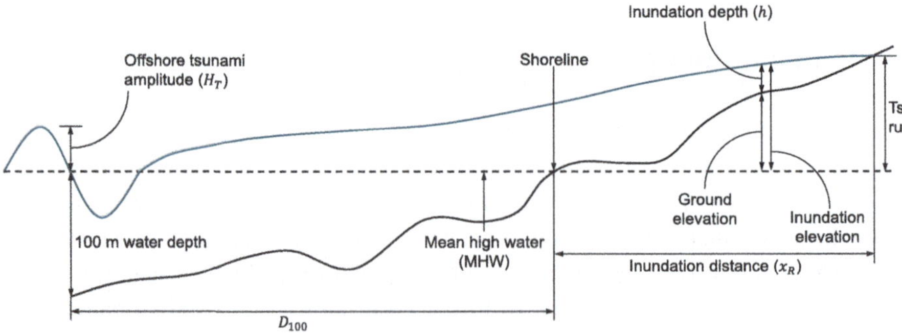

Fig. 14.3 Tsunami transect showing inundation distance and runup

The maximum tsunami runup R is a function of the wave steepness (wave height to wavelength ratio) and bathymetric slope. To account for such effects, the surf similarity parameter (ξ_{100}) is adopted:

$$\xi_{100} = \frac{T_{tsu}}{\cot\Phi}\sqrt{\frac{g}{2\pi H_T}} \tag{14.1}$$

where H_T and T_{tsu} respectively denote the offshore tsunami amplitude (meters) and period (seconds) at a water depth of 100 m. Φ is the average slope angle of the seabed from the location where H_T and T_{tsu} were taken to the MHW shoreline. This angle may be approximated using:

$$\Phi = \arctan\left(\frac{100}{D_{100}}\right) \tag{14.2}$$

with D_{100} being the horizontal distance from the shoreline to 100 m deep water (Fig. 14.3). The surf similarity parameter is also referred to as the Iribarren number which may also be used to classify breaking waves into different categories. After obtaining ξ_{100}, the ratio between maximum runup and offshore tsunami amplitude (R/H_T) is calculated via:

$$\frac{R}{H_T} = \begin{cases} 1.5, & \xi_{100} \leq 0.6 \\ 2.5\log_{10}(\xi_{100}) + 2.05, & 0.6 < \xi_{100} \leq 6 \\ 4.0, & 6 < \xi_{100} \leq 20 \\ -2.15\log_{10}(\xi_{100}) + 6.80, & 20 < \xi_{100} \leq 100 \\ 2.5, & \xi_{100} > 100 \end{cases} \tag{14.3}$$

which was determined based on an extensive number of field observations in addition to physical and numerical studies involving wave runup on slopes. Equation (14.3) is plotted in Fig. 14.4. Note that potential sea level rise due to climate change should also be factored into the maximum tsunami runup R.

For coastal regions associated with the US states of Alaska, Hawaii, Washington, Oregon, and California (all vulnerable to tsunamis), H_T and T_{tsu} values in 100 m deep water may be obtained from the ASCE Tsunami Design Geodatabase (http://asce7tsunami.online). This online database may also be used to obtain nearshore bathymetries and inundation limits that define the tsunami design zone. Note that when undertaking tsunami design for regions covered by the ASCE Tsunami Design Geodatabase, runup and inundation zones may be applied directly without using Eqs. (14.1)–(14.3). Additionally, values of H_T and T_{tsu} given by the database pertain to 2500-year tsunami events (i.e., expected to occur once every 2500 years on average). This return period was selected to maintain consistency with ASCE seismic design provisions.

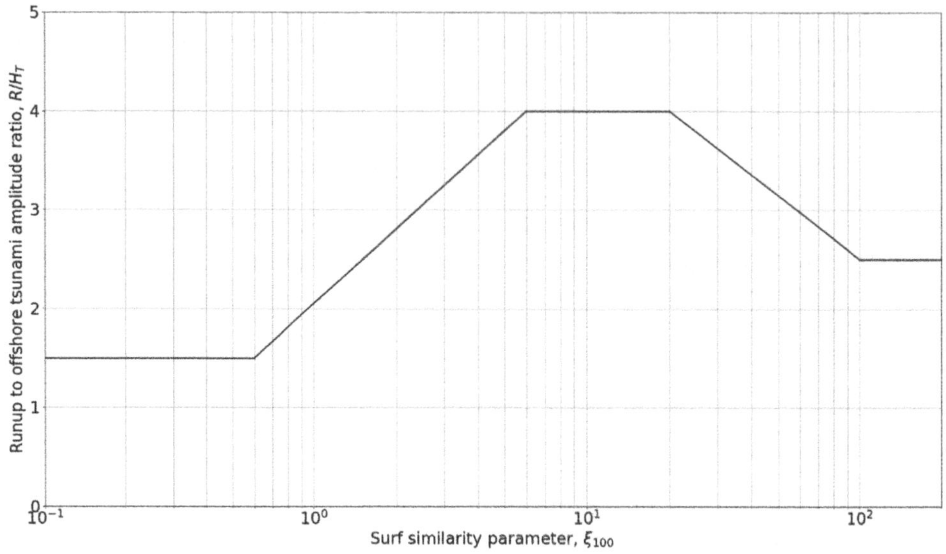

Fig. 14.4 Relationship between runup and surf similarity parameter

14.2 Tsunami Inundation Depth and Flow Velocity

Once the extent of runup has been determined, we can now estimate the expected tsunami inundation depth (h) and water velocity (u) at any location x between the shoreline ($x = 0$) and maximum distance of inundation ($x = x_R$) along a given transect. This will enable us to calculate the tsunami force exerted upon a structure located at x. The transect must pass through the location of interest (i.e., where the structure is located) and travel perpendicular to the shoreline. In addition, ASCE 7-22 [2] requires two additional transects 22.5° clockwise and counterclockwise from this center transect (see Fig. 14.5) to also be investigated such that the maximum tsunami force can be identified. This requirement accounts for the fact that tsunami waves may impact the structure from a non-perpendicular direction relative to the shoreline.

The inundation depth and water velocity at a given location i along a transect is determined using the Energy Grade Line Analysis (ELGA) method presented by Kriebel et al. [7]. This is an analytical method which separates the total energy associated with tsunami flow into kinetic (E_k) and potential (E_p) components given by:

$$E_k = \frac{1}{2}mu^2 \tag{14.4}$$

$$E_p = mgh \tag{14.5}$$

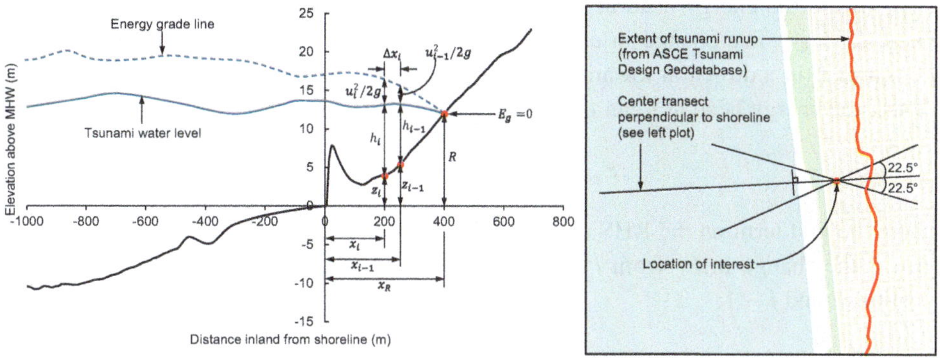

Fig. 14.5 Transect for Energy Grade Line Analysis. Satellite image courtesy of USGS

where $m = \rho h$ is the total mass of a column of water with unit area and height h. Note that the kinetic energy given by Eq. (14.4) assumes that the horizontal fluid velocity is constant down the entire water column. The total energy is therefore:

$$E_k + E_p = \frac{1}{2}\rho h u^2 + \rho g h^2 \tag{14.6}$$

Subsequently dividing both sides by $\rho g h$ yields:

$$\frac{E_k + E_p}{\rho g h} = \frac{u^2}{2g} + h = E_g \tag{14.7}$$

which is referred to as the energy head (E_g). The energy head at location x_i inland from the shoreline is thus:

$$E_{g,i} = \frac{u_i^2}{2g} + h_i \tag{14.8}$$

where u_i and h_i denote the flow velocity and inundation depth at x_i, respectively (Fig. 14.5). Equation (14.8) can also be expressed in terms of the Froude number $F_{r,i}$ via:

$$E_{g,i} = \frac{u_i^2}{2g} + h_i = h_i\left(1 + 0.5F_{r,i}^2\right) \tag{14.9}$$

The Froude number is a dimensionless parameter that expresses the ratio between inertia and gravity forces and is given as:

$$F_{r,i} = \frac{u_i}{\sqrt{g h_i}} \tag{14.10}$$

when treating the tsunami as flow in an open channel with infinite width.

To obtain the energy head at i, we must start at the location of maximum runup (R) where $E_g = 0$. The change in energy head is then calculated at regular intervals along the transect (in a direction towards the shoreline) until the location of interest is reached. Hence, the energy head at i can only be determined after first solving for $E_{g,i-1}$, namely:

$$E_{g,i} = E_{g,i-1} + (\varphi_i + s_i)\Delta x_i \tag{14.11}$$

where the first term on the RHS denotes the energy head at location $i - 1$ and the second term is the change in E_g from $i - 1$ to i. φ_i is the average slope of the terrain between locations i and $i - 1$:

$$\varphi_i = \frac{\Delta z_i}{\Delta x_i} = \frac{z_{i-1} - z_i}{x_{i-1} - x_i} \tag{14.12}$$

with z representing the vertical elevation above the MHW datum and x is the horizontal shoreward distance from the shoreline (see Fig. 14.5). s_i is the friction slope of the energy grade line between locations i and $i - 1$ which accounts for energy losses due to tsunami flow over frictional terrain:

$$s_i = \frac{gF_{r,i}^2}{\left(\frac{1}{n_i}\right)^2 h_{i-1}^{1/3}} \tag{14.13}$$

Equation (14.13) is also known as Manning's Equation which is an empirical formula used to determine the velocity of open channel flows while accounting for frictional effects. Here, $F_{r,i}$ denotes the Froude number at i which is assumed to decay from a value of 1 at the shoreline to 0 at the location of maximum runup (R). Since there is no theoretical basis capable of describing how the Froude number decays over a given topology, an empirical decay function following a square root form is adopted by ASCE 7-22:

$$F_{r,i} = \alpha\sqrt{1 - \frac{x_i}{x_R}} \tag{14.14}$$

where x_R is the maximum horizontal inundation distance from the shoreline and α is a term accounting for the presence of tsunami bores. Bores are highly turbulent wavefronts that form when a tsunami wave breaks as it approaches land. They are most commonly observed when the bathymetric slope is shallower than 1:100 or if there are discontinuities (e.g., shallow reefs) present within the nearshore bathymetry. ASCE 7-22 stipulates the presence of tsunami bores if any of the following conditions are satisfied:

- The nearshore bathymetric slope is 1:100 or milder.
- Existence of shallow reefs or other discontinuities in the bathymetric profile.
- Where historically documented or described in the literature.
- As determined by a site-specific inundation analysis.

Depending on the presence of bores, the parameter α takes a value of:

$$\alpha = \begin{cases} 1.0, & \text{No bores present} \\ 1.3, & \text{Tsunami bores present} \end{cases} \tag{14.15}$$

From Eq. (14.13), we also notice the presence of a parameter n. This is known as Manning's coefficient and is used to denote the macroroughness of the terrain. A value of n_i must be determined for each slope segment between i and $i - 1$ for substitution into Eq. (14.13) to calculate s_i. Table 14.1 provides a list of Manning coefficients applicable for various types of terrain. For areas largely consisting of residential housing which offer little resistance to tsunami flows, the category for "all other cases" should be used ($n = 0.03$). However, for highly built-up commercial centers adopting concrete or structural steel construction, n may be more accurately estimated using:

$$n = \sqrt{n_a^2 + \frac{C_D}{2gw} \times \frac{\Theta}{100 - \Theta} \times D^{4/3}} \leq 0.05 \tag{14.16}$$

where $n_a = 0.025$ and $C_D = 1.5$. w is the width of the buildings, D is the inundation depth, and Θ is the percentage of building coverage. Note that even in heavily urbanized areas, n need not exceed 0.05. Evidently, a higher Manning's coefficient will result in a more conservative assessment of tsunami loading. Finally, the Δx_i term in Eq. (14.11) is the horizontal distance between i and $i - 1$:

$$\Delta x_i = x_{i-1} - x_i \tag{14.17}$$

which should not be taken as greater than 30.5 m. This limit is related to the size of computational grids adopted for the modeling of tsunami inundation used to create the ASCE Tsunami Design Geodatabase.

After solving Eq. (14.11) to obtain the energy head at location i, the corresponding inundation depth can be determined via rearranging (14.9) in terms of h_i:

$$h_i = \frac{E_{g,i}}{1 + 0.5F_{r,i}^2} \tag{14.18}$$

Table 14.1 Manning's coefficient for EGLA per ASCE 7-22

Terrain type	Manning's coefficient, n
Coastal water	0.025–0.03
Open land	0.025
All other cases	0.03
Urbanized areas	0.04

Table 14.2 Tsunami design conditions for hydrodynamic analysis

Design condition	Design inundation depth (h)	Design flow velocity (u)
Maximum velocity	$h = \left(\frac{2}{3}\right)h_{max}$	$u = u_{max}$
Maximum inundation	$h = h_{max}$	$u = \left(\frac{1}{3}\right)u_{max}$

Likewise, the flow velocity at location i can be directly determined from $F_{r,i}$ and h_i by rearranging (14.10) in terms of u_i:

$$u_i = F_{r,i}\sqrt{gh_i} \tag{14.19}$$

where $F_{r,i}$ is per Eq. (14.14). Equations (14.11) and (14.18) must be solved for all points i along the transect (each separated by a horizontal distance Δx_i) from the location of maximum runup (R) to the site of interest. Once this site is reached, Eqs. (14.18) and (14.19) will yield the maximum design inundation depth and flow velocity denoted as h_{max} and u_{max}, respectively. In addition, the following constraints are placed on u_{max} by ASCE 7-22 to avoid excessively small or large flow velocities:

$$3.0 \leq u_{max} \leq \min\left[15.2,\ 1.5\sqrt{gh_{max}}\right] \tag{14.20}$$

with units of m s^{-1}. Note that h_{max} and u_{max} only refer to the maximum inundation depth and flow velocity at the site which are unlikely to occur simultaneously. Hence, ASCE 7-22 requires only the two design conditions corresponding to when the velocity is at a maximum and when the inundation depth is at a maximum when assigning hydrodynamic forces to structures impacted by tsunamis. The design inundation depth (h) and flow velocity (u) for these two conditions are given in Table 14.2.

14.3 Tsunami Pressure

As a tsunami with depth h and velocity u flows past overland structures, the hydrodynamic pressure below the free surface due to drag forces acting on a structure is:

$$p_d = \begin{cases} \frac{1}{2}\rho_{tsu} I_{tsu} C_d u^2, & \text{No bores present} \\ \frac{1}{2}C_{bore}\rho_{tsu} I_{tsu} C_d u^2, & \text{Tsunami bores present and } B/h \geq 3.0 \end{cases} \tag{14.21}$$

Unlike the forces produced by wind waves on piles (see Chap. 13), inertial components will not be present since we are assuming a constant flow velocity u (i.e., no acceleration). An explanation of the terms featured in Eq. (14.21) is presented in this section.

14.3.1 Flow Conditions

The flow velocity u and inundation depth h for each design condition are given in Table 14.2. Both conditions must be checked to obtain the most severe hydrodynamic pressure.

14.3.2 Water Density

The density of water associated with tsunami flows (ρ_{tsu}) must first be established when discussing the treatment of hydrostatic and/or hydrodynamic forces. During a tsunami, it is expected that incoming water from the ocean will accumulate significant amounts of sediment and debris which will increase its overall density. ASCE 7-22 assumes an equivalent sediment concentration of 7% which corresponds to a 10% increase in the overall seawater density (ρ_{sw}). This increase in density comes from the assumption that the suspended sediment has an average specific gravity of 2.5 (i.e., the sediment grains are 2.5 times heavier than water). The seawater density during a tsunami (ρ_{tsu}) is:

$$\rho_{tsu} = k_s \rho_{sw} = 1.1 \times 1025 = 1128\,\mathrm{kg\,m^{-3}} \tag{14.22}$$

where $k_s = 1.1$ is the fluid density factor.

14.3.3 Tsunami Importance Factor

I_{tsu} is the tsunami importance factor given by ASCE 7-22 as presented in Table 14.3. The risk categories for tsunami loading follow standard ASCE 7-22 risk categories. However, certain structures designated as risk category IV may be excluded from this category if they are not critical for post disaster response after a tsunami event.

Table 14.3 Tsunami importance factors per ASCE 7-22

Tsunami risk category	Importance factor, I_{tsu}
II	1.0
III	1.25
IV[a]	1.25

[a] Includes tsunami evacuation shelters but excludes facilities that are not essential to post-tsunami disaster management

14.3.4 Drag Coefficient

The parameter C_d in Eq. (14.21) is the drag coefficient which depends on the B/h ratio when applied to structures with total width B (e.g., an entire building):

$$C_d = \begin{cases} 1.25, & B/h \leq 12 \\ 1.3, & B/h = 16 \\ 1.4, & B/h = 26 \\ 1.5, & B/h = 36 \\ 1.75, & B/h = 60 \\ 1.8, & B/h = 100 \\ 2.0, & B/h \geq 120 \end{cases} \tag{14.23}$$

However, different drag coefficients are defined for flows interacting with individual structural components (e.g., beams, columns, and walls) based on their shape and orientation. Figure 14.6 summarizes the tsunami drag coefficients for structures and structural components recommended by ASCE 7-22.

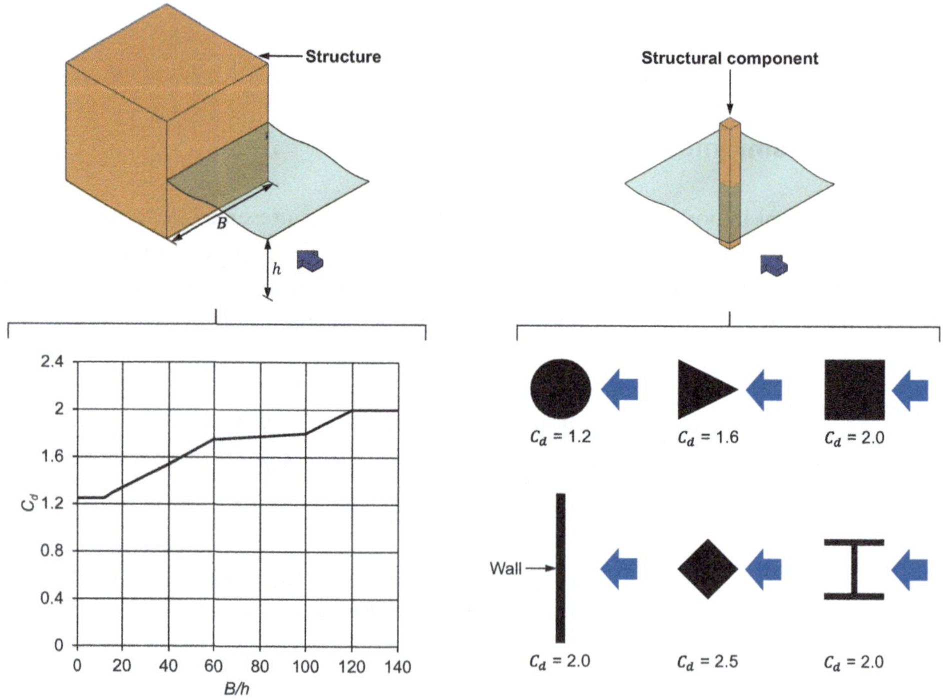

Fig. 14.6 Drag coefficients for structures (left) and structural components (right) from ASCE 7-22

14.3.5 Amplification Factor for Tsunami Bores

If tsunami bores are expected to occur, an amplification factor of $C_{bore} = 1.5$ (determined from experimental testing) must be applied to the drag force as shown in Eq. (14.21). However, this effect needs only to be applied for structures with an overall width to inundation depth ratio exceeding 3, namely, $B/h \geq 3.0$.

14.4 Simplified Procedure for Tsunami Pressure

As an alternative to Eq. (14.21), the hydrodynamic pressure may also be estimated via a simplified procedure that considers the worst possible loading scenario. When assuming the presence of tsunami bores and an inundation depth of h, the force per unit width on a structure becomes:

$$f_d = \frac{1}{2} C_{bore} \rho_{tsu} I_{tsu} C_d C_{cx} h u^2 \tag{14.24}$$

The maximum hydrodynamic forcing is assumed to occur when $h = (2/3)h_{max}$ and the Froude number is $\sqrt{2}$. This yields:

$$h = \frac{2}{3} h_{max}, \quad u = C_{amp} \sqrt{2} \sqrt{g \frac{2}{3} h_{max}} \tag{14.25}$$

where u is from Eq. (14.19) with the inclusion of a velocity amplification factor C_{amp} to account for potential obstructions upstream that may provide a funneling effect to increase the localized flow speed at the structure under consideration. Experimental observations into flow amplification reveal a maximum C_{amp} of 1.25. A closure ratio of $C_{cx} = 0.7$ and drag coefficient of 2.0 are next assumed. Equation (14.24) thus becomes:

$$\begin{aligned} f_d &= 1.5 \times \frac{1}{2} \rho_{tsu} I_{tsu} \times 2 \times 0.7 \times \left(\frac{2}{3} h_{max} \right) \left(1.25 \sqrt{2} \sqrt{g \frac{2}{3} h_{max}} \right)^2 \\ &= 1.458 \rho_{tsu} g I_{tsu} h_{max}^2 \end{aligned} \tag{14.26}$$

This drag force f_d is next further increased by 10% to account for any potential uncertainties yielding:

$$f_d = 1.1 \times 1.458 \rho_{tsu} g I_{tsu} h_{max}^2 = 1.6 \rho_{tsu} g I_{tsu} h_{max}^2 \tag{14.27}$$

To accommodate any water build-up at the face of the structure, f_d is treated as a rectangular pressure distribution up to a maximum height of $1.3h_{max}$. This constant hydrodynamic pressure (p_d) is therefore:

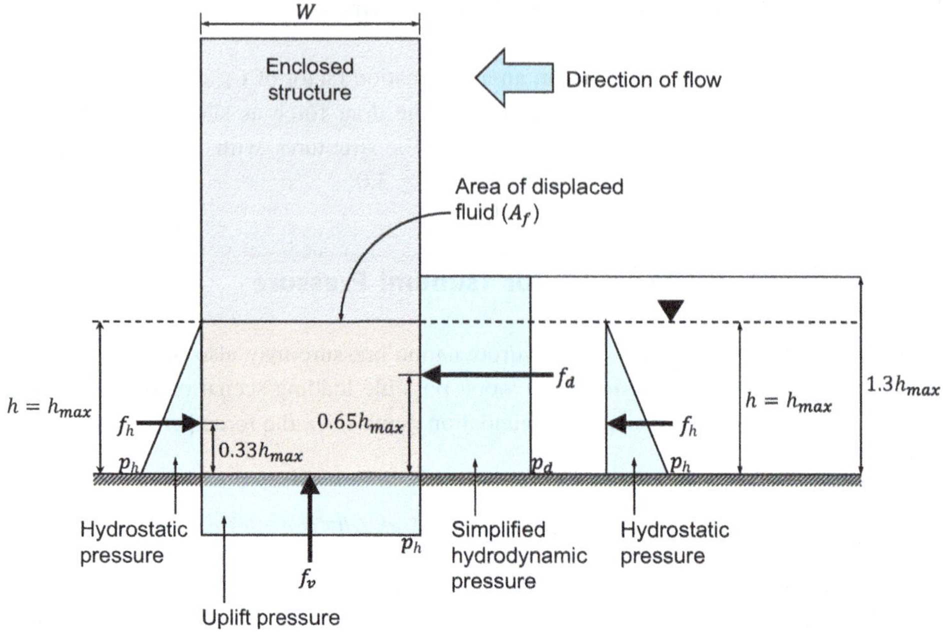

Fig. 14.7 Simplified hydrodynamic and hydrostatic tsunami pressures on an enclosed structure

$$p_d = \frac{1.6\rho_{tsu} g I_{tsu} h_{max}^2}{1.3h_{max}} \approx 1.25\rho_{tsu} g I_{tsu} h_{max} \tag{14.28}$$

which may be adopted in lieu of Eq. (14.21) to provide a conservative estimate of the hydrodynamic tsunami pressure on a structure as illustrated in Fig. 14.7. However, the simplified method may result in significant overprediction of hydrodynamic loads. As an example, Doyle et al. [5] revealed that Eq. (14.28) can produce forces up to 15 times greater than that observed experimentally.

14.5 Hydrostatic Pressure

In addition to the hydrodynamic pressure p_d calculated using either (14.21) or (14.28), hydrostatic pressures must also be considered which increase from 0 at the free surface to p_h at the base of the structure (at depth h):

$$p_h = \rho_{tsu} g h \tag{14.29}$$

The total lateral hydrostatic force per unit width acting on one side of an inundated structure is therefore:

$$f_h = \frac{1}{2}\rho_{tsu} g h^2 \tag{14.30}$$

Due to buoyancy, the hydrostatic pressure at the base of the structure (p_h) will also cause a vertical uplift force per unit width equal to:

$$f_v = \rho_{tsu} g A_f \tag{14.31}$$

where $A_f = Wh$ is the total submerged volume per unit width for a structure with length W as shown in Fig. 14.7. This buoyancy force will only be present for fully enclosed structures with no water intrusion.

14.6 Tsunami Forces

The hydrodynamic pressure p_d from Eqs. (14.21) or (14.28) can be converted into a uniformly distributed line load (f_d) over structural elements such as beams and columns to facilitate the design of a structure subject to tsunami loading. This may be achieved by first converting the pressure into a force (F_d) acting over the tributary area associated with a particular component (A_w):

$$F_d = p_d A_w \tag{14.32}$$

where:

$$A_w = C_{cx} B h_s \tag{14.33}$$

is the tributary area with width B and height h_s as shown in Fig. 14.8. The C_{cx} parameter is the closure ratio and represents the percentage of given area comprising of structural elements (i.e., beams, columns, and walls) that are expected to resist tsunami forces. The closure ratio is calculated for any given tributary area as follows:

$$C_{cx} = \frac{\sum(A_{col} + A_{wall}) + 1.5\sum A_{beam}}{Bh_s}, \quad 0.7 \leq C_{cx} \leq 1.0 \tag{14.34}$$

where A_{col}, A_{wall}, and A_{beam} respectively denote the total projected areas of the columns, walls, and beams covered by the tributary area as shown in Fig. 14.8. Note that ASCE 7-22 imposes a minimum closure ratio of 0.7. This minimum threshold is designed to account for waterborne debris that may accumulate against the side of a structure thus providing additional blockage. On the other hand, C_{cx} need not be taken as greater than 1. Evidently, only areas of a building below the inundation depth (h) need to be considered when obtaining h_s and C_{cx} as illustrated via Fig. 14.8. The uniformly distributed line load f_d over an element (with units of N/m) can now be obtained by dividing the total force F_d (with units of N) over the length of the element:

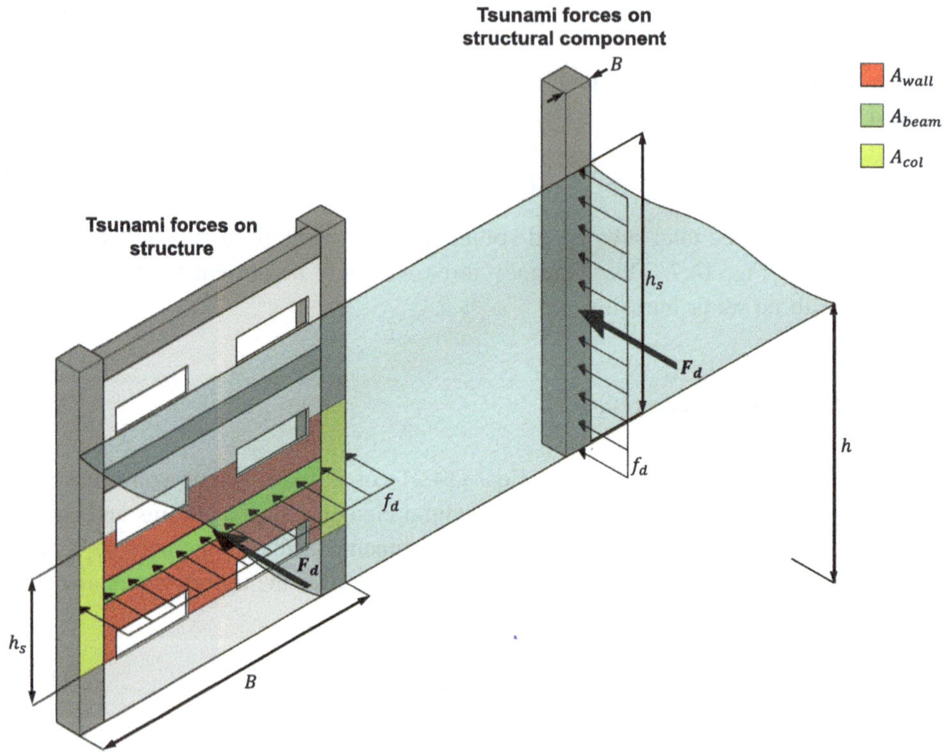

Fig. 14.8 Tsunami forces on a beam as part of a structure and an isolated structural component

$$f_d = \begin{cases} \frac{F_d}{B}, & \text{Horizontal elements} \\ \frac{F_d}{h_s}, & \text{Vertical elements} \end{cases} \tag{14.35}$$

and is demonstrated in Fig. 14.8.

14.7 Example: Tsunami Forces on a Frame Building

The determination of tsunami forces on a structure located within the inundation zone will be demonstrated via an example. Figure 14.9 shows the transect of a coastal profile in the Sunset District of San Francisco where tsunami forces are to be evaluated. Note that although two additional transects oriented 22.5° clockwise and counterclockwise also should be examined (see Fig. 14.5), only the center transect will be considered for this example. Using the ASCE Tsunami Design Geodatabase (http://asce7tsunami.online), the maximum runup is approximately $R = 12$ m above MHW which corresponds to an approximate horizontal distance of $x_R = 400$ m from the shoreline. A structure

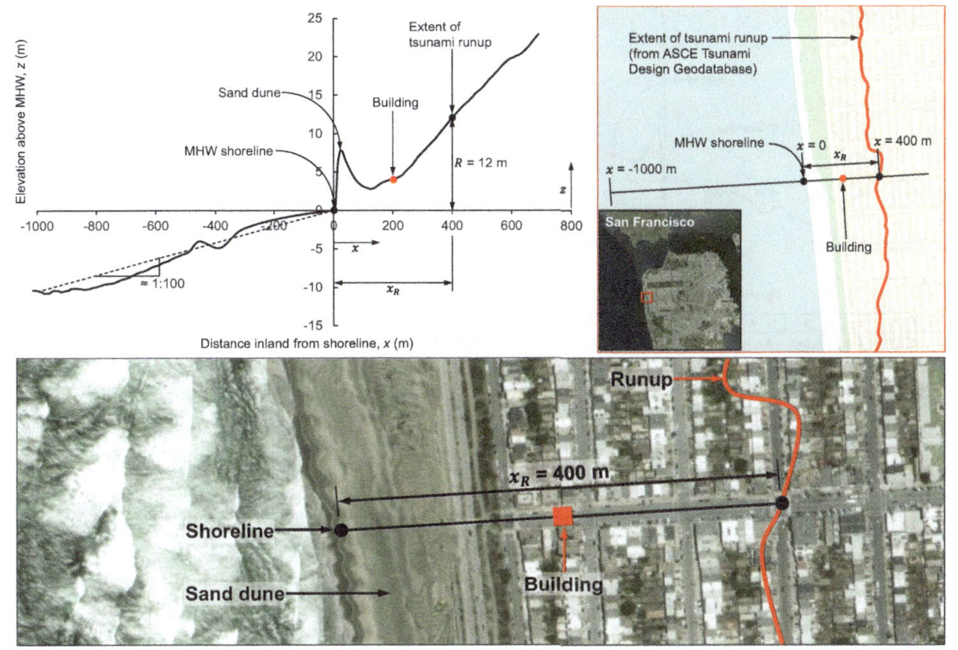

Fig. 14.9 Location of hypothetical building in the Sunset District of San Francisco for tsunami analysis. Satellite images courtesy of USGS

placed at $x = 200\,\text{m}$ as indicated via Fig. 14.9 will be examined against tsunami forces at this location.

14.7.1 Maximum Inundation and Velocity

We begin by plotting the transect elevation profile between the location of maximum runup ($x = x_R = 400$ m) and the location of our structure ($x = 200\,\text{m}$) as shown in Fig. 14.10. The profile is then discretized into 10 individual segments and 11 nodes (represented by $i = \{0, 1, 2, \ldots, 10\}$ in column 1 of the table in Fig. 14.10) where the horizontal distance between each node (Δx_i) is approximately 20 m. This satisfies the $\Delta x_i \leq 30.5\,\text{m}$ requirement. The horizontal (x_i) and vertical (z_i) distance from the shoreline used to plot each node were obtained directly from the ASCE Tsunami Design Geodatabase and summarized in columns 2 and 3, respectively. The maximum inundation depth (h_i) and maximum flow velocity (u_i) at each node are next computed from $i = 1$ to $i = 10$ (location of structure).

The procedure described in Sect. 14.2 is subsequently demonstrated to obtain h_i and u_i for node $i = 1$ boxed in Fig. 14.10. The slope φ_1 is first calculated via Eq. (14.12):

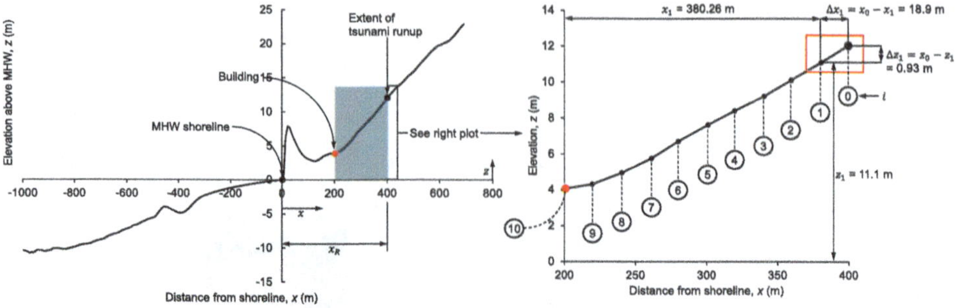

1	2	3	4	5	6	7	8	9	10	11
i	x_i (m)	z_i (m)	Δx_i (m)	n_i	φ_i	$F_{r,i}$	s_i	$E_{g,i}$ (m)	h_i (m)	u_i (ms^{-1})
0	399.16	12.03						0	0.01	0
1	380.26	11.1	18.9	0.04	0.049	0.283	0.006	1.040	1.000	0.886
2	359.46	10.11	20.8	0.04	0.048	0.410	0.003	2.085	1.923	1.781
3	340.56	9.21	18.9	0.04	0.048	0.498	0.003	3.044	2.708	2.567
4	319.77	8.4	20.79	0.04	0.039	0.580	0.004	3.933	3.367	3.332
5	300.87	7.62	18.9	0.04	0.041	0.645	0.004	4.795	3.969	4.026
6	280.07	6.71	20.8	0.04	0.044	0.710	0.005	5.809	4.640	4.791
7	261.17	5.76	18.9	0.04	0.050	0.764	0.005	6.863	5.312	5.517
8	240.38	4.94	20.79	0.04	0.039	0.820	0.006	7.809	5.844	6.208
9	219.59	4.3	20.79	0.04	0.031	0.872	0.007	8.587	6.222	6.812
10	200.68	4.05	18.91	0.04	0.013	0.917	0.007	8.972	6.318	7.217
									h_{max}	u_{max}

Column number

Refers to boxed segment in plot

Data required as initial input

Constants applicable for all profiles

h_{max} and u_{max} at building location (final output used for design)

Fig. 14.10 Calculation procedure for h_{max} and u_{max} at building location

$$\varphi_1 = \frac{\Delta z_1}{\Delta x_1} = \frac{z_0 - z_1}{x_0 - x_1} = \frac{12.03 - 11.1}{399.16 - 380.26} = 0.049$$

and is placed in column 6 of the table. The Froude number at $i = 1$ is next determined per (14.14):

$$F_{r,1} = \alpha\sqrt{1 - \frac{x_1}{x_R}} = 1.3\sqrt{1 - \frac{380.26}{399.16}} = 0.283$$

and placed in column 7. Here, $\alpha = 1.3$ is taken since tsunami bores are assumed to occur along this transect per Eq. (14.15). Recall that tsunami bores may occur if the nearshore profile has a slope shallower than 1:100. As indicated in Fig. 14.9, the bathymetry leading up to the shoreline has an average slope of almost exactly 1:100. Hence, the bore condition is selected to enable a conservative assessment of tsunami forces. The Manning's coefficient (column 5) for segment $i = 1$ is next selected to be $n_1 = 0.04$ per Table 14.1 due to its urbanized location as evidenced in Fig. 14.9. The friction slope of the energy grade line at $i = 1$ can now be determined using Eq. (14.13):

$$s_1 = \frac{gF_{r,1}^2}{\left(\frac{1}{n_1}\right)^2 h_0^{1/3}} = \frac{9.81 \times 0.283^2}{\left(\frac{1}{0.04}\right)^2 \times 0.01^{1/3}} = 0.006$$

which is entered into column 8. Note that when computing s_1, a nominally small value of 0.01 m is adopted for h_0 (i.e., the inundation at the maximum runup point). While h_0 should technically equal zero, a small non-zero value must be used instead to avoid division by zero when determining s_1 as seen above. After solving for Δx_1 (column 4) per Eq. (14.17):

$$\Delta x_1 = x_0 - x_1 = 399.16 - 380.26 = 18.9\,\text{m}$$

the energy head at $i = 1$ can be obtained using Eq. (14.11):

$$E_{g,1} = E_{g,0} + (\varphi_1 + s_1)\Delta x_1 = 0 + (0.049 + 0.006) \times 18.9 = 1.04\,\text{m}$$

and placed into column 9. Note that $E_{g,0} = 0$ by definition, since this location represents the maximum runup point with no inundation. The tsunami inundation depth at $i = 1$ can finally be determined via (14.18):

$$h_1 = \frac{E_{g,1}}{1 + 0.5F_{r,1}^2} = \frac{1.04}{1 + 0.5 \times 0.283^2} = 1.0\,\text{m}$$

and entered into column 10. The flow velocity at $i = 1$ can also be obtained through Eq. (14.19):

$$u_1 = F_{r,1}\sqrt{gh_1} = 0.283 \times \sqrt{9.81 \times 1.0} = 0.886\,\text{m}\,\text{s}^{-1}$$

then placed into column 11. This procedure must now be repeated for $i = 2, 3, 4, 5, 6, 7, 8, 9$, and 10 to ultimately arrive at h_{10} and u_{10}, which respectively correspond to the maximum inundation and flow velocity at the building h_{max} and u_{max}. From the table in Fig. 14.10, $h_{max} = 6.3\,\text{m}$ and $u_{max} = 7.2\,\text{m}\,\text{s}^{-1}$.

14.7.2 Hydrodynamic Pressure and Forces

The maximum inundation and flow velocity from Fig. 14.10 will be used to determine the hydrodynamic tsunami forces acting on the front wall of the concrete frame structure shown in Fig. 14.11. A tsunami risk category of II is assumed for the building ($I_{tsu} = 1.0$ from Table 14.3). From Table 14.2, two design conditions (maximum velocity and maximum inundation) must be considered. The hydrodynamic pressure (p_d) calculated via Eq. (14.21) for the two cases are summarized in Table 14.4. Note that the drag coefficient $C_d = 1.25$ is from Fig. 14.6 when adopting $B/h < 12$ for both cases, since $B/h = 5$ and 3.33, where $B = 21\,\text{m}$ is the total width of the building.

The total drag force (F_d) acting over the entire submerged surface of the building for each case can now be determined by multiplying p_d with the projected wall area per Eq. (14.32):

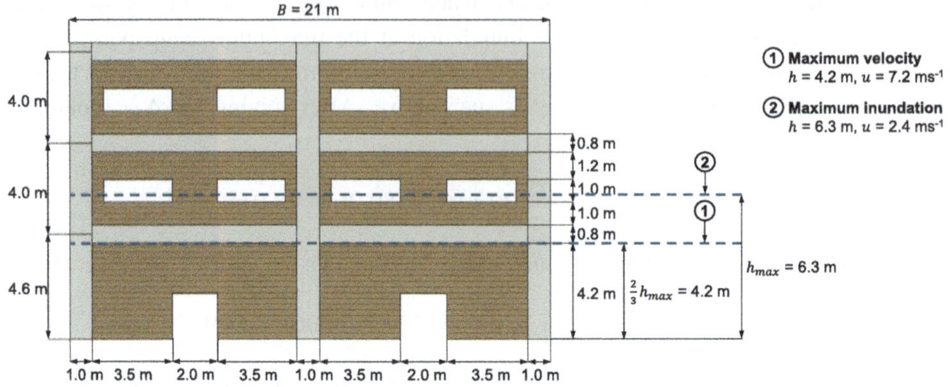

Fig. 14.11 Front wall of building subject to tsunami flows

Table 14.4 Hydrodynamic pressures for maximum velocity and maximum inundation cases

Design condition	Design inundation depth (h) (m)	Design flow velocity (u) (m s^{-1})	Hydrodynamic pressure (p_d) (Pa)
Maximum velocity	$h = \left(\frac{2}{3}\right) h_{max}$ $= 4.2$	$u = u_{max}$ $= 7.2$	$p_d = \frac{1}{2} C_{bore} \rho_{tsu} I_{tsu} C_d u^2$ $= \frac{1}{2} \times 1.5 \times 1128 \times 1 \times 1.25 \times 7.2^2$ $= 54{,}820$
Maximum inundation	$h = h_{max}$ $= 6.3$	$u = \left(\frac{1}{3}\right) u_{max}$ $= 2.4$	$p_d = \frac{1}{2} C_{bore} \rho_{tsu} I_{tsu} C_d u^2$ $= \frac{1}{2} \times 1.5 \times 1128 \times 1 \times 1.25 \times 2.4^2$ $= 6090$

Note $h_{max} = 6.3$ m and $u_{max} = 7.2$ m s^{-1} were determined via the EGLA procedure in Fig. 14.10

$$F_d = p_d A_w = \begin{cases} 54{,}820 \times 21 \times 4.2 = 4835\,\text{kN}, & \text{Maximum velocity} \\ 6090 \times 21 \times 6.3 = 806\,\text{kN}, & \text{Maximum inundation} \end{cases}$$

which gives the total hydrodynamic base shear on the structure. Here, A_w is from Eq. (14.33) with the closure ratio set to $C_{cx} = 1$ to conservatively ignore all openings. It is seen that the maximum velocity scenario ($h = 4.2$ m and $u = 7.2$ m s^{-1}) will govern the performance of the structure. If the hydrodynamic force on the first level beam is to be determined, the closure ratio for the tributary area shown in Fig. 14.12 must first be computed using Eq. (14.34):

$$C_{cx} = \frac{\sum(A_{col} + A_{wall}) + 1.5\sum A_{beam}}{Bh_s} = \frac{(3 \times 2.1 + 2 \times 18.9) + 1.5 \times 0}{21 \times 2.1} = 1$$

which is the expected result since there are no openings associated with this tributary area as evidenced in Fig. 14.12. In addition, the projected area of the individual columns, walls, and beams are respectively $A_{col} = 2.1\,\text{m}^2$, $A_{wall} = 18.9\,\text{m}^2$, and $A_{beam} = 0$. Note that the projected area of beams is not included in this calculation since the water level is below the first level beam. The total drag force acting on the level 1 beam is thus:

$$\begin{aligned} F_d &= p_d A_w = p_d C_{cx} B h_s \\ &= 54{,}820 \times 1 \times 21 \times 2.1 \\ &= 2418 \times 10^3\,\text{N} = 2418\,\text{kN} \end{aligned}$$

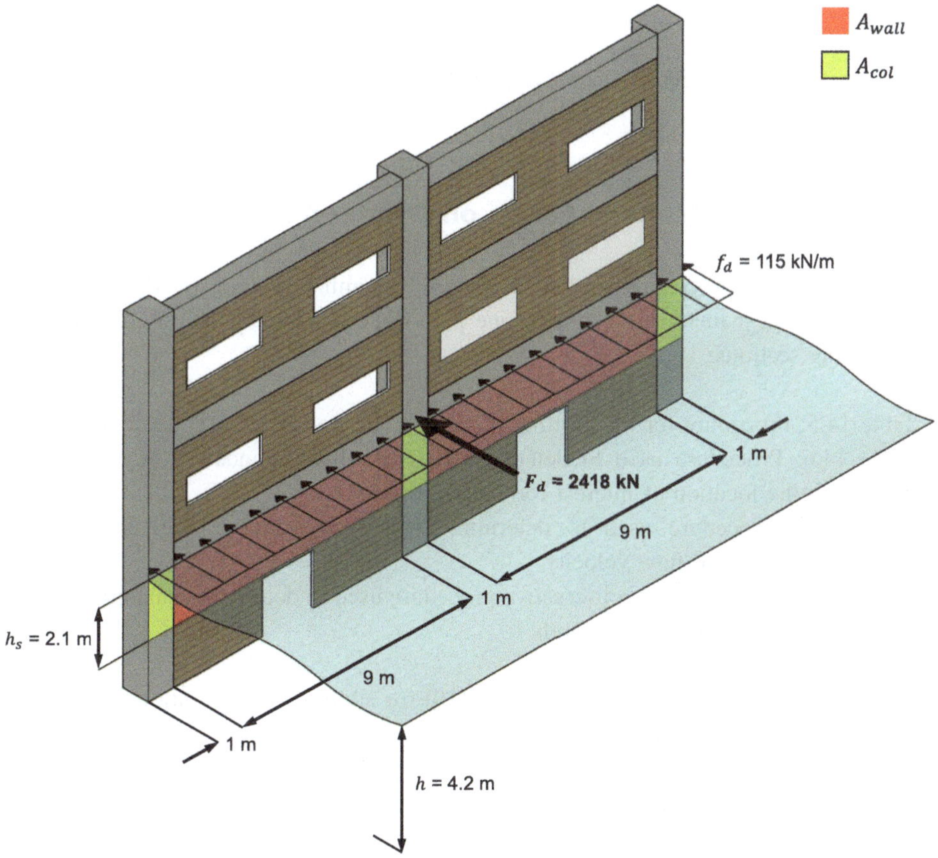

Fig. 14.12 Total (F_d) and distributed (f_d) hydrodynamic force on first level beam

Using Eq. (14.35), the equivalent line load is:

$$f_d = \frac{F_d}{B} = \frac{2418}{21} = 115\,\text{kN/m}$$

If the simplified procedure is instead adopted to obtain the hydrodynamic pressure via Eq. (14.28), p_d becomes:

$$\begin{aligned} p_d &= 1.25\rho_{tsu} g I_{tsu} h_{\max} \\ &= 1.25 \times 1128 \times 9.81 \times 1 \times 6.3 \\ &= 87{,}140\,\text{Pa} = 87.14\,\text{kPa} \end{aligned}$$

which is approximately 60% greater than the maximum p_d obtained using the more precise approach in Table 14.4. Applying this pressure over a height of $1.3h_{max}$ (and ignoring all openings) will therefore yield a total hydrodynamic base shear of:

$$F_d = 87.14 \times 21 \times (1.3 \times 6.3) = 14{,}990\,\text{kN}$$

which is over 3 times greater than the base shear of 4835 kN obtained using the more precise approach.

14.8 Summary: Tsunami Forces on Overland Structures

A summary of the procedure used to determine tsunami forces on overland structures located within the tsunami inundation zone per ASCE 7-22 guidelines is presented via the following sections:

- Table 14.5: Procedure for the plotting of appropriate transects for tsunami analysis.
- Table 14.6: Procedure used to determine the maximum inundation depth and flow velocity at the location of interest for each transect.
- Table 14.7: Procedure used to determine tsunami forces for a given maximum inundation depth and flow velocity.
- Table 14.8: Simplified but conservative procedure used to determine tsunami forces for a given maximum inundation depth.

Interested readers are encouraged to consult [10] for additional information pertaining to tsunami design practices when adopting the guidelines from ASCE 7-22.

Table 14.5 Procedure used to plot transects for EGLA

Step 1	Plot the following three transects (terrain elevation profile) all intersecting at the location of the desired structure: • A transect oriented perpendicular to the shoreline. This is referred to as the center transect • A transect oriented 22.5° clockwise from the center transect • A transect oriented 22.5° counterclockwise from the center transect
Step 2	Identify the maximum tsunami runup (R) and inundation distance from the MHW shoreline (x_R) for the transects plotted in Step 1 using one of the following methods: **Method 1** Use Eq. (14.3) to obtain R based on the tsunami amplitude (H_T) and period (T_{tsu}) at 100 m depth: $\frac{R}{H_T} = \begin{cases} 1.5, & \xi_{100} \leq 0.6 \\ 2.5\log_{10}(\xi_{100}) + 2.05, & 0.6 < \xi_{100} \leq 6 \\ 4.0, & 6 < \xi_{100} \leq 20 \\ -2.15\log_{10}(\xi_{100}) + 6.80, & 20 < \xi_{100} \leq 100 \\ 2.5, & \xi_{100} > 100 \end{cases}$ where ξ_{100} is the surf similarity parameter per Eq. (14.1): $\xi_{100} = \frac{T_{tsu}}{\cot\Phi}\sqrt{\frac{g}{2\pi H_T}}, \quad \Phi = \arctan\left(\frac{100}{D_{100}}\right)$ and D_{100} is the distance from 100 m depth to the shoreline **Method 2** Directly obtain R and x_R from the ASCE Tsunami Design Geodatabase (http://asce7tsunami.online)
Step 3	Plot the transects in Step 1 from the point of maximum runup to the structure's location for EGLA
Step 4	For each transect, identify if tsunami bores are present: • The nearshore bathymetric slope is 1:100 or milder • Existence of shallow reefs or other discontinuities in the bathymetric profile • Where historically documented or described in the literature • As determined by a site-specific inundation analysis

14.9 Impact Forces Due to Floating Debris

The tsunami forces summarized in Sect. 14.8 are commonly referred to as "damming forces" due to their relatively slow rise time and long duration. Yet, tsunamis often pick up and transport large quantities of debris during their propagation over land (Fig. 14.13). When floating debris collide with land-based structures, they exert a short-duration but high-intensity impact force which may result in failure. The failure of one structure will likely result in the release of additional waterborne debris that increases the susceptibility

Table 14.6 EGLA procedure used to determine maximum inundation depth and flow velocity

Step 1	Divide each transect plotted in Step 3 of Table 14.5 into individual nodes i with the distance of each segment Δx_i not exceeding 30.5 m. The runup location is denoted $i = 0$
Step 2	For each transect, determine the slope of segment $i = 1$ via Eq. (14.12): $\varphi_i = \frac{\Delta z_i}{\Delta x_i} = \frac{z_{i-1}-z_i}{x_{i-1}-x_i}$, x_i = Horizontal distance from shoreline to node i, z_i = Vertial distance from shoreline to node i
Step 3	Determine the Froude number of segment $i = 1$ via Eq. (14.14): $F_{r,i} = \alpha\sqrt{1-\frac{x_i}{x_R}}$, $\alpha = \begin{cases} 1.0, & \text{No bores present} \\ 1.3, & \text{Tsunami bores present} \end{cases}$
Step 4	Determine Manning's coefficient n_i for segment $i = 1$ from Table 14.1
Step 5	Determine the friction slope for segment $i = 1$ via Eq. (14.13): $s_i = \frac{gF_{r,i}^2}{\left(\frac{1}{n_i}\right)^2 h_{i-1}^{1/3}}$, $h_0 = 0.01$ m
Step 6	Determine the energy head at $i = 1$ via Eq. (14.11): $E_{g,i} = E_{g,i-1} + (\varphi_i + s_i)\Delta x_i$, $E_{g,0} = 0$
Step 7	Determine the inundation depth and flow velocity at $i = 1$ via Eqs. (14.18) and (14.19) using $E_{g,1}$ from Step 6 and $F_{r,1}$ from Step 3: $h_i = \frac{E_{g,i}}{1+0.5F_{r,i}^2}$, $u_i = F_{r,i}\sqrt{gh_i}$
Step 8	Repeat Steps 2–7 for $i = 2, 3, 4, \ldots$ until the location of interest is reached. At this location, $h_i = h_{max}$ and $u_i = u_{max}$. Ensure that u_{max} satisfies $3.0 \leq u_{max} \leq \min\left[15.2, 1.5\sqrt{gh_{max}}\right]$ per Eq. (14.20)
Step 9	Determine hydrodynamic pressures and forces using Tables 14.7 or 14.8

of downstream structures to debris impact, thereby leading to cascading hazards. In addition to the damming forces induced by the moving water itself, it is of great importance to predict the maximum impact force exerted by different types of waterborne debris as they make contact with coastal structures.

The type of debris to be considered during tsunami design depends on the nature of the surrounding environment (e.g., shipping containers should be accounted for if the structure is near a port facility) and the depth of inundation (h). As indicated in ASCE 7-22 via Table 14.9, larger and heavier debris types require a higher level of inundation before they can be carried by the tsunami. A detailed discussion into the mobilization of different debris types and their transport pathways following the 2011 Tohoku earthquake and tsunami is given by Naito et al. [8].

Table 14.7 Procedure to determine tsunami forces on structure

Step 1	Determine the design inundation (h) and flow velocity (u) for the following two cases specified in Table 14.2: Maximum velocity $\rightarrow h = \left(\frac{2}{3}\right)h_{max}, \quad u = u_{max}$ Maximum inundation $\rightarrow h = h_{max}, \quad u = \left(\frac{1}{3}\right)u_{max}$
Step 2	For a structure with width B, obtain the hydrodynamic pressure for the cases in Step 1 via Eq. (14.21): $p_d = \begin{cases} \frac{1}{2}\rho_{tsu}I_{tsu}C_d u^2, & \text{No bores present or } B/h < 3.0 \\ \frac{1}{2}C_{bore}\rho_{tsu}I_{tsu}C_d u^2, & \text{Tsunami bores present and } B/h \geq 3.0 \end{cases}$ where $\rho_{tsu} = 1128\,\text{kg m}^{-3}$, I_{tsu} is from Table 14.3, and C_d is from Fig. 14.6. Alternatively, refer to Table 14.8 for a simplified but conservative procedure to determine p_u
Step 3	Determine the total hydrodynamic force on structural elements with tributary area A_w via Eqs. (14.32) and (14.33): $F_d = p_d A_w, \quad A_w = C_{cx}Bh_s$ where h_s is the tributary height of the element below the inundation level and C_{cx} is the closure coefficient per Eq. (14.34): $C_{cx} = \frac{\sum(A_{col}+A_{wall})+1.5\sum A_{beam}}{Bh_s}, \quad 0.7 \leq C_{cx} \leq 1.0$ with A_{col}, A_{wall}, and A_{beam} denoting the projected area of all columns, walls, and beams
Step 4	If necessary, covert F_d obtained in Step 3 to a uniformly distributed line load (f_d) acting on linear elements via Eq. (14.35): $f_d = \begin{cases} \frac{F_d}{B}, & \text{horizontal elements} \\ \frac{F_d}{h_s}, & \text{vertical elements} \end{cases}$
Step 5	Apply additional hydrostatic lateral and/or buoyancy forces for structural analysis if necessary

Table 14.8 Simplified procedure to determine hydrodynamic pressure

Step 1	For h_{max} obtained in Table 14.6, determine the hydrodynamic pressure: $p_d = 1.25\rho_{tsu}gI_{tsu}h_{max}$ where $\rho_{tsu} = 1128$ kg m^{-3} and I_{tsu} is from Table 14.3
Step 2	Apply p_d from Step 1 as a uniform lateral pressure over a height of $1.3h_{max}$ on the structure
Step 3	Carry out Steps 3–5 in Table 14.7

Fig. 14.13 Debris transported during the 2011 Tohoku tsunami (courtesy of USGS)

Table 14.9 Minimum inundation depths to mobilize varies types of floating debris per ASCE 7-22

Debris type	Inundation threshold (h) required for mobilization (m)
Poles, logs, and passenger vehicles	0.9
Shipping containers[a]	0.9
Boulders and concrete debris	1.8
Ships or barges[a]	3.6

[a] *Note* Shipping containers and ship collisions need only apply for structures located near ports

When compared with drag-induced forces from moving water alone, the modeling of debris impact with stationary structures is significantly less understood. Yet, numerous analytical solutions have been proposed to parameterize the maximum impact force F_I as a function of the debris' mass (m_d) and velocity (u_d). In addition to m_d and u_d, a third parameter is also required which depends on the particular solution adopted. These solutions largely fall under the following three approaches:

- Contact-stiffness
- Impulse-momentum

- Work-energy.

When adopting the contact-stiffness approach [6], the impact force is expressed as:

$$F_I = u_d\sqrt{k_d m_d} \tag{14.36}$$

where k_d denotes the debris stiffness. The impulse-momentum approach [1] yields:

$$F_I = \frac{2u_d m_d}{t_d} \tag{14.37}$$

where t_d is the impact duration. Finally, the work-energy approach [4] is:

$$F_I = \frac{u_d^2 m_d}{x_d} \tag{14.38}$$

where x_d is the stopping distance of the debris. Of these three approaches, the most widely used is the contact-stiffness method since the debris stiffness is generally easier to estimate than the impact duration or stopping distance [9]. For this reason, the contact-stiffness approach is adopted by ASCE 7-22 for the calculation of debris forces and will be the primary focus of this section.

14.9.1 Dynamic Force from Debris Impact

To derive Eq. (14.36), we assume that floating debris will travel at the velocity of tsunami flow (u), a phenomenon that has been experimentally demonstrated. In addition, we know that Newton's second law states:

$$\text{Force} = \frac{\text{Change in momentum}}{\text{Change in time}} = \frac{m_d\left(u_{final} - u_{initial}\right)}{t_d} \tag{14.39}$$

where m_d is the mass of an object with its velocity before and after the collision represented by $u_{initial}$ and u_{final}, respectively, and t_d is the duration of impact. If we assume that the collision is completely elastic (i.e., no kinetic energy is lost), the object's velocity before and after impact must be equal in magnitude but opposite in direction (see Fig. 14.14). Hence, if the object has an initial velocity of $u_{initial} = -u_{max}$, Eq. (14.39) becomes:

$$F_I = \frac{m_d[u_{max} - (-u_{max})]}{t_d} = \frac{2m_d u_{max}}{t_d} \tag{14.40}$$

where F_I denotes the force of impact. We also notice that this result essentially reproduces the impulse-momentum equation given by (14.37). As the object contacts a rigid boundary, a compressive stress wave will be generated at the impact location which will propagate through the object along its entire length (L_d) and back again towards the impact end (Fig. 14.14). During the propagation of this stress wave, the object will be in constant contact with the boundary over a total time of t_d as shown in Fig. 14.14. If the object has a length L_d, the impact time is therefore the time required for the stress wave to travel twice its length:

$$t_d = \frac{2L_d}{c_0} = \frac{2L_d}{\sqrt{E_d/\rho_d}} \tag{14.41}$$

where $c_0 = \sqrt{E_d/\rho_d}$ is the speed of the wave through a solid medium with Young's modulus E_d and density ρ_d. Substituting Eq. (14.41) into (14.40) yields:

$$F_I = \frac{m_d u_{max}}{L_d}\sqrt{\frac{E_d}{\rho_d}} \tag{14.42}$$

as the impact force. However, we also know that if the object has a cross-sectional area of A_d, its stiffness (k_d) and mass (m_d) would be:

$$k_d = \frac{E_d A_d}{L_d}, \quad m_d = \rho_d A_d L_d \tag{14.43}$$

Substituting Eq. (14.43) into (14.42) thus gives another expression for the impact force associated with a velocity u_{max} only in terms of the object's mass and stiffness:

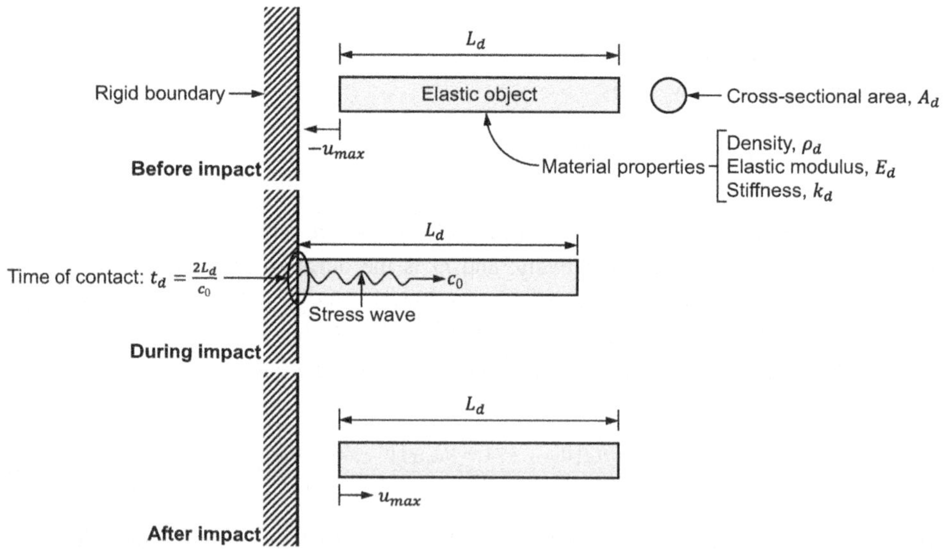

Fig. 14.14 Description of elastic debris collision

$$F_I = u_{max}\sqrt{k_d m_d} \tag{14.44}$$

where u_{max} is the maximum flow velocity at the location of interest determined using the EGLA procedure described in Sect. 14.2. Note that impact forces do not need to be applied concurrently with tsunami hydrostatic and hydrodynamic forces for purposes of structural analysis and design. The impact time is:

$$t_d = \frac{2u_{max}m_d}{F_I} \tag{14.45}$$

which demonstrates that the impact force F_I must be treated as a dynamic load occurring over a very short period of time t_d.

14.9.2 Equivalent Static Force from Debris Impact

The dynamic force from Eq. (14.44) may be converted into an equivalent static load if multiplied with a dynamic response factor R_{max}. This dynamic response factor is influenced by the ratio between the duration of impact (t_d) and the natural period of the structural component (T_n). The impact force when treated as an equivalent static load (F_S) is then:

$$F_S = F_I R_{max} \tag{14.46}$$

As an example, a structural component with plan dimensions B_c and W_c and length L_c (fixed at one end and free/fixed at the other end) is shown in Fig. 14.15. The component has a mass density of ρ_c and elastic modulus E_c. For fixed-free or fixed-fixed boundary conditions, the natural periods of vibration are:

$$T_n = \begin{cases} \frac{2\pi L_c^2}{3.52}\sqrt{\frac{\rho_c B_c W_c}{E_c I_c}}, & \text{fixed-free} \\ \frac{2\pi L_c^2}{22.37}\sqrt{\frac{\rho_c B_c W_c}{E_c I_c}}, & \text{fixed-fixed} \end{cases} \tag{14.47}$$

where:

$$I_c = \frac{B_c W_c^3}{12} \tag{14.48}$$

is the second moment of area in the direction of impact. The dynamic response factor (R_{max}) as a function of t_d/T_n given by ASCE 7-22 is also plotted in Fig. 14.15. If the impacted structure is assumed to be fully rigid such that $T_d \approx 0$, a dynamic response factor of $R_{max} = 1.5$ can be adopted.

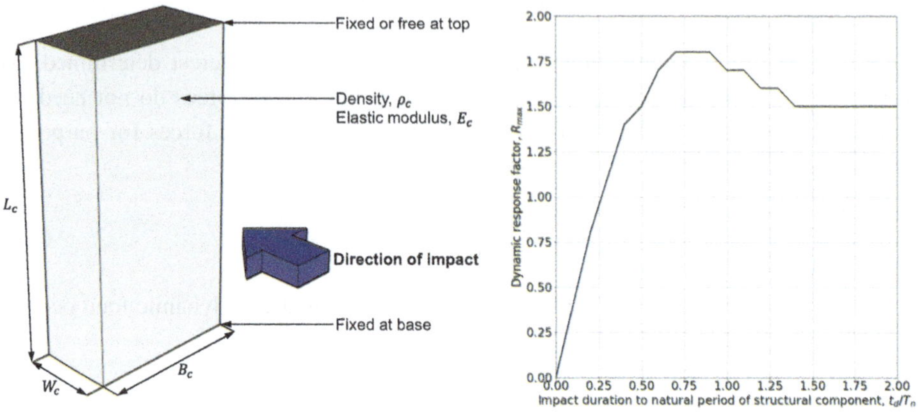

Fig. 14.15 Structural element impacted by debris and its dynamic response factor as a function of t_d/T_n (adapted from ASCE 7-22)

14.9.3 Example of Debris Impact

To illustrate the application of Eqs. (14.44) and (14.46), consider the example illustrated in Fig. 14.16. Here, a 0.5 m diameter wooden pole with a length of $L_d = 8\,\mathrm{m}$ is impacting a 1 m × 0.8 m concrete column fixed at both ends. The maximum velocity of $u_{max} = 7.2\,\mathrm{m\,s^{-1}}$ and its associated inundation depth of $h = 4.2\,\mathrm{m}$ were taken from Table 14.4, which correspond to critical flow conditions obtained from the example described in Sect. 14.7. The density and Young's modulus of the debris are $\rho_d = 700\,\mathrm{kg\,m^{-3}}$ and $E_d = 9000\,\mathrm{MPa}$, respectively. For the concrete column, its material properties are taken as $\rho_s = 2400\,\mathrm{kg\,m^{-3}}$ and $E_s = 25{,}000\,\mathrm{MPa}$. As shown in Fig. 14.16, the debris is impacting the column along a direction that is perpendicular to its long (1 m) side.

To determine the impact force F_I according to the contact-stiffness approach, the stiffness of the debris must be known. This is given as:

$$k_d = \frac{E_d A_d}{L_d} = \frac{9000 \times 10^6 \times 0.2}{8} = 225 \times 10^6\,\mathrm{N/m}$$

where:

$$A_d = \frac{\pi \times 0.5^2}{4} = 0.20\,\mathrm{m}^2$$

is the cross-sectional area of the debris. The total debris mass is then:

$$m_d = \rho_d A_d L_d = 700 \times 0.2 \times 8 = 1120\,\mathrm{kg}$$

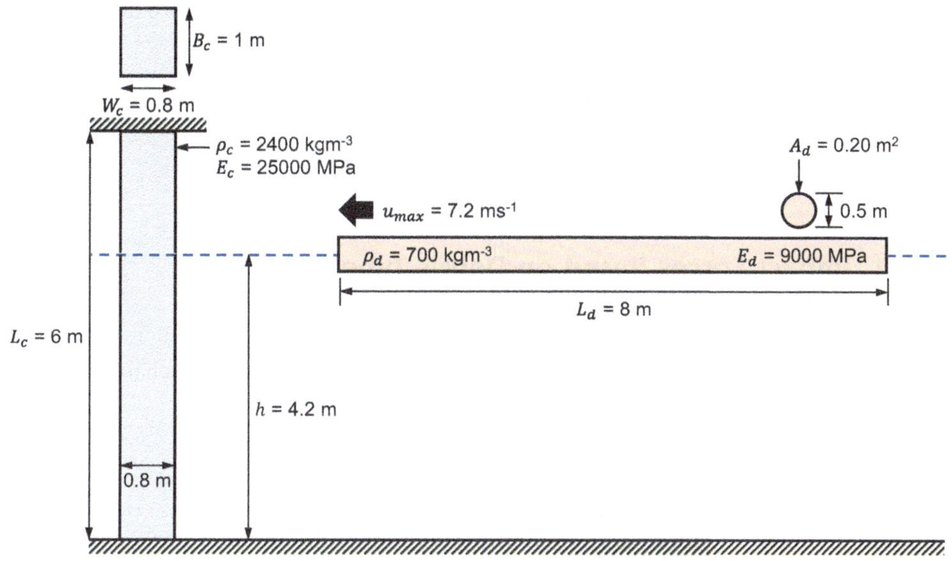

Fig. 14.16 Debris impact with concrete column

We can now determine the maximum impact force from (14.44):

$$F_I = u_{max}\sqrt{k_d m_d} = 7.2\sqrt{225 \times 10^6 \times 1120} = 3614 \times 10^3\,\text{N}$$

In order to convert this dynamic force into an equivalent static load, we must determine the duration of impact:

$$t_d = \frac{2u_{max}m_d}{F_I} = \frac{2 \times 7.2 \times 1120}{3614 \times 10^3} = 0.0045\,\text{s}$$

Now we must determine the natural period of vibration of the impacted column T_n. The second moment of area along the direction of impact is:

$$I_c = \frac{B_c W_c^3}{12} = \frac{1 \times 0.8^3}{12} = 0.0427\,\text{m}^4$$

For a fixed-fixed column, the natural period of vibration is then:

$$T_n = \frac{2\pi L_c^2}{22.37}\sqrt{\frac{\rho_c B_c W_c}{E_c I_c}} = \frac{2\pi \times 6^2}{22.37}\sqrt{\frac{2400 \times 1 \times 0.8}{25{,}000 \times 10^6 \times 0.0427}} = 0.0136\,\text{s}$$

The ratio between the impact duration and natural period is thus:

$$\frac{t_d}{T_n} = \frac{0.0045}{0.0136} = 0.34$$

From Fig. 14.15, this ratio results in a dynamic response factor of approximately R_{max} = 1.25. The equivalent static load on the column is then:

$$F_S = F_I R_{max} = 3614 \times 1.25 = 4520\,\text{kN}$$

which can be applied as a horizontal load 4.2 m above the base.

14.10 Impact Forces Based on Debris Type Per ASCE 7 Guidelines

ASCE 7-22 has provided special guidelines for the assessment of impact forces arising from the various types of tsunami debris summarized in Table 14.9.

14.10.1 Wooden Logs and Poles

ASCE 7-22 stipulates that logs and poles must have a minimum mass of $m_d = 454\,\text{kg}$ and stiffness $k_d = 61.3 \times 10^6\,\text{N/m}$ when using Eq. (14.44) to determine the force of impact. In addition, this force is further modified by the tsunami importance factor (I_{tsu}) and the orientation coefficient (C_o):

$$F_I = I_{tsu} C_o u_{max} \sqrt{k_d m_d} \tag{14.49}$$

or:

$$F_S = I_{tsu} C_o R_{max} u_{max} \sqrt{k_d m_d} \tag{14.50}$$

when treated as an equivalent static load. Since the force given by (14.44) assumes a direct longitudinal collision (the worst case) which is not very likely to occur, the orientation coefficient accounts for potential glancing blows and takes a value $C_o = 0.65$ based on experimental observations.

14.10.2 Boulders and Concrete Debris

Boulders and concrete debris are treated as an equivalent static force by ASCE 7-22 with a magnitude of:

$$F_S = 36 I_{tsu}\,(\text{kN}) \tag{14.51}$$

and applied 0.61 m above ground level. This assumes a 2270 kg boulder tumbling at 4 m s^{-1} and a dynamic response factor of 2.

Table 14.10 Minimum mass and stiffness values of standard shipping containers per ASCE 7-22

Type of container	Mass (m_d) (kg)	Stiffness (k_d) (N/m)
20-ft (6.1 m) standard shipping container	Empty: 2270 Loaded: 13,150	42.9×10^6
40-ft (12.2 m) standard shipping container	Empty: 3810 Loaded: 17,240	29.8×10^6

14.10.3 Passenger Vehicles

ASCE 7-22 stipulates an equivalent static force of:

$$F_S = 130 I_{tsu}\ (\text{kN}) \tag{14.52}$$

applied 0.91 m or greater above ground level. This assumes a 1090 kg vehicle traveling at 4 m s^{-1}.

14.10.4 Shipping Containers

The dynamic and equivalent static force of impact from shipping containers may be determined using Eqs. (14.49) and (14.50), respectively, with $C_o = 0.65$. When determining F_I from (14.44), the minimum mass (m_d) and stiffness (k_d) which can be used is given in Table 14.10. In addition, the dynamic force F_I need not be greater than 980 kN.

References

1. Aghl, P.P., Naito, C.J., Riggs, H.R., 2014. Full-Scale Experimental Study of Impact Demands Resulting from High Mass, Low Velocity Debris. Journal of Structural Engineering 140.
2. ASCE/SEI, 2022. ASCE 7-22. Minimum design loads and associated criteria for buildings and other structures. Reston, Virginia 20191.
3. ASCE/SEI, 2017. ASCE 7-16. Minimum design loads and associated criteria for buildings and other structures. Reston, Virginia 20191.
4. Blok, J.J., Brozius, L.H., Dekker, J.N., 1983. The Impact Loads of Ships Colliding With Fixed Structures. Presented at the Offshore Technology Conference, OnePetro, Houston, Texas.
5. Doyle, K., Koh, M.J., Jayasekara, J.R., Cox, D., Park, H., Kameshwar, S., Lomonaco, P., 2025. Tsunami Debris Damming Forces and Associated Coefficients for Elevated Coastal Structure Columns: Experimental Comparison to ASCE 7-22 Minimum Design Loads. Journal of Structural Engineering 151, 04024200.
6. Haehnel, R.B., Daly, S.F., 2004. Maximum Impact Force of Woody Debris on Floodplain Structures. Journal of Hydraulic Engineering 130.

7. Kriebel, D.L., Lynett, P.J., Cox, D.T., Petroff, C.M., Robertson, I.N., Chock, G.Y.K., 2017. Energy Method for Approximating Overland Tsunami Flows. Journal of Waterway, Port, Coastal, and Ocean Engineering 143.
8. Naito, C., Cercone, C., Riggs, H.R., Cox, D., 2014. Procedure for Site Assessment of the Potential for Tsunami Debris Impact. J. Waterway, Port, Coastal, and Ocean Engineering 140, 223–232.
9. Riggs, H.R., Cox, D.T., Naito, C.J., Kobayashi, M.H., Piran Aghl, P., Ko, H.T.-S., Khowitar, E., 2014. Experimental and Analytical Study of Water-Driven Debris Impact Forces on Structures. Journal of Offshore Mechanics and Arctic Engineering 136.
10. Robertson, I.N., 2020. Tsunami Loads and Effects: Guide to the Tsunami Design Provisions of ASCE 7–16. American Society of Civil Engineers, Reston, Virginia 20191.

Wave Attenuation Over Vegetated Terrain 15

The increasing susceptibility of coastal infrastructure to natural hazards (e.g., tropical cyclones) exacerbated by climate change necessitates a rethinking of adaptation solutions beyond traditional hard structures like seawalls. In the twenty-first century, there has been an increasing push towards the use of natural ecosystems, referred to as natural and nature-based features (NNBFs), for coastal protection rather than solely relying on human-made defenses [4]. One of the most common examples of NNBFs is that of coastal vegetation. In particular, vegetation such as mangroves or salt marshes (e.g., Fig. 15.1) can be highly effective at protecting communities from coastal hazards like tsunamis or extreme wave impact [8] while maintaining natural processes and conserving ecological systems [6]. Even oyster reefs have been demonstrated to be effective at shielding shorelines from large waves [9]. This chapter will discuss how reductions in wave height can be predicted based on the properties of submerged vegetation.

Unlike artificial defenses, the feasibility of adopting certain NNBF solutions will be heavily influenced by the local geography. As an example, Fig. 15.2 shows that mangrove forests are largely concentrated within the tropics (region between $\pm 23.4^\circ$ latitude), while salt marshes are mainly found in the subtropics and temperate zones. The objective of this chapter is to demonstrate the mechanisms by which coastal vegetation interacts with and attenuate surface gravity waves. This knowledge will help us predict the type and/or density of vegetation required to achieve the wave conditions desired at any given site.

The original version of the chapter has been revised. A correction to this chapter can be found at https://doi.org/10.1007/978-3-032-18199-2_22

S. Wang, *Applied Ocean Mechanics for Structural Engineering and Climate Adaptation*, Synthesis Lectures on Ocean Systems Engineering,
https://doi.org/10.1007/978-3-032-18199-2_15

Fig. 15.1 Left: Mangrove forest in southern Florida. Right: Salt marsh in northern Florida. Photographs courtesy of USGS

15.1 Overview of Wave Attenuation

Wave attenuation refers to the reduction in wave height during the propagation of surface gravity waves over frictional terrain. Oftentimes, this friction is provided by submerged vegetation. As sea levels rise or storm surges inundate coastal land, the presence of vegetation could serve to mitigate storm damage by acting as a non-intrusive buffer in dissipating wave energy and reducing incoming wave forces acting on nearshore structures. A theory for the attenuation of linear waves in 2D by submerged vegetation based on the work of Kobayashi et al. [7] will be presented in this chapter. This approach is founded on the assumption that real-world vegetation fields can be approximated as a collection of uniformly spaced rigid rods, like that shown in Fig. 15.3, when determining their damping characteristics.

As linear waves with initial height H_0 and frequency $\omega = 2\pi f$ propagate into a field of submerged vegetation strands of width b_v, height h_v, with N strands per unit area (Fig. 15.3), we wish to determine how the wave height H changes as a function of the horizontal distance x into the field. As shown in Fig. 15.4, the total water depth is assumed to be constant and is separated into two distinct regions above (region 1) and below (region 2) the interface at the tip of the vegetation strands. Subscripts "1" and "2" are used denote the regions above and below this interface. η_1 describes the free surface elevation above the still water level (SWL) and η_2 is the vertical displacement at the interface between regions 1 and 2.

At the SWL ($z = 0$) relative to the horizontal position where waves first enter the submerged vegetation field ($x = 0$), the free surface displacement η_1 takes the form:

$$\eta_1 = \frac{H_0}{2} \exp\big(-k_j x\big) \cos(kx - \omega t) \tag{15.1}$$

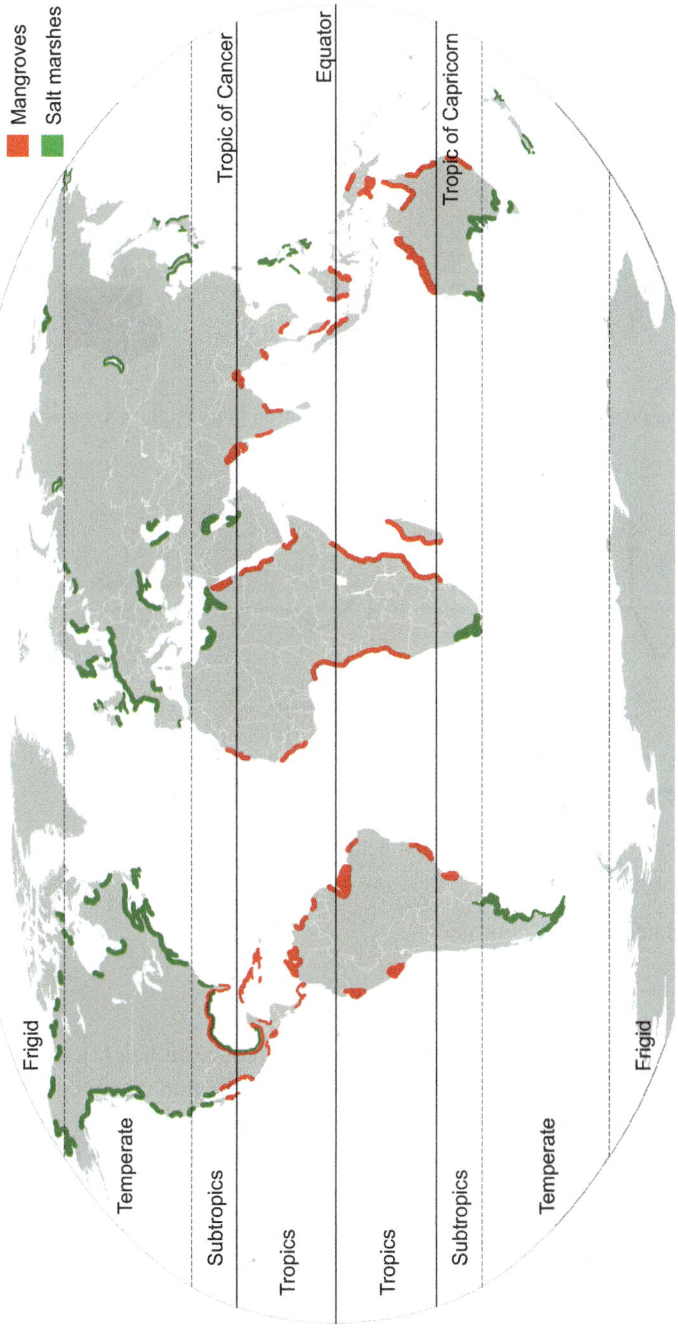

Fig. 15.2 Distribution of mangroves and salt marshes in the early twenty-first century per D'Odorico et al. [5]

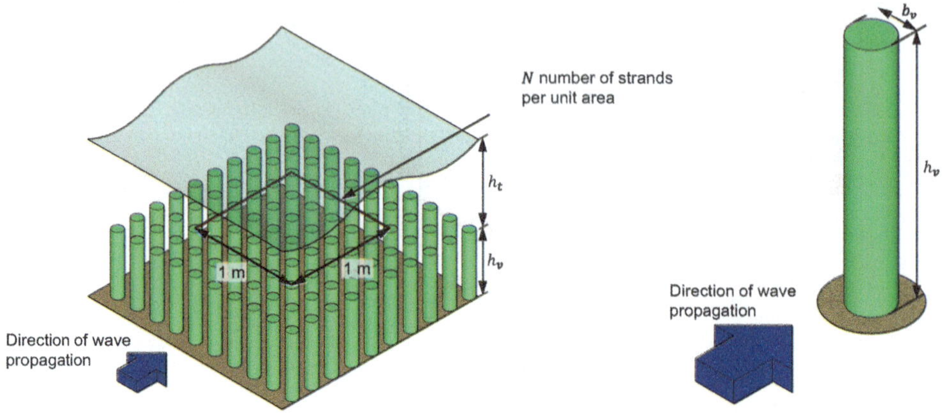

Fig. 15.3 Left: Wave propagation over a submerged vegetation field. Right: Dimensions of a single strand

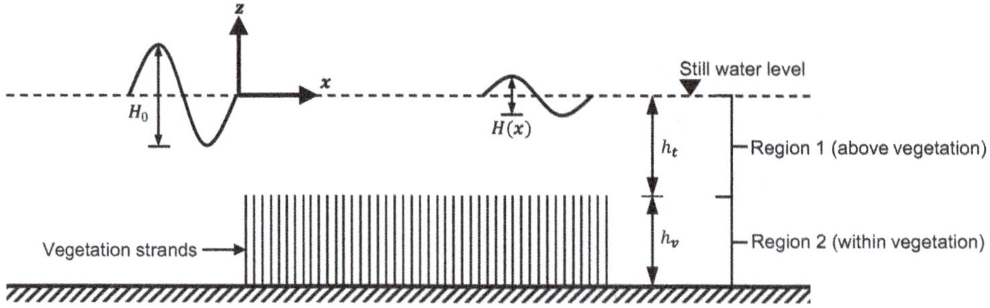

Fig. 15.4 Decay of wave height into vegetation field

whereas the attenuated wave height is:

$$H = H_0 \exp\left(-k_j x\right) \tag{15.2}$$

As indicated in (15.2), we assume waves attenuate (reduce in height) along x following an exponential relationship, where k_j denotes the exponential decay coefficient. This model is consistent with experimental observations by Asano et al. [3]. For a given incoming wave height (H_0) and frequency (ω), we therefore wish to determine the wavenumber k and decay coefficient k_j such that we can compute the attenuated wave properties at some distance x into the field. The wavenumber will inform the change in wavelength while the decay coefficient governs the reduction in wave height.

15.2 Navier–Stokes Equations and Boundary Conditions

15.2.1 Navier–Stokes Equations

In region 1 above the vegetation where $(\eta_2 - h_t) < z < \eta_1$, the 2D Navier–Stokes continuity equation is (see Table 2.3):

$$\frac{\partial u_1}{\partial x} + \frac{\partial w_1}{\partial z} = 0 \tag{15.3}$$

Similarly, the linearized 2D Navier–Stokes momentum equations in region 1 are:

$$\rho \frac{\partial u_1}{\partial t} = -\frac{\partial p_1}{\partial x} \tag{15.4}$$

$$\rho \frac{\partial w_1}{\partial t} = -\frac{\partial p_1}{\partial z} \tag{15.5}$$

In region 2 within the vegetated zone, the continuity equation is:

$$\frac{\partial u_2}{\partial x} + \frac{\partial w_2}{\partial z} = 0 \tag{15.6}$$

The momentum equations for region 2 must include a drag force (F_x and F_z for horizontal and vertical components, respectively) induced by the vegetation on fluid flow:

$$\rho \frac{\partial u_2}{\partial t} = -\frac{\partial p_2}{\partial x} - F_x \tag{15.7}$$

$$\rho \frac{\partial w_2}{\partial t} = -\frac{\partial p_2}{\partial z} - F_z \tag{15.8}$$

where F_x and F_z denote the horizontal and vertical drag forces acting on the vegetation strands per unit volume. We assume the vertical component is negligible ($F_z = 0$) and the horizontal component can be approximated by the Morison equation for flow around slender piles. When summing up the drag force associated with each individual strand within a unit area, F_x is given by:

$$F_x = \frac{1}{2} \rho C_D b_v N |u_2| u_2 \tag{15.9}$$

where C_D is the drag coefficient. Since Eq. (15.9) is nonlinear, we assume that F_x can be further approximated by a linear equation that takes the following form:

$$F_x \approx \rho D u_2 \tag{15.10}$$

to which D is an unknown linearized drag coefficient to be determined later.

15.2.2 Boundary Conditions

At the free surface ($z = 0$), the boundary conditions are:

$$\frac{\partial \eta_1}{\partial t} = w_1 \quad \text{at } z = 0 \tag{15.11}$$

$$p_1 = \rho g \eta_1 \quad \text{at } z = 0 \tag{15.12}$$

At the boundary between regions 1 and 2, the boundary conditions are:

$$\frac{\partial \eta_2}{\partial t} = w_1 = w_2 \quad \text{at } z = -h_t \tag{15.13}$$

$$p_1 = p_2 \quad \text{at } z = -h_t \tag{15.14}$$

while the boundary condition at the seabed is:

$$w_2 = 0 \quad \text{at } z = -(h_t + h_v) \tag{15.15}$$

This chapter will demonstrate how these boundary conditions can be used in conjunction with the linearized Navier–Stokes equations to obtain solutions that describe wave attenuation due to submerged vegetation.

15.3 Fluid Dynamics Above Vegetated Zone

In order to obtain the wavenumber (k) and decay coefficient (k_j) of surface waves above a vegetation field, we first re-express k and k_j as a complex wave number k_c defined as follows:

$$k_c = k + ik_j \tag{15.16}$$

where $i = \sqrt{-1}$ is the imaginary number. Equation (15.1) can therefore be rewritten as:

$$\eta_1 = \text{Re}\left\{ \frac{H_0}{2} \exp[i(k_c x - \omega t)] \right\} \tag{15.17}$$

to which Re is the real part of the complex expression (the "Re" term will not be explicitly stated henceforth for brevity). To confirm that (15.17) indeed is equivalent to (15.1):

$$\begin{aligned} \exp[i(k_c x - \omega t)] &= \exp(ik_c x) \times \exp(-i\omega t) \\ &= \exp\left[i\left(k + ik_j\right)x\right] \times \exp(-i\omega t) \\ &= \exp\left(ikx - k_j x\right) \times \exp(-i\omega t) \end{aligned}$$

$$
\begin{aligned}
&= \exp(ikx - k_j x - i\omega t) \\
&= \exp[i(kx - \omega t)] \times \exp(-k_j x) \\
&= \exp(-k_j x)[\cos(kx - \omega t) + i \sin(kx - \omega t)] \\
&= \exp(-k_j x) \cos(kx - \omega t) \\
&\quad + i \exp(-k_j x) \sin(kx - \omega t)
\end{aligned}
$$

$$
\therefore \mathrm{Re}\left\{ \begin{array}{l} \exp(-k_j x) \cos(kx - \omega t) \\ \quad +i \exp(-k_j x) \sin(kx - \omega t) \end{array} \right\} = \exp(-k_j x) \cos(kx - \omega t) \tag{15.18}
$$

which when multiplied with $H_0/2$ yields (15.1). Equation (15.17) states that η_1 is proportional to $\exp[i(k_c x - \omega t)]$. We can extend this relationship and assume that u_1, u_2, w_1, w_2, p_1, and p_2 are also proportional to $\exp[i(k_c x - \omega t)]$. When considering region 1, u_1, w_1, and p_1 therefore takes the form:

$$
\begin{aligned}
u_1 &= U_1 \exp[i(k_c x - \omega t)] \\
w_1 &= W_1 \exp[i(k_c x - \omega t)] \\
p_1 &= P_1 \exp[i(k_c x - \omega t)]
\end{aligned} \tag{15.19}
$$

where U_1, W_1, and P_1 are proportionality terms not dependent on x or t. We wish to use the linearized momentum equations (15.4) and (15.5) to express u_1 and w_1 in terms of p_1. Differentiating u_1 and w_1 with respect to time t, and p_1 with respect to x yields:

$$
\frac{\partial u_1}{\partial t} = -i\omega U_1 \exp[i(k_c x - \omega t)] = -i\omega u_1 \tag{15.20}
$$

$$
\frac{\partial p_1}{\partial x} = ik_c P_1 \exp[i(k_c x - \omega t)] = ik_c p_1 \tag{15.21}
$$

$$
\frac{\partial w_1}{\partial t} = -i\omega W_1 \exp[i(k_c x - \omega t)] = -i\omega w_1 \tag{15.22}
$$

Substituting (15.20) and (15.21) into (15.4) and (15.22) into (15.5) thus gives:

$$
u_1 = \frac{k_c}{\rho\omega} p_1 \tag{15.23}
$$

$$
w_1 = -\frac{i}{\rho\omega} \frac{\partial p_1}{\partial z} \tag{15.24}
$$

Substituting (15.17) and (15.24) into (15.11) yields:

$$
-\frac{i\omega H_0}{2} \exp[i(k_c x - \omega t)] = -\frac{i}{\rho\omega} \frac{\partial p_1}{\partial z} \quad \text{at } z = 0 \tag{15.25}
$$

By substituting (15.17) into (15.12):

$$p_1 = \rho g \frac{H_0}{2} \exp[i(k_c x - \omega t)] \quad \text{at } z = 0 \tag{15.26}$$

Equation (15.26) suggests that p_1 is not dependent on z at $z = 0$ (this is a trivial observation). Hence, we need to generalize (15.26) to account for the depth component z in order to make (15.26) applicable at all depths within region 1. This can be achieved by incorporating a function $R_1(z)$ into (15.26) with the constraint that $R_1(z) = 1$ at $z = 0$:

$$\begin{gathered} p_1 = \rho g \frac{H_0}{2} R_1(z) \exp[i(k_c x - \omega t)] \\ \text{where } R_1(z) = 1 \text{ at } z = 0 \end{gathered} \tag{15.27}$$

Differentiating (15.27) with respect to z and substituting into (15.25) gives:

$$\begin{aligned} & -\frac{i\omega H_0}{2} \exp[i(k_c x - \omega t)] \\ & = -\frac{i}{\rho\omega} \rho g \frac{H_0}{2} \frac{\partial R_1(z)}{\partial z} \exp[i(k_c x - \omega t)] \quad \text{at } z = 0 \end{aligned} \tag{15.28}$$

which simplifies to:

$$\frac{\omega^2}{g} = \frac{\partial R_1(z)}{\partial z} \quad \text{at } z = 0 \tag{15.29}$$

From (15.27) and (15.29), we therefore need to search for a function $R(z)$ that satisfies the following conditions:

$$\text{Find } R_1(z) \text{ such that } \begin{cases} R_1(z) = 1, & z = 0 \\ \frac{\partial R_1(z)}{\partial z} = \frac{\omega^2}{g}, & z = 0 \end{cases} \tag{15.30}$$

A function that satisfies (15.30) is:

$$R_1(z) = \cosh(k_c z) + \frac{\omega^2}{k_c g} \sinh(k_c z) \tag{15.31}$$

Substituting (15.31) into (15.27) subsequently yields the expression for pressure within region 1, p_1:

$$p_1 = \rho g \frac{H_0}{2} \left[\cosh(k_c z) + \frac{\omega^2}{k_c g} \sinh(k_c z) \right] \exp[i(k_c x - \omega t)] \tag{15.32}$$

To confirm that (15.31) is indeed the correct function for $R_1(z)$, Eqs. (15.23) and (15.24) can be substituted into the continuity equation (15.3), which is then shown to satisfy (15.3) when adopting (15.32) for p_1.

15.4 Fluid Dynamics Within Vegetated Zone

We now shift our attention to region 2. By assuming that $\eta_2 \propto \exp[i(k_c x - \omega t)]$, we can determine the expression for η_2 via Eq. (15.13):

$$\frac{\partial \eta_2}{\partial t} = -i\omega\eta_2 = w_1 = -\frac{i}{\rho\omega}\frac{\partial p_1}{\partial z} \quad \text{at } z = -h_t \tag{15.33}$$

after substituting in (15.32), we obtain:

$$\eta_2 = \frac{1}{\rho\omega^2}\frac{\partial p_1}{\partial z}\bigg|_{z=-h_t} = \frac{k_c g H_0}{2\omega^2}\left[-\sinh(k_c h_t) + \frac{\omega^2}{k_c g}\cosh(k_c h_t)\right]\exp[i(k_c x - \omega t)] \tag{15.34}$$

which simplifies to:

$$\eta_2 = \frac{H_0}{2}\left[\cosh(k_c h_t) - \frac{g k_c}{\omega^2}\sinh(k_c h_t)\right]\exp[i(k_c x - \omega t)] \tag{15.35}$$

When considering region 2 inside the vegetation strands, Eqs. (15.7) and (15.8) can be expressed as:

$$\rho(-i\omega u_2) = -(ik_c p_2) - \rho D u_2 \tag{15.36}$$

$$\rho(-i\omega w_2) = -\frac{\partial p_2}{\partial z} \tag{15.37}$$

which is rearranged to produce:

$$u_2 = \frac{k_c}{\rho(\omega + iD)} p_2 \tag{15.38}$$

$$w_2 = -\frac{i}{\rho\omega}\frac{\partial p_2}{\partial z} \tag{15.39}$$

To obtain an expression for p_2, Eqs. (15.38) and (15.39) are substituted into continuity equation for region 2 (15.6):

$$\frac{k_c}{\rho(\omega + iD)}\frac{\partial p_2}{\partial x} = \frac{i}{\rho\omega}\frac{\partial^2 p_2}{\partial z^2} \tag{15.40}$$

Furthermore, the condition at the boundary between regions 1 and 2 depicted via (15.14) states that $p_1 = p_2$ when $z = -h_t$. p_2 can therefore take the form of p_1 given in (15.32) multiplied by a function $R_2(z)$ which must equal 1 at $z = -h_t$ in order to satisfy (15.14):

$$\begin{aligned} p_2 &= \rho g \frac{H_0}{2}\left[\cosh(k_c h_t) - \frac{\omega^2}{k_c g}\sinh(k_c h_t)\right] R_2(z)\exp[i(k_c x - \omega t)] \\ &= \left(p_1|_{z=-h}\right) R_2(z) \end{aligned} \tag{15.41}$$

Substituting (15.41) into (15.40) gives:

$$\frac{\omega k_c^2}{\omega + iD} i\left(p_1|_{z=-h}\right) R_2(z) = i\left(p_1|_{z=-h}\right)\frac{\partial^2 R_2(z)}{\partial z^2} \tag{15.42}$$

which simplifies to:

$$\alpha^2 R_2(z) = \frac{\partial^2 R_2(z)}{\partial z^2} \tag{15.43}$$

where:

$$\alpha^2 = \frac{\omega k_c^2}{\omega + iD} \tag{15.44}$$

Combining (15.15) and (15.39) therefore implies:

$$-\frac{i}{\rho\omega}\frac{\partial p_2}{\partial z} = 0 \to \frac{\partial R_2(z)}{\partial z} = 0, \quad z = -(h_t + h_v) \tag{15.45}$$

The function $R_2(z)$ must therefore satisfy the following conditions:

$$\text{Find } R_2(z) \text{ such that } \begin{cases} R_2(z) = 1, & z = -h_t \\ \frac{\partial R_2(z)}{\partial z} = 0, & z = -(h_t + h_v) \\ \frac{\partial^2 R_2(z)}{\partial z^2} = \alpha^2 R_2(z), & -(h_t + h_v) < z < (\eta_2 - h_t) \end{cases} \tag{15.46}$$

where it is found that $R_2(z)$ can be satisfied via:

$$R_2(z) = \frac{\cosh[\alpha(z + h_t + h_v)]}{\cosh(\alpha h_v)} \tag{15.47}$$

Substituting (15.47) into (15.41) thus yields the expression for p_2:

$$p_2 = \rho g \frac{H_0}{2}\left[\cosh(k_c h_t) - \frac{\omega^2}{k_c g}\sinh(k_c h_t)\right]\frac{\cosh[\alpha(z + h_t + h_v)]}{\cosh(\alpha h_v)}\exp[i(k_c x - \omega t)] \tag{15.48}$$

which represents the dynamic pressure field within the vegetated zone.

15.5 Dispersion Relation and Decay Coefficient

The dispersion relation relating the wave frequency ω to the complex wavenumber k_c can now be determined from the conditions stipulated in (15.13) and (15.39):

$$\frac{\partial \eta_2}{\partial t} = w_1 = w_2 = -\frac{i}{\rho\omega}\frac{\partial p_2}{\partial z} \quad \text{at } z = -h_t \tag{15.49}$$

After substituting in η_2 from (15.35), we obtain:

$$-\frac{i\omega H_0}{2}\left[\cosh(k_c h_t) - \frac{g k_c}{\omega^2}\sinh(k_c h_t)\right]\exp[i(k_c x - \omega t)] = -\frac{i}{\rho\omega}\frac{\partial p_2}{\partial z}\bigg|_{z=-h_t} \tag{15.50}$$

Next, we can incorporate p_2 from (15.48) to yield:

$$\begin{aligned} &-\frac{i\omega H_0}{2}\left[\cosh(k_c h_t) - \frac{g k_c}{\omega^2}\sinh(k_c h_t)\right]\exp[i(k_c x - \omega t)] \\ &= -\frac{i}{\rho\omega}\rho g\frac{H_0}{2}\left[\cosh(k_c h_t) - \frac{\omega^2}{k_c g}\sinh(k_c h_t)\right]\alpha\tanh(\alpha h_v)\exp[i(k_c x - \omega t)] \end{aligned} \tag{15.51}$$

By rearranging (15.51) for ω^2 and simplifying, the dispersion relation governing wave attenuation is produced:

$$\omega^2 = k_c g\frac{k_c\tanh(k_c h_t) + \alpha\tanh(\alpha h_v)}{k_c + \alpha\tanh(\alpha h_v)\tanh(k_c h_t)} \tag{15.52}$$

Equation (15.52) may be solved numerically to yield $k_c = k + ik_j$ for given values of ω, h_v, h_t, and D. By determining k_c, we obtain both the wavenumber k and the exponential decay coefficient k_j which enables the use of Eq. (15.2) to calculate the wave height at some distance x into the vegetated zone. However, (15.52) is fairly complicated and a simplified equation could be found for cases of weak damping. For this purpose, we introduce the dimensionless damping coefficient ε:

$$\varepsilon = \frac{D}{2\omega} \tag{15.53}$$

The expression for α^2 in (15.44) may now be reformulated through substitution of (15.53):

$$\alpha^2 = \frac{\omega k_c^2}{\omega + iD} = \frac{\omega k_c^2}{\omega(1 + 2i\varepsilon)} = \frac{k_c^2}{1 + 2i\varepsilon} = \frac{k_c^2(1 - 2i\varepsilon)}{(1 + 2i\varepsilon)(1 - 2i\varepsilon)} = \frac{k_c^2(1 - 2i\varepsilon)}{\left(1 - 4i^2\varepsilon^2\right)} \tag{15.54}$$

For small damping, we can assume that $\varepsilon^2 \approx 0$:

$$\alpha^2 \approx k_c^2(1 - 2i\varepsilon) \approx k_c^2\left(1 - 2i\varepsilon + i^2\varepsilon^2\right) \approx k_c^2(1 - i\varepsilon)^2 \tag{15.55}$$

leading to:

$$\alpha \approx k_c(1 - i\varepsilon) \tag{15.56}$$

Substituting (15.56) and (15.16) into (15.52) ultimately yields the simplified dispersion relation which can be used to compute the wavenumber k and decay coefficient k_j:

$$\omega^2 \approx gk \tanh[k(h_v + h_t)] \tag{15.57}$$

$$k_j \approx \varepsilon k \frac{2kh_v + \sinh(2kh_v)}{2k(h_v + h_t) + \sinh[2k(h_v + h_t)]} \tag{15.58}$$

Note that (15.58) is valid only for $\varepsilon \lesssim 0.1$. We also notice that Eq. (15.57) is identical to the dispersion relation governing linear waves derived in Chap. 3. This result implies that for small damping, the wavenumber is not affected by the presence of vegetation. Hence, linear waves will maintain their wavelength while propagating over vegetated terrain and will only experience a reduction in wave height. In fact, experimental observations suggest that (15.57) might be valid even for larger damping values.

15.6 Damping Coefficient

In order to apply (15.52) or (15.58), we must determine an expression for the linear damping coefficient D featured within (15.53). For this purpose, we first define the rate of energy dissipation (D_d) within the vegetated zone (region 2) per unit horizontal area due to the drag force F_x as:

$$D_d = \int_{-(h_v+h_t)}^{-h_t} F_x u_2 dz \tag{15.59}$$

Here, D_d represents the rate at which wave energy is being dissipated by a unit area of vegetation extending the entire depth of region 2 from $z = -h_t$ to $z = -(h_v + h_t)$. Note that $F_x u_2$ denotes the horizontal drag force multiplied by the fluid velocity in region 2. Since work = force × distance and velocity = distance/time, force × velocity therefore returns the rate of energy loss (i.e., work done per unit time, otherwise known as power) associated with vegetation damping. From (15.9) and (15.10), we see that F_x can be expressed in two ways—the nonlinear Morison drag Eq. (15.9) and its linearized approximation (15.10). When substituted into (15.59), the dissipation rate becomes:

$$D_d = \frac{1}{2}\rho C_D b_v N \int_{-(h_v+h_t)}^{-h_t} |u_2| u_2^2 dz \tag{15.60}$$

$$D_d \approx \rho D \int_{-(h_v+h_t)}^{-h_t} u_2^2 dz \tag{15.61}$$

We wish to find an expression for the linear damping coefficient D such that (15.60) is approximately equal to (15.61) for the case of small damping where $\varepsilon \ll 1$. The pressure in region 2 (p_2) stated via (15.48) when adopting $k_c = k(1 + i\delta)$ and $\alpha = k[1 + i(\delta - \varepsilon)]$, where $\delta = k_j/k$ is the ratio between the decay coefficient and wavenumber, can be expressed as follows:

$$p_2 \approx \rho g \frac{H}{2} \frac{\cosh[k(h_v + h_t + z)]}{\cosh[k(h_v + h_t)]} [1 + i\delta c_2 + i(\delta - \varepsilon)c_3] \exp[i(kx - \omega t)] \tag{15.62}$$

where:

$$c_2 = \frac{\sinh(kh_t)\sinh[k(h_t + h_v)]}{\cosh(kh_v)} - kh_t \tanh(kh_v)$$
$$c_3 = k(h_v + h_t + z)\tanh[k(h_v + h_t + z)] - kh_v \tanh(kh_v) \tag{15.63}$$

Equation (15.62) can then be substituted into (15.38) to produce an expression for u_2, which is subsequently substituted into (15.60) and (15.61):

$$D_d = \frac{2}{3\pi} \rho C_D b_v N \left(\frac{kgH}{2\omega} \right)^3 \frac{\sinh^3(kh_v) + 3\sinh(kh_v)}{3k\cosh^3[k(h_v + h_t)]} \tag{15.64}$$

$$D_d \approx D \left(\frac{\rho g H^2}{8} \right) \frac{2kh_v + \sinh(2kh_v)}{2\sinh[2k(h_v + h_t)]} \tag{15.65}$$

It is seen that in (15.64), D_d is proportional to H^3 whereas in (15.65), D_d is proportional to H^2. Since we are assuming small damping ($\varepsilon \ll 1$), we can take the approximation that $H^3 \approx H^2 H_0$ for substitution into (15.64). Subsequently equating (15.64) with (15.65) finally allows us to derive an approximate expression for the linear damping coefficient D:

$$D \approx \frac{2\omega}{9\pi} C_D b_v N H_0 \frac{\sinh(3kh_v) + 9\sinh(kh_v)}{[2kh_v + \sinh(2kh_v)]\sinh[k(h_v + h_t)]} \tag{15.66}$$

which can be substituted into (15.53) to determine ε. Note that while D was derived assuming that $\varepsilon \ll 1$, it may also be applied for cases where ε is in the order of unity, i.e., $\varepsilon \lesssim 1$.

15.7 Drag Coefficient

From Eq. (15.66), we notice that the linear damping coefficient D is also dependent upon the drag coefficient C_D. This term is highly variable and is based on the type of vegetation considered which must be determined through experimental testing [2]. Nevertheless, the drag coefficient C_D generally takes the form:

$$C_D = \left(\frac{C_\alpha}{R_e}\right)^{C_\beta} + C_\gamma \tag{15.67}$$

where C_α, C_β, and C_γ are constants dependent on the vegetation properties. R_e is the Reynolds number defined as:

$$R_e = \frac{b_v u_c}{\nu_k} \tag{15.68}$$

where $\nu_k = 10^{-6}$ m^2s^{-1} is the kinematic viscosity of water and u_c is the characteristic fluid velocity acting on the vegetation:

$$u_c = \frac{gkH_0}{2\omega}\frac{\cosh(kh_v)}{\cosh[k(h_v + h_t)]} \tag{15.69}$$

The constants C_α, C_β, and C_γ for two types of coastal vegetation, saltmarsh cordgrass and kelp (Fig. 15.5), are presented in Table 15.1.

Fig. 15.5 Left: A salt marsh in Georgia, United States (courtesy of Graham D. Schuster CC BY-SA 3.0). Right: An underwater kelp forest (courtesy of NOAA)

Table 15.1 Drag coefficient constants for saltmarsh cordgrass and kelp

Vegetation type	C_α	C_β	C_γ	Reference
Saltmarsh cordgrass	744	1.27	0.76	Anderson and Smith [1]
Kelp	2200	2.4	0.08	Kobayashi et al. [7]

15.8 Example: Attenuation of Regular Waves Over Submerged Vegetation

An example is adopted to highlight how the equations of wave attenuation can be used to describe the damping of surface waves as they travel over submerged vegetation. Consider waves with an initial wave height of $H_0 = 1$ m and period $T = 5$ s propagating in water with a total depth of $h = 7$ m. The waves then enter a field of submerged kelp $h_v = 2$ m tall with a stem density of $N = 200$ stems/m^2 (see Fig. 15.6). Each vegetation strand has a diameter of $b_v = 0.05$ m. Note that the wave profile technically falls into the nonlinear regime when assessed via Le Méhauté's abacus, while the equations by Kobayashi et al. [7] were derived assuming linear wave theory. Nevertheless, they can still serve as a useful approximation when determining the decay rate.

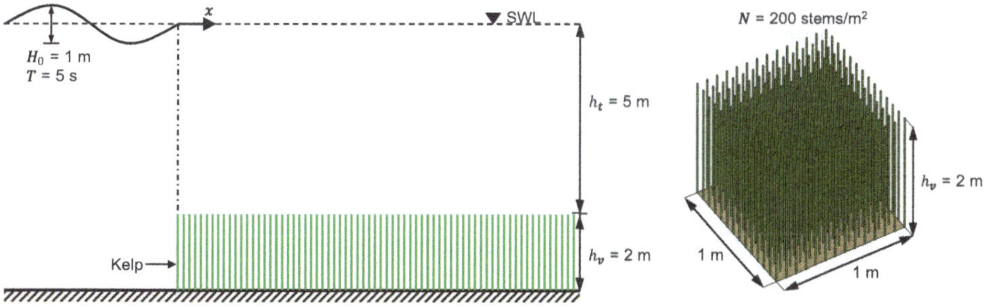

Fig. 15.6 Wave propagation over an idealized kelp field

The wavenumber k is first determined via the dispersion relation:

$$\omega^2 = \left(\frac{2\pi}{T}\right)^2 \approx gk \tanh[k(h_v + h_t)]$$

where $g = 9.81\,\mathrm{ms}^{-2}$, $T = 5$ s, $h_v = 2$ m, and $h_t = 5$ m. Solving for k numerically yields $k = 0.186\,\mathrm{m}^{-1}$. However, if the approximate method is used to obtain k:

$$\frac{2\pi}{k} \approx \frac{gT^2}{2\pi}\sqrt{\tanh\left(\frac{4\pi^2 h}{T^2 g}\right)} = \frac{9.81 \times 5^2}{2\pi}\sqrt{\tanh\left(\frac{4\pi^2 \times 7}{5^2 \times 9.81}\right)}$$

we get $k = 0.179\,\mathrm{m}^{-1}$ which has an error of less than 4%. To determine the drag coefficient C_D, we must first compute the Reynolds number R_e as a function of the characteristic fluid velocity:

$$u_c = \frac{gkH_0}{2\omega}\frac{\cosh(kh_v)}{\cosh[k(h_v + h_t)]} = \frac{9.81 \times 0.186 \times 1}{2 \times 1.257} \times \frac{\cosh(0.186 \times 2)}{\cosh[0.186(5+2)]} = 0.393\ \mathrm{ms}^{-1}$$

where $\omega = 2\pi/T = 1.257\,\mathrm{rads}^{-1}$. This gives a Reynolds number of:

$$R_e = \frac{b_v u_c}{\nu_k} = \frac{0.05 \times 0.393}{10^{-6}} = 19670$$

The drag coefficient is thus:

$$C_D = \left(\frac{C_\alpha}{R_e}\right)^{C_\beta} + C_\gamma = \left(\frac{2200}{19670}\right)^{2.4} + 0.08 = 0.0852$$

where $C_\alpha = 2200$, $C_\beta = 2.4$, and $C_\gamma = 0.08$ are from Table 15.1 for kelp. The linear damping coefficient D is then:

$$
\begin{aligned}
D &\approx \frac{2\omega}{9\pi} C_D b_v N H_0 \frac{\sinh(3kh_v) + 9\sinh(kh_v)}{[2kh_v + \sinh(2kh_v)]\sinh[k(h_v + h_t)]} \\
&= \frac{2 \times 1.257}{9\pi} \times 0.0852 \times 0.05 \times 200 \times 1 \\
&\quad \times \frac{\sinh(3 \times 0.186 \times 2) + 9\sinh(0.186 \times 2)}{[2 \times 0.186 \times 2 + \sinh(2 \times 0.186 \times 2)]\sinh[0.186(2+5)]} \\
&= 0.136\ \text{rads}^{-1}
\end{aligned}
$$

and the dimensionless damping coefficient ε is next determined:

$$
\varepsilon = \frac{D}{2\omega} = \frac{0.136}{2 \times 1.257} = 0.0542
$$

We notice that $\varepsilon < 0.1$ so Eq. (15.58) can be used directly to determine the exponential decay coefficient k_j. However, if ε was substantially larger than 0.1, then Eq. (15.52) should be solved numerically to determine k_c, where k_j is given by the complex part of k_c. Solving Eq. (15.58) subsequently yields:

$$
\begin{aligned}
k_j &\approx \varepsilon k \frac{2kh_v + \sinh(2kh_v)}{2k(h_v + h_t) + \sinh[2k(h_v + h_t)]} \\
&= 0.0542 \times 0.186 \times \frac{2 \times 0.186 \times 2 + \sinh(2 \times 0.186 \times 2)}{2 \times 0.186(2+5) + \sinh[2 \times 0.186(2+5)]} = 0.00168\ \text{m}^{-1}
\end{aligned}
$$

The equation describing the attenuation of wave height (H) as the wave propagates a distance x meters into the submerged vegetation field is thus:

$$
H = H_0 \exp(-k_j x) = 1 \times \exp(-0.00168x)
$$

The wave height from $x = 0$ to $x = 1000$ m is shown via Fig. 15.7.

15.9 Summary: Wave Attenuation Over Rigid Vegetation for Regular Waves

For regular waves propagating with frequency $\omega = 2\pi f = 2\pi/T$ and initial wave height H_0, the process to determine the wavenumber k and decay coefficient k_j for rigid vegetation using the procedure provided by Kobayashi et al. [7] is summarized in Table 15.2. Parameters required as input are as follows:

- The C_α, C_β, and C_γ parameters based on vegetation type.
- Height of vegetation h_v.
- Water depth between top of vegetation and still water level h_t.
- Vegetation width normal to the direction of wave propagation b_v.
- Number of vegetation strands per unit area (m^2) N.

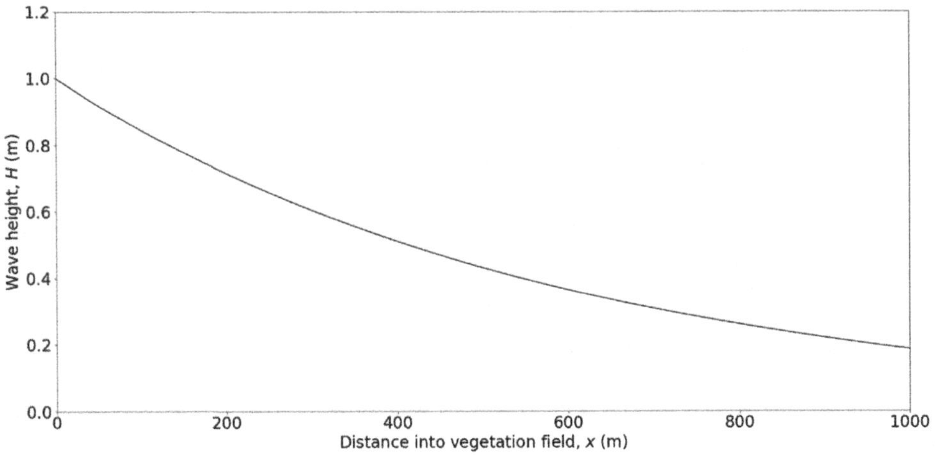

Fig. 15.7 Decay of wave height (H) as a function of distance (x) into the vegetation field

Table 15.2 Wave attenuation calculations assuming rigid vegetation per Kobayashi et al. [7]

Step 1	Determine k via (15.57)	$\omega^2 \approx gk\tanh[k(h_v + h_t)]$
Step 2	Determine C_D via (15.67)	$C_D = \left(\frac{C_\alpha}{R_e}\right)^{C_\beta} + C_\gamma$
Step 3	Determine D via (15.66)	$D \approx \frac{2\omega}{9\pi} C_D b_v N H_0 \frac{\sinh(3kh_v)+9\sinh(kh_v)}{[2kh_v+\sinh(2kh_v)]\sinh[k(h_v+h_t)]}$
Step 4	Determine ε via (15.53)	$\varepsilon = \frac{D}{2\omega}$
Step 5	Check if $\varepsilon \leq 0.1$	**If YES**, determine k_j from (15.58): $k_j \approx \varepsilon k \frac{2kh_v+\sinh(2kh_v)}{2k(h_v+h_t)+\sinh[2k(h_v+h_t)]}$ **If NO**, solve for k_c numerically using (15.52): $\omega^2 = k_c g \frac{k_c\tanh(k_c h_t)+\alpha\tanh(\alpha h_v)}{k_c+\alpha\tanh(\alpha h_v)\tanh(k_c h_t)}$ where α is given by (15.44): $\alpha^2 = \frac{\omega k_c^2}{\omega+iD}$ The solution for k_c will consist of a real part and a complex part. k_j will be given by the complex part
Step 6	Determine H at distance x via (15.2) using k_j calculated in Step 5	$H = H_0\exp(-k_j x)$
Step 7	Determine the wavelength L from the wavenumber k calculated in Step 1	$L = \frac{2\pi}{k}$
Step 8	Use H and L from Steps 6 and 7 as design wave parameters (e.g., within the Goda or Morison Equation)	

References

1. Anderson, M.E., Smith, J.M., 2014. Wave attenuation by flexible, idealized salt marsh vegetation. Coastal Engineering 83, 82–92.
2. Anderson, M.E., Smith, J.M., McKay, S.K., 2011. Wave dissipation by vegetation (No. ERDC/CHL CHETN-I-82). US Army Corps of Engineers.
3. Asano, T., Tsutsui, S., Sakai, T., 1988. Wave damping characteristics due to seaweed, in: Proceedings of the 35th Coastal Engineering Conference in Japan. JSCE, pp. 138–142.
4. Chang, C.-W., Mori, N., 2021. Green infrastructure for the reduction of coastal disasters: A review of the protective role of coastal forests against tsunami, storm surge, and wind waves. Coastal Engineering Journal 63, 370–385.
5. D'Odorico, P., He, Y., Collins, S., De Wekker, S.F.J., Engel, V., Fuentes, J.D., 2013. Vegetation–microclimate feedbacks in woodland–grassland ecotones. Global Ecology and Biogeography 22, 364–379.
6. Guannel, G., Arkema, K., Ruggiero, P., Verutes, G., 2016. The Power of Three: Coral Reefs, Seagrasses and Mangroves Protect Coastal Regions and Increase Their Resilience. PLoS ONE 11.
7. Kobayashi, N., Raichle, A.W., Asano, T., 1993. Wave attenuation by vegetation. Journal of Waterway, Port, Coastal, and Ocean Engineering 119, 30–48.
8. Wang, S., Chang, C.-W., 2025. SPH Simulations to Investigate the Influence of Realistic Mangroves in Reducing Breaking Wave Forces on Coastal Structures. Ocean Engineering 316, 120014.
9. Wiberg, P.L., Taube, S.R., Ferguson, A.E., Kremer, M.R., Reidenbach, M.A., 2019. Wave attenuation by oyster reefs in shallow coastal bays. Estuaries and Coasts 42, 331–347.

Extra Topics on Wave Attenuation Over Vegetated Terrain

16

It is important to understand that the procedure presented by Kobayashi et al. [3], described in Chap. 15, is limited to monochromatic waves propagating over rigid vegetation stems on flat bathymetry. Evidently, these idealized conditions may not always apply in the real world. In this chapter, we will explore a few additional topics pertaining to wave attenuation under more varied wave, bathymetric, and vegetation conditions with a focus on the following:

- Attenuation of shallow-water regular waves via rigid vegetation on sloping bathymetry.
- Attenuation of irregular waves via rigid vegetation on flat bathymetry.
- Attenuation of regular waves via non-rigid vegetation on flat bathymetry.

16.1 Attenuation of Shallow-Water Waves on Sloping Bathymetry

In Chap. 9, we discussed the notion that waves propagating onto sloping bathymetry will increase in height due to shoaling, thereby counteracting the effects of drag-induced attenuation provided by the vegetation. Using linear wave theory, Mendez and Losada [6] developed a formulation to account for both shoaling and attenuation effects for regular waves traveling onto a sloping beach. Figure 16.1 presents an illustration of this problem where we are interested in determining the wave height H at some distance x into a vegetation field founded on the beach.

To derive such a wave transformation model, we begin with the understanding that the rate at which wave energy is being lost over a given distance must be equal to the rate

S. Wang, *Applied Ocean Mechanics for Structural Engineering and Climate Adaptation*, Synthesis Lectures on Ocean Systems Engineering,
https://doi.org/10.1007/978-3-032-18199-2_16

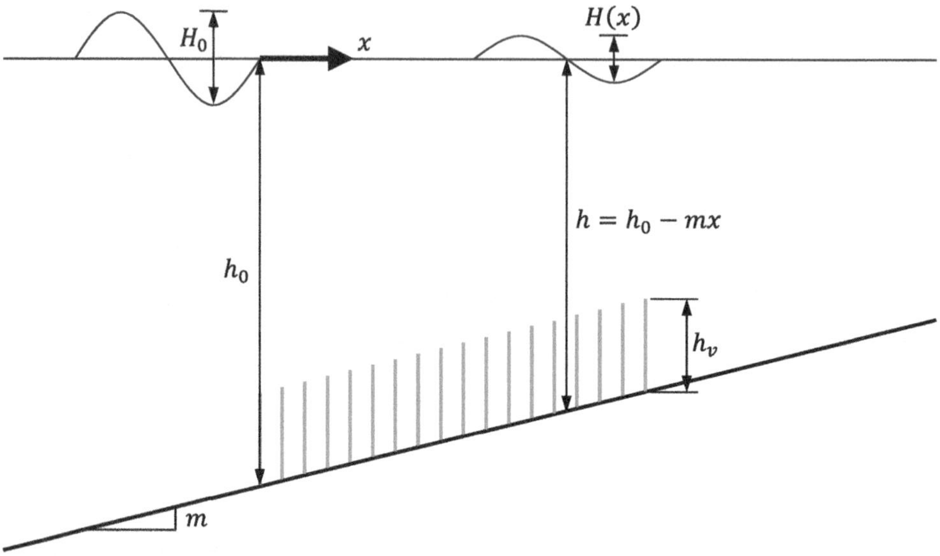

Fig. 16.1 Wave attenuation over a sloping beach with idealized vegetation strands

of energy dissipation provided by the vegetation. This means that the divergence of the horizontal energy flux, as given by Eq. (4.46), must be equal to the time-averaged rate of energy dissipation per unit area. This concept holds true irrespective of the bathymetric configuration is mathematically expressed as:

$$\frac{\partial E c_g}{\partial x} = -\varepsilon_v \tag{16.1}$$

where E is the wave energy per unit area (i.e., energy density) and c_g is the group velocity. From Chap. 4 on wave energetics, we know that these quantities are:

$$E = \frac{\rho g H^2}{8} \tag{16.2}$$

$$c_g = \frac{g}{2\omega}\left[\tanh(kh) + \frac{kh}{\cosh^2(kh)}\right] \tag{16.3}$$

with $h = h_v + h_t$ being the total water depth. The time-averaged energy dissipation rate due to the presence of vegetation is analogous to Eq. (15.59), and is expressed by Mendez and Losada [6] as:

$$\varepsilon_v = \overline{\int_{-h}^{-h+\alpha h} F_x u_2 dz} \tag{16.4}$$

where F_x is the Morison drag force per Eq. (15.9) and u_2 is the horizontal fluid velocity within the vegetated zone. The overbar over Eq. (16.4) implies time-averaging over a wave period. Note that unlike the derivations by Kobayashi et al. [3], the vegetation height is now expressed in terms of the total water depth where:

$$\alpha = \frac{h_v}{h_t + h_v} = \frac{h_v}{h} \leq 1 \tag{16.5}$$

is the vegetation height-to-water depth ratio. This ratio must be less than 1 for submerged vegetation or equal to 1 in the case of stems protruding out of the water. While Kobayashi et al. [3] employed a linear approximation to F_x in the form of Eq. (15.10), a more general solution to (16.4) was obtained by Dalrymple et al. [1] by considering the nonlinear drag force:

$$\varepsilon_v = \frac{2}{3\pi}\rho C_D b_v N \left(\frac{gk}{2\omega}\right)^3 \frac{\sinh^3(k\alpha h) + 3\sinh(k\alpha h)}{3k\cosh^3(kh)} H^3 \tag{16.6}$$

Like before, C_D is the vegetation drag coefficient from Eq. (15.67), b_v is width of each vegetation strand, and N denotes the number of strands per unit area (e.g., per m^2). To arrive at an analytical solution for the attenuation problem, we must adopt the shallow water assumption with $kh \ll 1$. The energy dissipation rate then becomes:

$$\varepsilon_v = \frac{2}{3\pi}\rho C_D b_v N \left(\frac{gk}{2k\sqrt{gh}}\right)^3 \frac{3k\alpha h}{3k} H^3 = \frac{2}{3\pi}\rho C_D b_v N \frac{g^3 \alpha h}{8(gh)^{3/2}} H^3 \tag{16.7}$$

since $\sinh(k\alpha h) \approx k\alpha h$, $\cosh(kh) \approx 1$, and $\omega \approx k\sqrt{gh}$ when $kh \ll 1$. The energy flux in shallow water is then:

$$Ec_g = \frac{\rho g H^2}{8}\sqrt{gh} \tag{16.8}$$

where $c_g = \sqrt{gh}$ for shallow-water waves. Equation (16.1) then becomes:

$$\frac{\partial}{\partial x}\left(\frac{\rho g H^2}{8}\sqrt{gh}\right) = -\frac{2}{3\pi}\rho C_D b_v N \frac{g^3 \alpha h}{8(gh)^{3/2}} H^3 \tag{16.9}$$

which can be simplified to:

$$\frac{\partial H^2 h^{1/2}}{\partial x} = -A_1 \frac{H^3}{h^{1/2}} \tag{16.10}$$

where:

$$A_1 = \frac{2C_D b_v N \alpha}{3\pi} \tag{16.11}$$

The water depth h for a beach with constant slope as a function of distance x (Fig. 16.1) can be described via the following equation:

$$h = h_0 - mx \tag{16.12}$$

where h_0 is the water depth at $x = 0$ and m describes the slope. For example, a 1:10 beach will have $m = 0.1$. Notice that when assuming a constant vegetation height (h_v), the A_1 parameter in Eq. (16.11) will not be constant if $m \neq 0$. This is because the α term will change as a function of x due to the change in water depth h as waves propagate along the slope. For simplicity, at any given location x, we can find an average value for α that encompasses the vegetation field extending a total distance x. If we also assume that the vegetation is completely submerged below water at the location x, Eq. (16.5) can be replaced with:

$$\alpha = \frac{1}{x}\int_0^x \frac{h_v}{h_0 - mx}dx = \frac{h_v}{mx}\ln\left(\frac{h_0}{h_0 - mx}\right), \quad h_v \leq h_0 - mx \tag{16.13}$$

Likewise, if vegetation is expected to protrude above the free surface or if the vegetation height h_v varies along x, the value of α used should correspond to an average value based on the particular situation.

We must now express Eq. (16.10) in a form that gives us the wave height H as a function of the initial wave height H_0 at a given distance x into a vegetation field. To do so, we first adopt the substitution:

$$y = H^2 h^{1/2} \tag{16.14}$$

which will allow us to express (16.10) as:

$$\frac{\partial y}{\partial x} = -A_1 \frac{y^{3/2}}{h^{5/4}} \tag{16.15}$$

since $H^3 = y^{3/2}/h^{3/4}$. We can now employ the chain rule to convert the left-hand side of (16.15) into $\partial y/\partial h$ via:

$$\frac{\partial y}{\partial h} = \frac{\partial y}{\partial x}\frac{\partial x}{\partial h} = \frac{A_1}{m}\frac{y^{3/2}}{h^{5/4}} \tag{16.16}$$

where $\partial x/\partial h = -1/m$ from Eq. (16.12). Replacing $y^{3/2}$ with $H^3 h^{3/4}$ gives:

$$\frac{\partial y}{\partial h} = \frac{A_1}{m}\frac{H^3 h^{3/4}}{h^{5/4}} \tag{16.17}$$

We can now integrate both sides of this expression with respect to h to yield:

$$y = \frac{2A_1 h^{1/2} H^3}{m} + C = \frac{2A_1 h^{1/2} y^{3/2}}{m h^{3/4}} + C \tag{16.18}$$

where C is the constant of integration. This then leads to:

$$\frac{1}{y^{1/2}} = \frac{2A_1}{m h^{1/4}} + C \tag{16.19}$$

Rearranging in terms of y finally gives:

$$y = \frac{1}{\left(\frac{2A_1}{m h^{1/4}} + C\right)^2} = H^2 h^{1/2} \tag{16.20}$$

To determine the constant of integration C, we prescribe our boundary condition that $h = h_0$ when $H = H_0$, i.e.:

$$\frac{1}{\left(\frac{2A_1}{m h_0^{1/4}} + C\right)^2} = H_0^2 h_0^{1/2} \tag{16.21}$$

Rearranging for C then produces:

$$C = \frac{1}{H_0 h_0^{1/4}} - \frac{2A_1}{m h_0^{1/4}} \tag{16.22}$$

which when substituted back into Eq. (16.20) yields:

$$H^2 h^{1/2} = \frac{1}{\left(\frac{2A_1}{m h^{1/4}} + \frac{1}{H_0 h_0^{1/4}} - \frac{2A_1}{m h_0^{1/4}}\right)^2} \tag{16.23}$$

Equation (16.23) can finally be rearranged for H to give:

$$H = K_s K_v H_0 \tag{16.24}$$

where:

$$K_s = \left(\frac{h_0}{h}\right)^{1/4} \tag{16.25}$$

$$K_v = \frac{1}{\frac{2A_1 H_0}{m}(K_s - 1) + 1}, \quad m \neq 0 \tag{16.26}$$

denote the changes in wave height due to shoaling and vegetation damping, respectively. Note that Eq. (16.26) cannot be applied for flat bathymetries when $m = 0$. Instead, we realize that:

$$\lim_{m\to 0}\frac{1}{m}(K_s-1)=\lim_{m\to 0}\frac{1}{m}\left[\frac{h_0^{1/4}}{(h_0-mx)^{1/4}}-1\right]=\frac{x}{4h_0} \tag{16.27}$$

which when substituted back into (16.26) gives:

$$K_v=\frac{1}{\frac{A_1H_0x}{2h_0}+1},\quad m=0 \tag{16.28}$$

Evidently, $K_s = 1$ when $m = 0$ due to the absence of shoaling. The shallow-water wave transformation equations for sloping bathymetry where the change in water depth follows $h = h_0 - mx$ are summarized as:

$$H(x)=\begin{cases}\left(\frac{h_0}{h_0-mx}\right)^{1/4}\frac{H_0}{\frac{2A_1H_0}{m}\left[\left(\frac{h_0}{h_0-mx}\right)^{1/4}-1\right]+1}, & m\neq 0\\ \frac{H_0}{\frac{A_1H_0x}{2h_0}+1}, & m=0\end{cases} \tag{16.29}$$

where:

$$A_1=\frac{2C_Db_vN\alpha}{3\pi} \tag{16.30}$$

for the general case, or:

$$A_1=\frac{2C_Db_vN}{3\pi}\frac{h_v}{mx}\ln\left(\frac{h_0}{h_0-mx}\right),\quad h_v\leq h_0-mx \tag{16.31}$$

if assuming a constant vegetation height h_v. The drag coefficient C_D can be determined via the Reynolds number R_e in the same manner as discussed in Sect. 15.7:

$$C_D=\left(\frac{C_\alpha}{R_e}\right)^{C_\beta}+C_\gamma,\quad R_e=\frac{b_vu_c}{\nu_k} \tag{16.32}$$

where $\nu_k = 10^{-6}\mathrm{m^2s^{-1}}$ and the C_α, C_β, and C_γ parameters can be found in Table 15.1. The characteristic fluid velocity acting on the vegetation is:

$$u_c=\frac{gkH_0}{2\omega}\frac{\cosh(kh_v)}{\cosh(kh)} \tag{16.33}$$

In shallow water, u_c can be approximated as:

$$u_c=\frac{H_0}{2}\sqrt{\frac{g}{h_0}} \tag{16.34}$$

Figure 16.2 illustrates the H/H_0 ratio as a function of h/h_0 for different values of A_1H_0/m obtained using Eq. (16.29) for $m \neq 0$. It is evident that an increase in damping (i.e., larger A_1) or a shallower slope (i.e., smaller m) results in increased wave attenuation

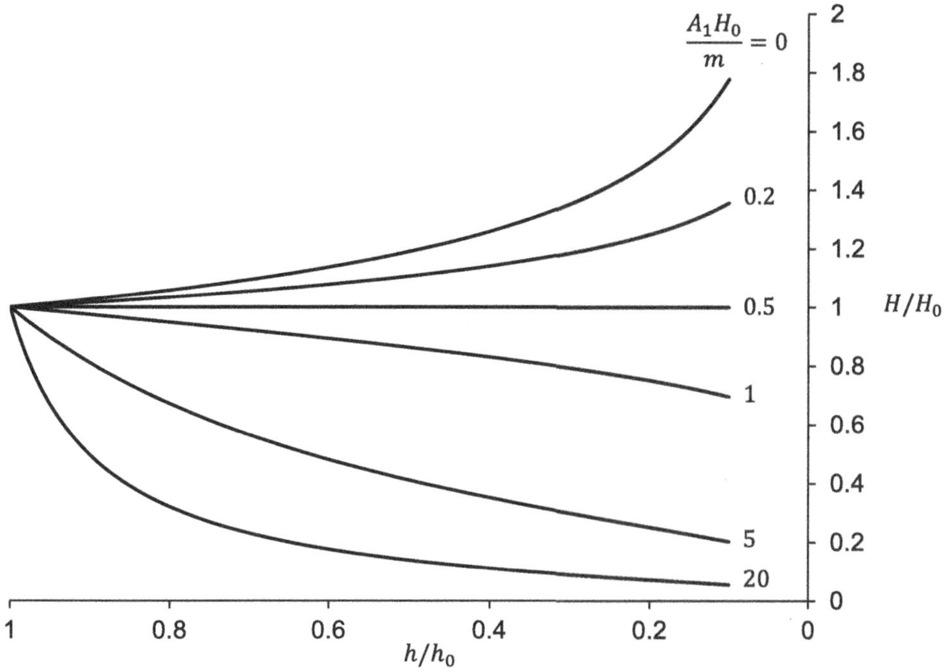

Fig. 16.2 Attenuation of shallow-water waves over sloped bathymetry for different A_1H_0/m (adapted from Mendez and Losada [6])

(i.e., smaller H/H_0) as the wave propagates up the slope (i.e., decreasing h/h_0). We also observe that when A_1H_0/m is exactly equal to 0.5, the wave growth due to shoaling is perfectly balanced by the attenuation due to vegetation. Assuming that breaking does not take place, this results in the wave maintaining its original height of H_0 as it travels up the beach. Conversely, $A_1H_0/m = 0$ implies the absence of vegetation ($K_v = 1$) such that wave transformation is purely due to shoaling. In addition, it is important to realize that Eq. (16.29) is valid only for shallow-water waves. This contrasts with the procedure described in Sect. 15.9 which exhibits no such limitation.

16.2 Attenuation of Irregular Waves

Since water waves in nature are rarely monochromatic (i.e., regular), Mendez and Losada [6] further developed a method to estimate the attenuation of irregular waves as they propagate over rigid vegetation founded on flat bathymetry. The model was validated against experiments performed by Dubi [2] and Lovas [5]. Just like with regular waves, the energy balance equation given by (16.1) must hold:

$$\frac{\partial Ec_g}{\partial x} = -\varepsilon_v \tag{16.35}$$

When assuming that waves follow a Rayleigh distribution, the height of irregular waves within a time series of N total waves will be expressed in terms of the root mean square (RMS) height, being:

$$H_{rms} = \sqrt{\frac{1}{N}\sum_{i=1}^{N} H_i^2} \approx \frac{H_s}{1.4} \tag{16.36}$$

which is approximately equal to the significant wave height (H_s) divided by 1.4. The horizontal energy flux from Eq. (16.35) then becomes:

$$\frac{\partial Ec_g}{\partial x} = \frac{\partial}{\partial x}\left\{\frac{\rho g H_{rms}^2}{8}\frac{g}{2\omega}\left[\tanh(kh) + \frac{kh}{\cosh^2(kh)}\right]\right\} \tag{16.37}$$

where we have replaced H from (16.2) with H_{rms}. Note that for the attenuation of irregular waves, the wavenumber $k = 2\pi/L_p$ is obtained from the wavelength L_p associated with the peak period T_p of the wave spectrum, and:

$$\omega = \frac{2\pi}{T_p} = \sqrt{gk\tanh(kh)} \tag{16.38}$$

is the peak frequency. We can also express the RMS wave height in terms of the Rayleigh probability density function (PDF) via:

$$H_{rms}^2 = \int_0^\infty H^2 f(H)dH \tag{16.39}$$

where:

$$f(H) = \frac{2H}{H_{rms}^2}\exp\left[-\left(\frac{H}{H_{rms}}\right)^2\right] \tag{16.40}$$

is the Rayleigh PDF. The energy dissipation rate ε_v from Eq. (16.6) can likewise be reformulated as:

$$\varepsilon_v = \frac{2}{3\pi}\rho\tilde{C}_D b_v N\left(\frac{gk}{2\omega}\right)^3 \frac{\sinh^3(k\alpha h) + 3\sinh(k\alpha h)}{3k\cosh^3(kh)} H_{rms}^3 \tag{16.41}$$

Since the drag coefficient is sensitive to wave height, an "average drag coefficient" $\tilde{C}_D$ is specified for irregular waves. From (16.39), Eq. (16.41) can be further expressed as:

$$\varepsilon_v = \frac{2}{3\pi}\rho \tilde{C}_D b_v N \left(\frac{gk}{2\omega}\right)^3 \frac{\sinh^3(k\alpha h) + 3\sinh(k\alpha h)}{3k\cosh^3(kh)} \int_0^\infty H^3 f(H) dH \tag{16.42}$$

where:

$$\int_0^\infty H^3 f(H) dH = \int_0^\infty \frac{2H^4}{H_{rms}^2} \exp\left[-\left(\frac{H}{H_{rms}}\right)^2\right] dH = \frac{3\sqrt{\pi}}{4} H_{rms}^3 \tag{16.43}$$

which when substituted back into (16.42) yields:

$$\varepsilon_v = \frac{1}{2\sqrt{\pi}}\rho \tilde{C}_D b_v N \frac{g^3 k^3}{8\omega^3} \frac{\sinh^3(k\alpha h) + 3\sinh(k\alpha h)}{3k\cosh^3(kh)} H_{rms}^3 \tag{16.44}$$

Substituting Eqs. (16.44) and (16.37) into (16.35) then gives:

$$\begin{aligned} &\frac{\partial}{\partial x}\frac{\rho g^2 H_{rms}^2}{16\omega}\left[\tanh(kh) + \frac{kh}{\cosh^2(kh)}\right] \\ &= -\frac{1}{2\sqrt{\pi}}\rho \tilde{C}_D b_v N \frac{g^3 k^3}{8\omega^3} \frac{\sinh^3(k\alpha h) + 3\sinh(k\alpha h)}{3k\cosh^3(kh)} H_{rms}^3 \end{aligned} \tag{16.45}$$

which can be simplified into:

$$\frac{\partial H_{rms}^2}{\partial x}\left[\frac{\tanh(kh)\cosh^2(kh) + kh}{\cosh^2(kh)}\right] = -\frac{1}{3\sqrt{\pi}} \tilde{C}_D b_v N k \frac{\sinh^3(k\alpha h) + 3\sinh(k\alpha h)}{\cosh^3(kh)\tanh(kh)} H_{rms}^3 \tag{16.46}$$

giving:

$$\begin{aligned} \frac{\partial H_{rms}^2}{\partial x} &= -\frac{1}{3\sqrt{\pi}} \tilde{C}_D b_v N k \frac{\sinh^3(k\alpha h) + 3\sinh(k\alpha h)}{\cosh(kh)\tanh(kh)\left[\tanh(kh)\cosh^2(kh) + kh\right]} H_{rms}^3 \\ &= -\frac{2}{3\sqrt{\pi}} \tilde{C}_D b_v N k \frac{\sinh^3(k\alpha h) + 3\sinh(k\alpha h)}{\sinh(kh)[\sinh(2kh) + 2kh]} H_{rms}^3 \end{aligned} \tag{16.47}$$

Equation (16.47) can now be expressed as:

$$\frac{\partial H_{rms}^2}{\partial x} = -B_0 H_{rms}^3 \tag{16.48}$$

with:

$$B_0 = \frac{2}{3\sqrt{\pi}} \tilde{C}_D b_v N k \frac{\sinh^3(k\alpha h) + 3\sinh(k\alpha h)}{\sinh(kh)[\sinh(2kh) + 2kh]} \tag{16.49}$$

Equation (16.48) is a first-order nonlinear ordinary differential equation (ODE) since H_{rms} is only a function of x. To solve (16.48), we rewrite its left-hand side as:

$$\frac{\partial H_{rms}^2}{\partial x} = \frac{\partial H_{rms} \cdot H_{rms}}{\partial x} = H_{rms}\frac{\partial H_{rms}}{\partial x} + H_{rms}\frac{\partial H_{rms}}{\partial x} = 2H_{rms}\frac{\partial H_{rms}}{\partial x} \tag{16.50}$$

which can be substituted back into (16.48) to yield:

$$\frac{\partial H_{rms}}{\partial x} = -\frac{B_0 H_{rms}^2}{2} \tag{16.51}$$

Employing separation of variables:

$$\frac{1}{H_{rms}^2}\partial H_{rms} = -\frac{B_0}{2}\partial x \tag{16.52}$$

then integrating both sides, we obtain:

$$\frac{1}{H_{rms}} = \frac{B_0}{2}x + C \tag{16.53}$$

To solve for the constant of integration C, we utilize the boundary condition that $H_{rms} = H_{rms,0}$ when $x = 0$, where $H_{rms,0}$ is the RMS wave height at the start of the vegetation field (i.e., when $x = 0$):

$$\frac{1}{H_{rms,0}} = C \tag{16.54}$$

Substituting C back into (16.53) gives:

$$\frac{1}{H_{rms}} = \frac{B_0}{2}x + \frac{1}{H_{rms,0}} \tag{16.55}$$

which can be rearranged for H_{rms} to model the transformation of the RMS wave height as a function of distance x into the vegetation field as shown in Fig. 16.3:

$$H_{rms} = \frac{H_{rms,0}}{1 + \tilde{\beta}x} \tag{16.56}$$

where $H_{rms,0}$ denotes the RMS wave height at $x = 0$. The parameter $\tilde{\beta}$ is:

$$\tilde{\beta} = \frac{B_0 H_{rms,0}}{2} = \frac{\tilde{C}_D b_v N H_{rms,0} k}{3\sqrt{\pi}} \frac{\sinh^3(kh_v) + 3\sinh(kh_v)}{[\sinh(2kh) + 2kh]\sinh(kh)} \tag{16.57}$$

As before, b_v is width of each vegetation strand and N denotes the number of strands per unit area. The wavenumber $k = 2\pi/L_p$ is obtained from the wavelength L_p associated with the peak period T_p of the wave spectrum and $h_v = \alpha h$ is the submerged height of the vegetation strands. Mendez and Losada [6] identified an ideal form for the average drag coefficient $\tilde{C}_D$, being:

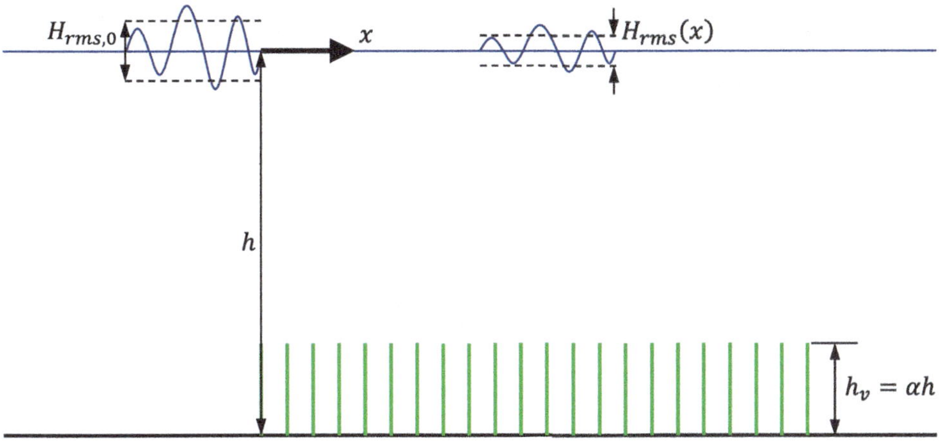

Fig. 16.3 Attenuation of irregular waves in terms of the RMS wave height

$$\tilde{C}_D = \frac{\exp(-0.0138Q)}{Q^{0.3}}, \quad 7 \leq Q \leq 172 \tag{16.58}$$

where:

$$Q = \frac{K_C}{\alpha^{0.76}}, \quad \alpha = \frac{h_v}{h} \leq 1 \tag{16.59}$$

is a function of the Keulegan-Carpenter number K_C:

$$K_C = \frac{u_c T_p}{b_v} \tag{16.60}$$

Here, the characteristic fluid velocity acting on the vegetation u_c is determined via the peak wave period and can be estimated from Eq. (16.33):

$$u_c = \frac{gkH_{rms,0}T_p}{4\pi} \frac{\cosh(kh_v)}{\cosh(kh)} \tag{16.61}$$

where we have replaced ω in (16.33) with $2\pi/T_p$.

16.3 Example: Attenuation of Irregular Waves Over Submerged Vegetation

We will now provide an example to illustrate the application of Eq. (16.56) for the modeling of wave height transformation for irregular waves. To carry out this example, let us consider the following wave and vegetation parameters (see Fig. 16.4):

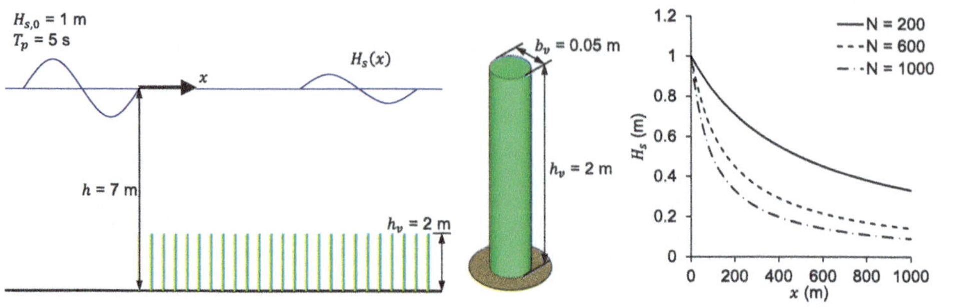

Fig. 16.4 Attenuation of irregular waves over vegetation on flat bathymetry over a distance of 1000 m

- Initial significant wave height $H_{s,0} = 1$ m
- Peak period $T_p = 5$ s
- Water depth $h = 7$ m
- Vegetation height $h_v = 2$ m
- Stem width $b_v = 0.05$ m
- Number of stems per unit area $N = 200$, 600, and 1000 m^{-2}.

We must first convert the initial significant wave height $H_{s,0}$ into the root mean square wave height $H_{rms,0}$:

$$H_{rms,0} = \frac{H_{s,0}}{1.4} = \frac{1}{1.4} = 0.714 \text{ m}$$

Next, the wavenumber k can be determined from the peak period T_p. First calculating the wavelength L_p via the approximate method in Table 3.2:

$$L_p = \frac{gT_p^2}{2\pi}\sqrt{\tanh\left(\frac{4\pi^2 h}{T_p^2 g}\right)} = \frac{9.81 \times 5^2}{2\pi}\sqrt{\tanh\left(\frac{4\pi^2 \times 7}{5^2 \times 9.81}\right)} = 35.1 \text{ m}$$

The wavenumber is then:

$$k = \frac{2\pi}{L_p} = \frac{2\pi}{35.1} = 0.179\ \mathrm{m}^{-1}$$

which can be used to calculate the characteristic fluid velocity u_c:

$$u_c = \frac{gkH_{rms,0}T_p}{4\pi}\frac{\cosh(kh_v)}{\cosh(kh)} = \frac{9.81 \times 0.179 \times 0.714 \times 5}{4\pi}\frac{\cosh(0.179 \times 2)}{\cosh(0.179 \times 7)} = 0.281\mathrm{ms}^{-1}$$

The Keulegan-Carpenter number K_C is then:

$$K_C = \frac{u_c T_p}{b_v} = \frac{0.281 \times 5}{0.05} = 28.1$$

thus giving:

$$Q = \frac{K_C}{\alpha^{0.76}} = \frac{28.1}{0.286^{0.76}} = 72.7$$

where $\alpha = h_v/h = 2/7 = 0.286$ is the relative stem height. Finally, the average drag coefficient $\tilde{C}_D$ is:

$$\tilde{C}_D = \frac{\exp(-0.0138Q)}{Q^{0.3}} = \frac{\exp(-0.0138 \times 72.7)}{72.7^{0.3}} = 0.1013$$

We now possess all the information required to determine the damping parameter $\tilde{\beta}$. For a stem density of $N = 200\,\mathrm{m}^{-2}$, this parameter is:

$$\begin{aligned}\tilde{\beta} &= \frac{\tilde{C}_D b_v N H_{rms,0} k}{3\sqrt{\pi}} \frac{\sinh^3(kh_v) + 3\sinh(kh_v)}{[\sinh(2kh) + 2kh]\sinh(kh)} \\ &= \frac{0.1013 \times 0.05 \times 200 \times 0.714 \times 0.179}{3\sqrt{\pi}} \\ &\quad \times \frac{\sinh^3(0.179 \times 2) + 3\sinh(0.179 \times 2)}{[\sinh(2 \times 0.179 \times 7) + 2 \times 0.179 \times 7]\sinh(0.179 \times 7)} \\ &= 0.01012\ \mathrm{m}^{-1}\end{aligned}$$

The RMS wave height as a function of x can now be found using Eq. (16.56). if we instead wish to determine the significant wave height, this equation can be multiplied with 1.4 giving:

$$H_s(x) = \frac{1.4H_{rms,0}}{1 + \tilde{\beta}x} = \frac{1}{1 + 0.01012x}$$

Figure 16.4 plots the decay in H_s as a function of distance x into the vegetation field for $N = 200$, 600, and 1000 m^{-2}. As expected, the rate of wave decay is greater for vegetation fields with higher stem densities.

16.4 Summary: Wave Attenuation Over Rigid Vegetation for Irregular Waves

Table 16.1 provides a summary of the calculation procedure to determine the attenuation of irregular waves as they propagate into a vegetation field on flat bathymetry. The following wave and vegetation parameters are required as input:

Wave parameters	**Vegetation parameters**
• Initial significant wave height $H_{s,0}$	• Vegetation height h_v
• Peak period T_p	• Stem width b_v
• Water depth h	• Number of stems per unit area N

Table 16.1 Summary of irregular wave attenuation calculations assuming rigid vegetation per [6]

Step 1	Determine the initial RMS wave height $H_{rms,0}$ from the initial significant wave height $H_{s,0}$ per (16.36)	$H_{rms,0} = \frac{H_{s,0}}{1.4}$
Step 2	Determine the wavelength L_p pertaining to the peak period T_p	$L_p = \frac{gT_p^2}{2\pi}\tanh\left(\frac{2\pi h}{L_p}\right)$ or $L_p \approx \frac{gT_p^2}{2\pi}\sqrt{\tanh\left(\frac{4\pi^2 h}{T_p^2 g}\right)}$
Step 3	Determine the wavenumber k corresponding to L_p	$k = \frac{2\pi}{L_p}$
Step 4	Determine the characteristic fluid velocity u_c via (16.61)	$u_c = \frac{gkH_{rms,0}T_p}{4\pi}\frac{\cosh(kh_v)}{\cosh(kh)}$
Step 5	Determine the Keulegan-Carpenter number K_C via (16.60)	$K_C = \frac{u_c T_p}{b_v}$
Step 6	Determine the Q parameter via (16.59)	$Q = \frac{K_C}{\alpha^{0.76}},\quad \alpha = \frac{h_v}{h} \le 1$
Step 7	Determine the average drag coefficient $\tilde{C}_D$ via (16.58)	$\tilde{C}_D = \frac{\exp(-0.0138Q)}{Q^{0.3}},\quad 7 \le Q \le 172$
Step 8	Determine the damping parameter $\tilde{\beta}$ via (16.57)	$\tilde{\beta} = \frac{\tilde{C}_D b_v N H_{rms,0} k}{3\sqrt{\pi}}\frac{\sinh^3(kh_v)+3\sinh(kh_v)}{[\sinh(2kh)+2kh]\sinh(kh)}$
Step 9	Determine H_{rms} or H_s at a distance of x into the vegetation field via (16.56)	$H_{rms} = \frac{H_{rms,0}}{1+\tilde{\beta}x},\quad H_s = \frac{H_{s,0}}{1+\tilde{\beta}x}$

16.5 Wave Attenuation Over Highly Flexible Vegetation

The previous methods treat the dissipation of wave energy as the summation of drag forces associated with each individual stem that are approximated as being rigid. For the attenuation of regular waves propagating over highly flexible vegetation, nonlinear stream function theory was employed by Lima et al. [4] to estimate the reduced wave height at some distance x from the beginning of the vegetation field. The derivation of such formulations is beyond the scope of this text hence only the final equations used to determine the attenuated wave height (H) will be presented. The transformation of H as a function of distance x into the vegetation field follows a similar form to Eq. (16.56):

$$H = \frac{H_0}{1 + \beta x} \tag{16.62}$$

where H_0 is the wave height at $x = 0$, and:

$$\beta = 7.916 \frac{L^2 \left(\frac{h_v}{h}\right)^4 b_v^{\frac{5}{2}} N^{\frac{5}{4}}}{g T^2 (4n - 1) h_v^2} \frac{\rho_v}{\rho} \tag{16.63}$$

Note that b_v and ρ_v refers to the diameter and density of each vegetation strand, respectively, while N is the number of strands per unit area. L and T are the wavelength and period, and:

$$n = 1 + \frac{2kh}{\sinh(2kh)} \tag{16.64}$$

Figure 16.5 compares the decay in wave height for rigid and highly flexible vegetation (pertaining to three different stem densities) computed using Eq. (16.62). The wave and vegetation profile were taken from the example in Sect. 15.8. It is observed that when using Kobayashi's approach assuming rigid vegetation, waves decay at a faster rate than that obtained for flexible vegetation strands at lower densities. Note that the formulation of Eqs. (16.63) and (16.64) were based upon waves propagating over nylon strings where entangling effects were also accounted for.

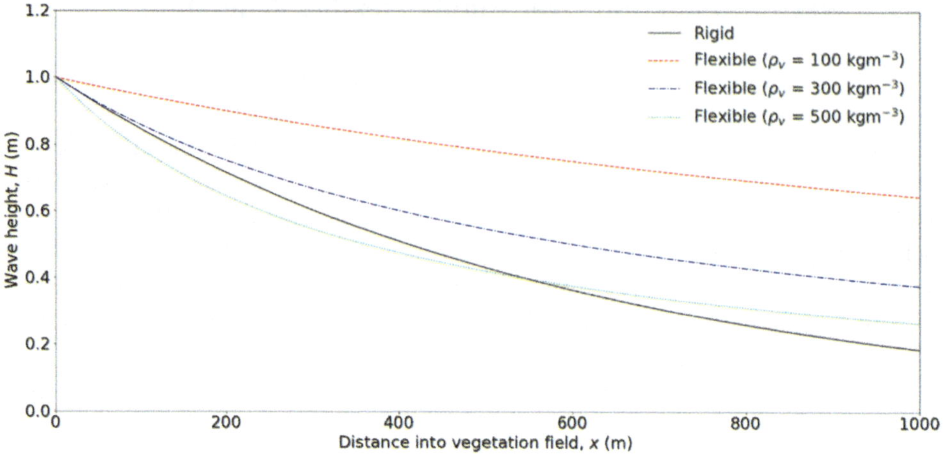

Fig. 16.5 Decay of wave height as a function of x for rigid and flexible vegetation with different stem densities

References

1. Dalrymple, R.A., Kirby, J.T., Hwang, P.A., 1984. Wave diffraction due to areas of energy dissipation. Journal of Waterway, Port, Coastal, and Ocean Engineering 110, 67–79.
2. Dubi, A., 1995. Damping of water waves by submerged vegetation: A case study on Laminaria hyperborea (Ph.D. Thesis). Norwegian University of Science and Technology, Trondheim, Norway.
3. Kobayashi, N., Raichle, A.W., Asano, T., 1993. Wave attenuation by vegetation. Journal of Waterway, Port, Coastal, and Ocean Engineering 119, 30–48.
4. Lima, S.F., Neves, C.F., Rosauro, N.M.L., 2007. Damping of gravity waves by fields of flexible vegetation, in: Coastal Engineering 2006. Presented at the Proceedings of the 30th International Conference, World Scientific Publishing Company, San Diego, California, USA, pp. 491–503.
5. Lovas, S.M., 2000. Hydro-physical conditions in kelp forests and the effect on wave damping and dune erosion: A case study on Laminaria hyperborea (Ph.D. Thesis). Norwegian University of Science and Technology, Trondheim, Norway.
6. Mendez, F.J., Losada, I.J., 2004. An empirical model to estimate the propagation of random breaking and nonbreaking waves over vegetation fields. Coastal Engineering 51, 103–118.

Part IV
Wave Interactions with Floating Structures

Static Analysis and Stability of Floating Structures

17

The melting of polar ice due to climate change coupled with land subsidence from the extraction of groundwater and subsurface hydrocarbons will likely see the continued rise of global sea-levels into the twenty-first century and beyond [13]. Yet, many coastal megacities continue to experience increased migration, largely due to the availability of economic opportunities, which will increase the number of people living in low-elevation coastal zones (LECZ) that are highly susceptible to flooding and sea-level rise [11]. This combination of land scarcity and sea-level rise has led to the United Nations (UN) convening its inaugural Round Table on Sustainable Floating Cities in 2019, where the UN Deputy Secretary-General Amina Mohammed proclaimed that "floating cities are a means of ensuring climate resilience, as buildings can rise along with the sea" [9]. From this chapter onwards, we will shift our analysis towards that of floating structures as a means for climate adaptation and coastal resilience. We will first introduce the conditions governing their hydrostatic stability in the absence of waves, before discussing the dynamics of floating structures under different sea states.

17.1 Floating Structures for Urban Development

The development of floating structures, communities, and even entire cities, can be thought of as an alternative form of land reclamation. Traditionally, land reclamation involves the creation of "artificial land" via the filling in of harbors, thereby forming new solid ground for urban development to take place. As an example, Fig. 17.1 shows the progress of land reclamation in Boston, Massachusetts, since 1630 [22]. Such practices can be extremely destructive to local ecosystems on the seabed while being economically

S. Wang, *Applied Ocean Mechanics for Structural Engineering and Climate Adaptation*, Synthesis Lectures on Ocean Systems Engineering,
https://doi.org/10.1007/978-3-032-18199-2_17

costly if the water depth is too deep. In this sense, floating structures can offer significant advantages over traditional land reclamation [17]. This is because floating structures are not limited by water depth and are inherently base isolated against earthquakes [21]. The use of modular platforms also allows them to be easily expanded upon or relocated with minimal impact on the seabed ecosystem [14]. In fact, floating structures can even enhance local marine activity by providing habitats for fish and other sessile organisms [4]. Most importantly, unlike structures built on reclaimed land, floating structures will be immune to flooding and sea-level rise from glacial discharge or extreme weather events. In addition to buildings, many other types of floating infrastructure can be integrated into urban communities. Examples include floating bridges or piers for transportation [8], floating farms or aquaculture facilities for food production [7], and floating wind, wave, or solar energy harvesters for power generation [16].

The basic concept of living on water is not a recent development, with small-scale floating homes being used for centuries in diverse cultures around the world [19]. Examples include the floating totora reed rafts on Lake Titicaca, Peru, and the floating fishing villages on the Mekong Delta in Vietnam [12]. More recent examples can be found in the floating neighborhoods on Puget Sound, Seattle, shown in Fig. 17.2 [1]. However, floating settlements capable of sustaining significant populations can only be realized through the implementation of pontoons that support high-density developments (e.g., multistory apartments or office buildings). These large-scale structures are often referred to as "Very Large Floating Structures" (VLFS) and are sometimes placed into different categories which are summarized in Table 17.1. Despite the advantages and promising potential of VLFS, few ideas have progressed beyond the realm of architectural speculation. As an

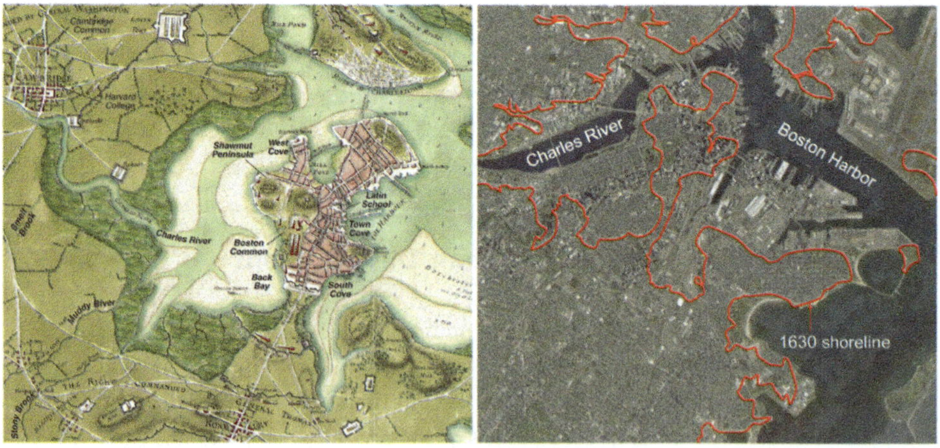

Fig. 17.1 Left: Boston in 1776 (map courtesy of the US Library of Congress). Right: Boston in 2024 with 1630 shoreline given by Weiskel et al. [22]. Satellite image courtesy of USGS

Fig. 17.2 Floating homes in Seattle, Washington (courtesy of Marilyn Johnson. Reproduced with permission)

example, Fig. 17.3 showcases a floating city concept proposed by OCEANIX in collaboration with the United Nations Human Settlements Program (UN-Habitat) that has yet to become reality [23]. The development of large-scale floating structures for permanent human habitation is further hampered by the fact that there are currently no guidelines accessible to structural engineers on their analysis and design. To address these limitations, this chapter will provide an introduction to the analysis of floating structures by first discussing the conditions for their stability under hydrostatic conditions. The following chapters will expand this discussion towards their dynamic behavior under regular and irregular waves.

17.2 Heel and Trim

For floating structures and vessels, the concept of stability reflects their capacity to resist overturning moments induced mainly by lateral forces such as wind. Furthermore, the structure must remain in a stable configuration during normal operations that may involve a slight shifting of its center of mass (e.g., when moving cargo on a floating platform). Floating bodies can exhibit motion about six degrees of freedom (three translational—x, y, z, and three rotational—θ_x, θ_y, θ_z) as shown in Fig. 17.4. The origin of this coordinate system is at the center of gravity (CoG) of the floating body, with the x-axis oriented along its longitudinal (lengthwise) direction.

Table. 17.1 Classification of VLFS reported in the literature [20]

Basis for classification	Classification	Description
Pontoon geometry [15]	Pontoon-type VLFS	A platform resting on the water surface
	Semisubmersible-type VLFS	A platform raised above the water surface via columns connected to pontoons located below the surface
Proximity to land [6]	Coastal VLFS	Floating structures designed for sheltered calm waters close to the coastline
	Offshore VLFS	Floating structures designed for unsheltered rough waters in the open sea subject to large waves
Pontoon rigidity [18]	Very Large Floating Structures (VLFS)	A single large mat-like platform where elastic deformations govern the structural response
	Modular Floating Structures (MFS)	Individual modular floating units governed by rigid-body motion working together to form a floating community

Fig. 17.3 Floating city concept proposed by OCEANIX (courtesy of OCEANIX/BIG-Bjarke Ingels Group. Reproduced with permission)

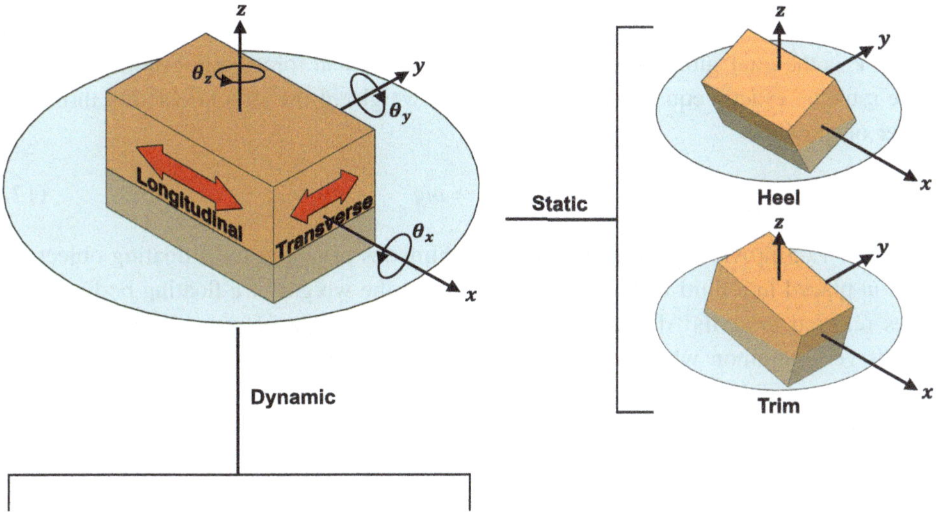

Translational motion		Rotational motion	
$x(t)$	Surge	$\theta_x(t)$	Roll
$y(t)$	Sway	$\theta_y(t)$	Pitch
$z(t)$	Heave	$\theta_z(t)$	Yaw

Fig. 17.4 Static and dynamic degrees of freedom for floating bodies

It is important to first familiarize ourselves with the terminology used to describe translational and rotational motions of a floating object. For static analysis, the terms "heel" and "trim" are used to denote a rotational offset about the x and y-axis (see Fig. 17.4), respectively. This tilting can be caused by a constant wind pressure applied laterally to the structure or by an asymmetric distribution of the structure's mass due to the placement of live loads. Hence, when assessing the static stability of floating structures, only the θ_x and $\theta_y$degrees of freedom (DoF) need to be considered. On the contrary, the dynamic (i.e., time varying) movements of a floating body induced by periodic waves encompass all six DoFs and will be discussed in the next chapter.

17.3 Principles of Buoyancy

We begin our discussion of hydrostatic stability by reintroducing Archimedes' principle which states that a body immersed in a fluid (either fully or partially) experiences an upwards force (referred to as the "buoyant force") that is equal to the weight of the fluid that the body displaces. This buoyant force acts vertically through the geometric centroid of the displaced fluid known as the "center of buoyancy" (CoB) and exhibits a magnitude:

$$F_B = \rho g V \tag{17.1}$$

where V is the total submerged volume. To satisfy vertical force equilibrium, the buoyant force must be exactly equal in magnitude to the weight of the structure acting through its center of gravity:

$$F_G = mg \tag{17.2}$$

Figure 17.5 provides an illustration of Archimedes principle for a floating object with mass m placed in a fluid with density ρ. Note that the weight of a floating body is sometimes referred to as its "displacement" to reflect the weight of water that it "displaces" (this is very common when describing the tonnage of ships). This text will not use this definition to avoid confusion with the conventional meaning of the term in the context of engineering mechanics. Given that $F_B = F_G$, we can thus determine the displaced volume:

$$V = \frac{m}{\rho} \tag{17.3}$$

Knowledge of V is crucial since it will allow us to obtain the draft D as shown in Fig. 17.5. For a rectangular pontoon with length C and width B, the draft is simply:

$$D = \frac{V}{BC} = \frac{m}{\rho BC} \tag{17.4}$$

which reflects the submerged height of the pontoon. The center of buoyancy would then be located a distance $D/2$ below the waterline.

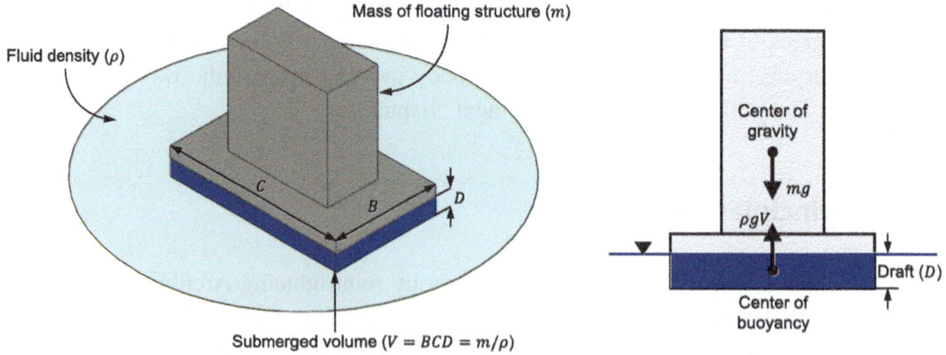

Fig. 17.5 Illustration of Archimedes' principle for a rectangular floating body

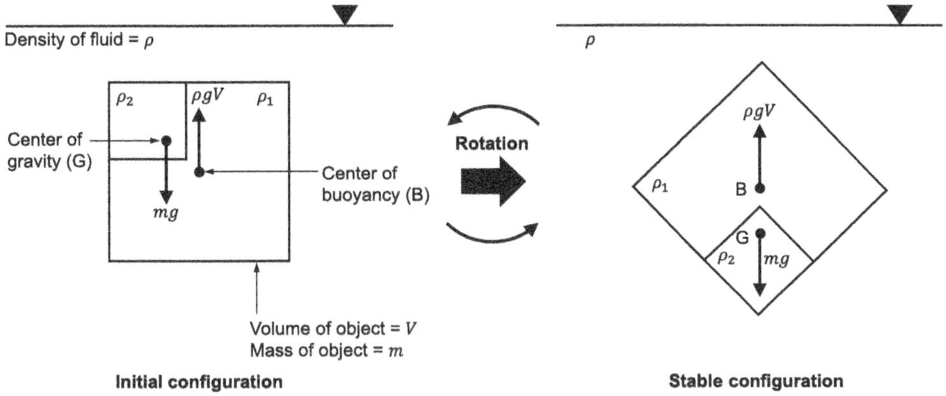

Fig. 17.6 Transition of a submerged body from an unstable configuration to its stable configuration

17.4 Stability of Submerged Bodies

For a fully submerged body of uniform density, the center of gravity is located at the same position as the center of buoyancy. If the weight of the body also equals the weight of the displaced fluid (i.e., its density matches the fluid density), the object will neither translate nor rotate and will be in static equilibrium. This state is known as neutral buoyancy. However, consider the submerged body shown in Fig. 17.6 composed of materials exhibiting two different densities where $\rho_2 > \rho_1$ so that its center of gravity (G) is shifted to the upper left of the center of buoyancy (B). To achieve stability, the body will rotate counterclockwise such that G is directly below B. In this configuration, the object is in moment equilibrium since any small rotation θ induced on the body (Fig. 17.7) will be resisted by a restoring moment equal to:

$$M_r = mg\overline{GB}\sin\theta \tag{17.5}$$

where m is the total mass of the object and $\overline{GB}$ is the distance between the centers of gravity and buoyancy. This example also illustrates that for a submerged body (e.g., a submarine) to be stable; the center of gravity must always be below its center of buoyancy.

17.5 Stability of Floating Bodies

Unlike submerged bodies, floating structures, such as that shown in Fig. 17.5, often exhibit centers of gravity located well above the center of buoyancy. Therefore, there must be an additional stabilizing component present within bodies that pierce the waterline. This additional restoring moment arises from changes in total wetted surface as the body heels or trims to one side such that the buoyant force associated with the newly submerged

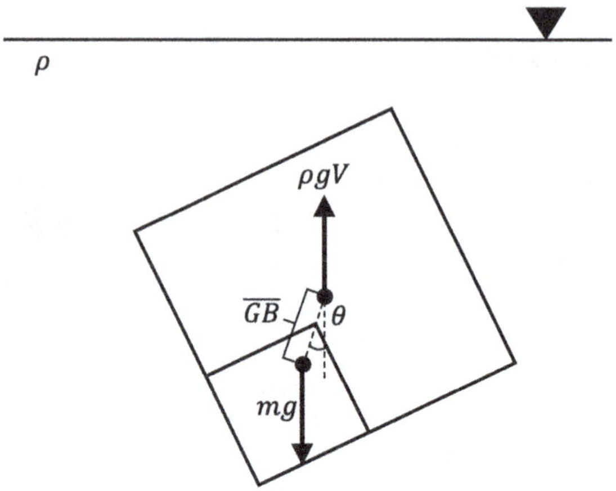

Fig. 17.7 Perturbation of a stable submerged body by a small angle θ

wedge provides a restoring moment which opposes the tendency of the structure to overturn under its own weight (see Fig. 17.8). This action will ultimately shift the center of buoyancy laterally towards the direction of rotation such that the total structural weight acting through the center of gravity can provide a restoring moment counteracting this rotation.

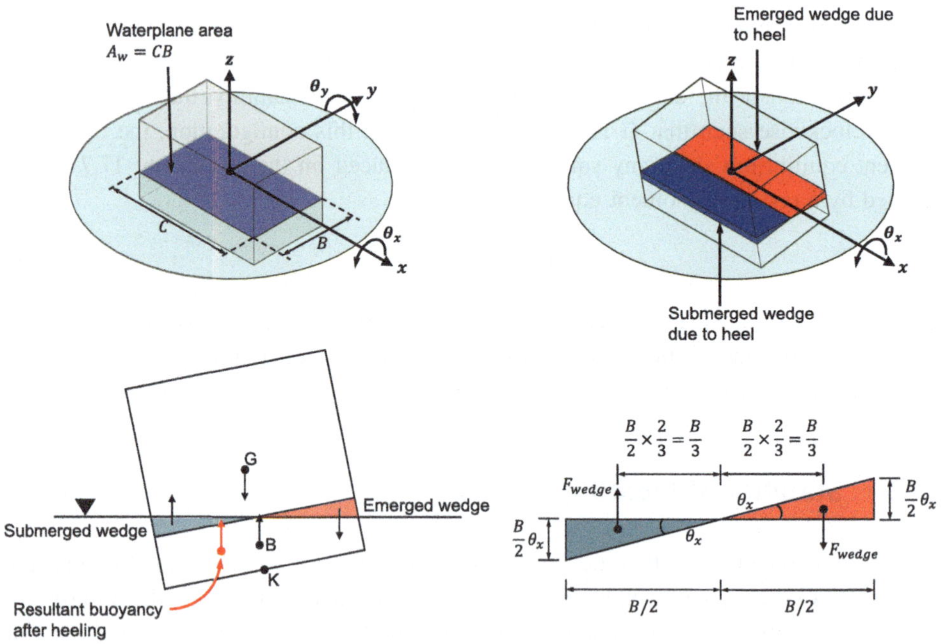

Fig. 17.8 Parameters of a rectangular floating body and its stability under heel

17.5.1 Derivation of Stability Requirements

To obtain an expression for the total restoring moment for rectangular floating bodies, we first consider only the restoring moment associated with the shaded wedges in Fig. 17.8 as shown for heel (θ_x). When assuming that the angle of heel is small, each wedge will have a total volume of:

$$V_{wedge} = C \times \frac{B}{2} \times \frac{B}{2}\theta_x \times \frac{1}{2} = \frac{CB^2}{8}\theta_x \tag{17.6}$$

The buoyancy force is therefore:

$$F_{wedge} = \rho g V_{wedge} = \rho g \frac{CB^2}{8}\theta_x \tag{17.7}$$

which results in a total restoring moment (associated with both the submerged and emerged wedge):

$$M_{rx} = 2 \times F_{wedge} \times \frac{B}{3} = \rho g \frac{CB^3}{12}\theta_x \tag{17.8}$$

We notice that $CB^3/12$ is the second moment of area of the rectangular waterplane (with length C and width B) about the x-axis. The restoring moment for heel and trim can therefore be expressed as:

$$\begin{aligned} M_{rx} &= \rho g I_x \theta_x \\ M_{ry} &= \rho g I_y \theta_y \end{aligned} \tag{17.9}$$

which are applicable to all geometries (not necessarily rectangular). Here, I_x and I_y respectively denote the second moment of area associated with the waterplane (A_w) about the x and y-axis relative to the origin (i.e., the CoG) determined using the parallel axis theorem:

$$\begin{aligned} I_x &= I_x' + A_w y_{cf}^2 \\ I_y &= I_{y\prime} + A_w x_{cf}^2 \end{aligned} \tag{17.10}$$

where x_{cf} and y_{cf} represent the horizontal distance from the CoG to the center of floatation (CoF), defined as the geometric centroid of the waterplane area, along the x and y axes, respectively. I_x' and I_y' are the second moments of area of the waterplane relative to the CoF:

$$\begin{aligned} I_x' &= \iint_{A_w} y^2 dA \\ I_y' &= \iint_{A_w} x^2 dA \end{aligned} \tag{17.11}$$

Figure 17.9 demonstrates an application of Eq. (17.10) for a C-shaped floating object where the center of floatation is offset from its center of gravity along the x-axis. In this situation, the waterplane second moment of area relative to the CoG about the y-axis (I_y) will not equal its corresponding second moment of area relative to the CoF (I'_y) since $x_{cf} \neq 0$. However, $I_x = I'_x$ still holds since $y_{cf} = 0$.

For the doubly symmetric rectangular object shown in Fig. 17.8 with width B and length C, the second moments of area are:

$$I_x = \int_{-\frac{C}{2}}^{\frac{C}{2}} \int_{-\frac{B}{2}}^{\frac{B}{2}} y^2 dydx + CB \times 0 = \frac{CB^3}{12}$$
$$I_y = \int_{-\frac{C}{2}}^{\frac{C}{2}} \int_{-\frac{B}{2}}^{\frac{B}{2}} x^2 dydx + CB \times 0 = \frac{BC^3}{12} \tag{17.12}$$

Unlike Fig. 17.9, the centers of gravity and flotation both lie on the same vertical axis, resulting in $x_{cf} = y_{cf} = 0$ in this instance.

Evidently, Eq. (17.9) assumes that the waterplane remains constant and is therefore only valid for small angles of rotation where $\sin\theta \approx \theta$. Combining (17.9) with (17.5) thus yields the total restoring moment for heel and trim, respectively:

$$M_{rx} = \rho g I_x \theta_x - mg\overline{BG}\theta_x$$
$$M_{ry} = \rho g I_y \theta_y - mg\overline{BG}\theta_y \tag{17.13}$$

Since G is above B, the distance $\overline{BG}$ will be positive. We can also rewrite (17.13) to express $\overline{BG}$ in terms of the keel location (K):

$$M_{rx} = \rho g I_x \theta_x - mg\left(\overline{KG} - \overline{KB}\right)\theta_x$$

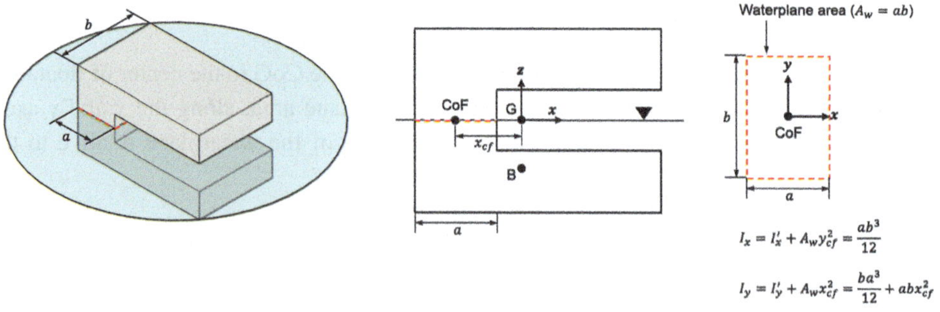

Fig. 17.9 Second moment of area of waterplane with its CoF offset from the CoG

$$M_{ry} = \rho g I_y \theta_y - mg\left(\overline{KG} - \overline{KB}\right)\theta_y \tag{17.14}$$

where $\overline{KG}$ and $\overline{KB}$ denote the distance from the keel to the centers of gravity and buoyancy, respectively. Subsequently dividing both sides of (17.14) by $mg\theta$ produces:

$$\begin{aligned} \frac{M_{rx}}{mg\theta_x} &= \frac{\rho I_x}{m} - \overline{KG} + \overline{KB} = \overline{KB} + \frac{I_x}{V} - \overline{KG} \\ \frac{M_{ry}}{mg\theta_y} &= \frac{\rho I_y}{m} - \overline{KG} + \overline{KB} = \overline{KB} + \frac{I_y}{V} - \overline{KG} \end{aligned} \tag{17.15}$$

since $\rho/m = 1/V$ where V is the displaced volume. It is seen that for the floating body to be statically stable, the right-hand side of (17.15) must be positive. As such, the following conditions must be true:

$$\begin{aligned} \overline{KB} + \frac{I_x}{V} &\geq \overline{KG} \\ \overline{KB} + \frac{I_y}{V} &\geq \overline{KG} \end{aligned} \tag{17.16}$$

The constants $\overline{KB} + I_x/V$ and $\overline{KB} + I_y/V$ represent the distance between the keel and a location called the "metacenter" and is denoted $\overline{KM}$. This metacenter is defined as the point where a vertical line crossing the tilted center of buoyancy intersects a line through the original vertical center of buoyancy (Fig. 17.10). As such, the metacenter can be interpreted as the instantaneous point of rotation as the structure is inclined through small angles of heel or trim. I_x/V and I_y/V therefore represent the distance from the center of buoyancy (B) to the metacenter (M), denoted $\overline{BM}_x$ and $\overline{BM}_y$, for heel and trim rotations, respectively. Equation (17.16) can now be expressed as:

$$\begin{aligned} \overline{KM}_x &= \overline{KB} + \overline{BM}_x \geq \overline{KG} \\ \overline{KM}_y &= \overline{KB} + \overline{BM}_y \geq \overline{KG} \end{aligned} \tag{17.17}$$

which implies that for floating bodies to remain stable, the metacenter (for both heel and trim) must be located above the center of gravity ($\overline{GM} > 0$). The distance from the center of gravity to the metacenter ($\overline{GM}$) is known as the "metacentric height" and may be different for heel and trim:

$$\begin{aligned} \overline{GM}_x &= \frac{I_x}{V} - \overline{BG} \\ \overline{GM}_y &= \frac{I_y}{V} - \overline{BG} \end{aligned} \tag{17.18}$$

The metacentric height is used to compute the righting arm ($\overline{GZ}$) defined as the horizontal distance between the center of buoyancy in the tilted configuration and the center of gravity. For small angles, this righting arm is given by:

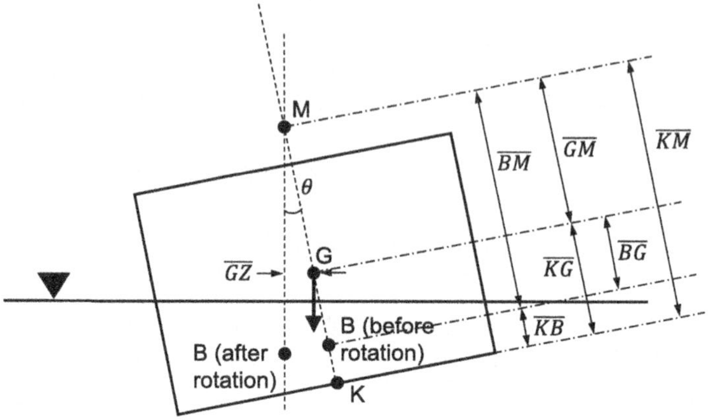

Fig. 17.10 Location of the metacenter and righting arm for a floating body

$$\begin{aligned}\overline{GZ}_x &= \overline{GM}_x \sin\theta_x \approx \overline{GM}_x\theta_x \\ \overline{GZ}_y &= \overline{GM}_y \sin\theta_y \approx \overline{GM}_y\theta_y\end{aligned} \tag{17.19}$$

A larger metacentric height thus infers greater stability against overturning due to the longer righting arm $\overline{GZ}$ for a constant rotation θ.

17.5.2 Angles of Heel and Trim

In addition to informing the extent of hydrostatic stability, by substituting (17.18) into (17.13), the metacentric height $\overline{GM}_x$ or $\overline{GM}_y$ can also be used to determine the angle of heel or trim produced by an externally applied moment about the x or y-axis:

$$\begin{aligned}\theta_x &= \frac{M_x}{mg\overline{GM}_x} \\ \theta_y &= \frac{M_y}{mg\overline{GM}_y}\end{aligned} \tag{17.20}$$

where θ_x and θ_y are given in radians and m is the total mass of the floating structure. M_x and M_y respectively denote the moment generated about the x and y-axis. For floating structures at sea, this moment is primarily generated by wind which is determined via:

$$M_{wind} = F_{wind} h_{wind} \tag{17.21}$$

where h_{wind} is the moment arm between the point of application of the wind force F_{wind} and the center of lateral resistance (CLR). For a floating structure without moorings, the CLR is typically located at the centroid of the submerged cross-sectional area. The total wind force F_{wind} is determined from:

$$F_{wind} = \frac{1}{2}\rho_a C_S C_H U_{10}^2 A_{wind} \tag{17.22}$$

where $\rho_a = 1.2\ \text{kg/m}^3$ is the air density, A_{wind} is the projected area of the surface exposed to wind, and U_{10} is the wind speed at 10 m above sea level. C_S is the shape coefficient which accounts for the geometry of the impacted surface [2]:

$$C_S = \begin{cases} 0.4, & \text{Spherical shape} \\ 0.5, & \text{Cylindrical shape} \\ 1.0, & \text{Flat surface} \\ 1.2, & \text{Wires} \\ 1.5, & \text{Isolated members (cranes, beams, etc.)} \end{cases} \tag{17.23}$$

and C_H is the height coefficient which adjusts the wind speed based on the elevation z above sea level. Barltrop [2] recommends the simple formula:

$$C_H = \left(\frac{z}{10}\right)^{0.25} \tag{17.24}$$

where z must be expressed with units of meters. Figure 17.11 illustrates the determination of wind-induced moments on a floating structure (with wind on the small freeboard ignored for simplicity). Note that in this example, the total wind force is assumed to act at the centroid of the impacted surface. For a more accurate estimate of the wind-induced moment, the distribution of wind force with height should be considered.

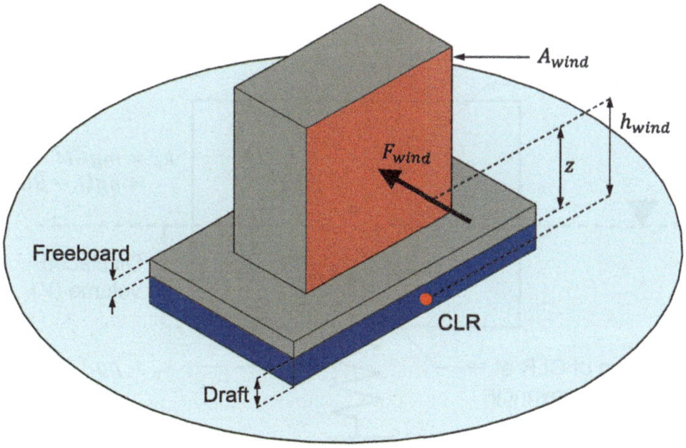

Fig. 17.11 Lateral wind force acting on a floating structure

17.5.3 Hydrostatic Stiffness

When examining (17.20), we see that the equations may be rearranged to exhibit the form $M = mg\overline{GM}\theta$. This is analogous to $M = K_\theta\theta$ where $K_\theta = mg\overline{GM}$ depicts the rotational stiffness. For heel and trim, the waterplane stiffness K_θ is therefore:

$$\begin{aligned} K_{\theta_x} &= mg\overline{GM}_x = \rho g\left(I_x - \overline{BG}V\right) \\ K_{\theta_y} &= mg\overline{GM}_y = \rho g\left(I_y - \overline{BG}V\right) \end{aligned} \tag{17.25}$$

Similarly, the vertical (heave) stiffness acting through the center of buoyancy is:

$$K_z = \rho g A_w \tag{17.26}$$

Using (17.25) and (17.26), the linear rigid body behavior of floating structures may be idealized via the spring model shown in Fig. 17.12. Note that the horizontal reaction force at the roller support is the force required to prevent lateral motion of the structure (not present for a freely floating structure). This reaction force could be provided by vertical mooring piles for example.

17.6 Stability Requirements for Small-Scale Floating Dwellings

Given the novelty of floating structures for permanent human habitation, there are few established guidelines that regulate their stability. However, some recommendations are published by the Danish Maritime Authority [3], which provide technical regulations on the stability and buoyancy of houseboats and other small-scale floating dwellings (e.g.,

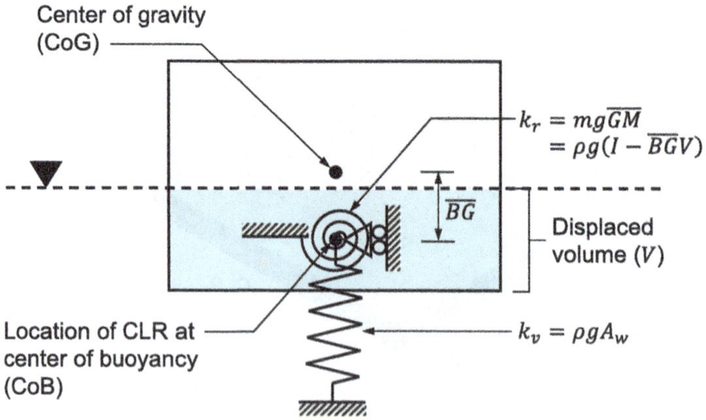

Fig. 17.12 Linear spring approximation for a floating structure with lateral restraints at the CLR

Fig. 17.2). These guidelines mandate that under a lateral wind pressure not less than 500 Pa, floating dwellings must satisfy both of the following requirements:

- The angles of heel (θ_x) and trim (θ_y) must not exceed 4°
- The freeboard of any given side must not be reduced by more than two thirds the original freeboard during heel or trim.

While the first condition is relatively straightforward, Fig. 17.13 illustrates the meaning behind the second condition. Here, the original freeboard height of D_F is reduced to D'_F due to the angle θ induced by either heel or trim. To satisfy stability requirements given by the Danish Maritime Authority [3], the amount that this freeboard is being reduced cannot exceed $2D_F/3$. In other words, the reduced freeboard D'_F must be greater than 1/3 of the original freeboard, i.e.:

$$D'_F \geq \frac{D_F}{3} \tag{17.27}$$

For small angles of θ, the reduced freeboard is approximately:

$$D'_F \approx D_F - \frac{B}{2}\theta \tag{17.28}$$

where θ is the angle of heel or trim in radians and B is the length of the pontoon parallel with the direction of rotation (see Fig. 17.13). While a minimum wind pressure of 500 Pa is mandated, the wind pressure used should be determined according to the relevant building standard (e.g., ASCE 7-22) for the location in question.

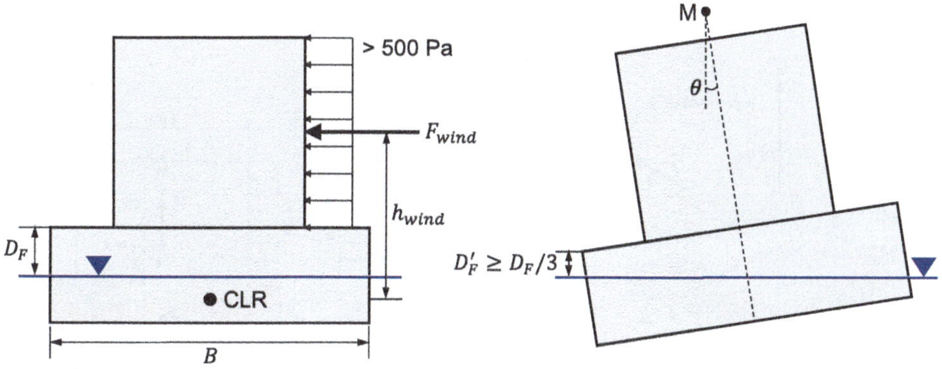

Fig. 17.13 Reduction of freeboard due to heel or trim per the Danish Maritime Authority [3] requirements

17.7 Example: Hydrostatic Stability of a Floating House

We will demonstrate the principles of hydrostatic stability via the analysis of a floating house shown in Fig. 17.14, which was inspired by Nakajima et al. [10]. The structure comprises a rectangular pontoon measuring $6.2 \times 5 \times 1.5$ m with a mass of $m_p = 8000$ kg. This pontoon supports a house measuring $4.5 \times 5 \times 3$ m with a mass of $m_b = 16500$ kg. The total mass of the entire structure is therefore $m = 8000 + 16500 = 24500$ kg. Let us assume that the mass of the pontoon and superstructure are uniformly distributed such that their centers of gravity correspond with their respective geometric centroids. In addition, suppose that a constant lateral wind pressure of 500 Pa is being exerted on the side walls of the superstructure as illustrated in Fig. 17.14. Technically speaking, the lateral wind pressure will also be acting on the exposed freeboard above the waterline. However, this additional wind pressure will be neglected for simplicity.

17.7.1 Metacentric Height

Since we know the geometry of the pontoon, we can determine the displaced volume V:

$$V = \frac{m}{\rho} = \frac{24500}{1025} = 23.9 \text{ m}^3$$

and the draft D:

$$D = \frac{V}{6.2 \times 5} = \frac{23.9}{6.2 \times 5} = 0.77 \text{ m}$$

The distance between the keel and center of buoyancy (CoB) is then:

$$\overline{KB} = \frac{D}{2} = \frac{0.77}{2} = 0.39 \text{ m}$$

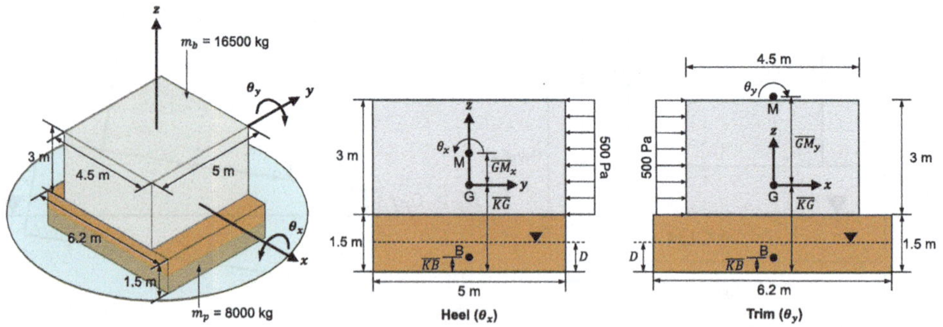

Fig. 17.14 Stability of a floating house in heel and trim

Next, we can calculate the height of center of gravity (CoG) of the entire structure above the keel ($\overline{KG}$):

$$\overline{KG} = \frac{8000 \times \left(\frac{1.5}{2}\right) + 16500 \times \left(1.5 + \frac{3}{2}\right)}{24500} = 2.27\text{m}$$

Now we can determine the distance between the CoB and CoG:

$$\overline{BG} = \overline{KG} - \overline{KB} = 2.27 - 0.39 = 1.9 \text{ m}$$

The second moments of area of the waterplane about the x and y axes are:

$$I_x = \frac{6.2 \times 5^3}{12} = 64.6 \text{ m}^4$$
$$I_y = \frac{5 \times 6.2^3}{12} = 99.3 \text{ m}^4$$

which yields a metacentric height in heel ($\overline{GM}_x$) and trim ($\overline{GM}_y$) of:

$$\overline{GM}_x = \frac{I_x}{V} - \overline{BG} = \frac{64.6}{23.9} - 1.9 = 0.8 \text{ m}$$
$$\overline{GM}_y = \frac{I_y}{V} - \overline{BG} = \frac{99.3}{23.9} - 1.9 = 2.3 \text{ m}$$

Since both $\overline{GM}_x$ and $\overline{GM}_y$ are positive, we know that the structure will be statically stable and will not overturn. In addition, $\overline{GM}_y > \overline{GM}_x$ implies that the structure will be more stable in trim (θ_y) than in heel (θ_x). This result makes intuitive sense since the pontoon is longer (6.2 m) than it is wide (5 m), making it more difficult to overturn for moments exerted about the y-axis.

17.7.2 Angles of Heel and Trim

To determine the angle of heel and trim, we must first compute the vertical distance between the resultant horizontal wind force (F_{wind}) and the center of lateral resistance (CLR) as shown in Fig. 17.15. When assuming a freely floating structure, the CLR will be located at the CoB. The moment lever arm is therefore:

$$h_{wind} = \left(1.5 + \frac{3}{2}\right) - \overline{KB} = \left(1.5 + \frac{3}{2}\right) - 0.39 = 2.61 \text{ m}$$

The resultant wind force in heel and trim are:

$$F_{wind,\theta_x} = 500 \times 3 \times 4.5 = 6750 \text{ N}$$
$$F_{wind,\theta_y} = 500 \times 3 \times 5 = 7500 \text{ N}$$

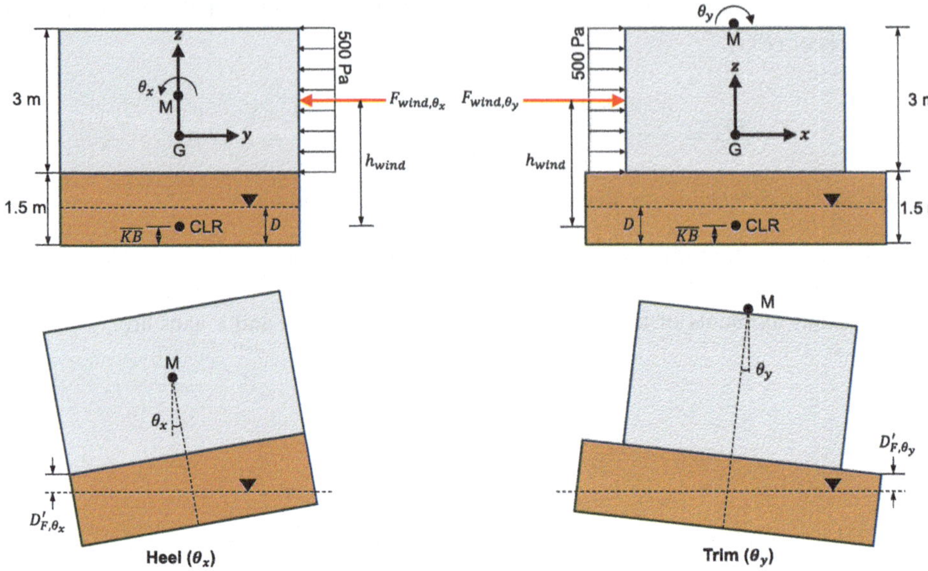

Fig. 17.15 Lateral wind force in heel and trim with induced rotations

and the overturning moments in heel and trim are:

$$M_x = F_{wind,\theta_x} h_{wind} = 6750 \times 2.61 = 17620 \text{ Nm}$$
$$M_y = F_{wind,\theta_y} h_{wind} = 7500 \times 2.61 = 19580 \text{ Nm}$$

The angles of rotation in heel (θ_x) are trim (θ_y) are therefore:

$$\theta_x = \frac{M_x}{mg\overline{GM}_x} = \frac{17620}{24500 \times 9.81 \times 0.8} = 0.092 \text{ rad} = 5.3^\circ$$
$$\theta_y = \frac{M_y}{mg\overline{GM}_y} = \frac{19580}{24500 \times 9.81 \times 2.3} = 0.035 \text{ rad} = 2.0^\circ$$

We notice that although the lateral wind force is greater in trim (due to the larger surface area), the structure experiences a much larger heel angle due to its smaller metacentric height $\overline{GM}_x$.

17.7.3 Freeboard After Rotation

The freeboard after heel (D'_{F,θ_x}) or trim (D'_{F,θ_y}) is:

$$D'_{F,\theta_x} = D_F - \frac{5}{2}\theta_x = 0.73 - \frac{5}{2} \times 0.092 = 0.5 \text{ m}$$

$$D'_{F,\theta_y} = D_F - \frac{6.2}{2}\theta_y = 0.73 - \frac{6.2}{2} \times 0.035 = 0.62 \text{ m}$$

where

$$D_F = 1.5 - D = 1.5 - 0.77 = 0.73 \text{ m}$$

is the original freeboard of the pontoon.

17.8 Summary: Analysis Procedure for Hydrostatic Stability

Table 17.2 provides a summary of the calculation procedure to determine the stability of a floating structure and its angle of heel or trim. The total mass (m), location of the CoG ($\overline{KG}$), and fluid density (ρ) are required as input. As an example, Fig. 17.16 illustrates the relevant parameters for a superstructure supported by a rectangular pontoon.

17.9 Structural Analysis of Floating Pontoons

We have thus far examined the hydrostatic stability of floating structures in response to overturning moments, such as that generated by lateral wind forces. In this section, we will introduce a simplified approach for the structural analysis of a conventional floating pontoon through a real-world example. A large floating platform to support human activities typically takes the form of a flat slab with planar dimensions much greater than its depth. Depending on its overall size, the platform may need to be assembled on-site by joining multiple smaller pontoons. The structure we will use to demonstrate this concept is the Marina Bay Floating Platform which operated in Singapore from 2007 until its demolition in 2023 (Fig. 17.17). Covering a total surface area of 120×83 m (9960 m^2), the platform was assembled from 15 rectangular steel pontoons 40 m long, 16.6 m wide, and 1.2 m deep arranged in a 3×5 grid [5]. As shown in Fig. 17.18, two longitudinal and two transverse bulkheads discretize each pontoon into nine watertight compartments. These structural elements provide flexural rigidity while ensuring that any perimeter damage (e.g., from ship collisions) would flood only a single compartment and can therefore be easily managed and repaired.

Table. 17.2 Stability calculation procedure for floating structures

Step 1	Determine the displaced volume V from the total mass m	$V = \frac{m}{\rho}$
Step 2	Determine the draft D and the elevation of the CoB from the keel ($\overline{KB}$)	For a rectangular pontoon with length C and width B: $D = \frac{V}{CB}$, $\overline{KB} = \frac{D}{2}$
Step 3	Determine the distance between the CoB and CoG ($\overline{BG}$)	$\overline{BG} = \overline{KG} - \overline{KB}$
Step 4	Determine the waterplane second moments of area in heel (I_x) and trim (I_y) via (17.10)	$I_x = I'_x + A_W y_{cf}^2$, $I_y = I'_y + A_W x_{cf}^2$ For a rectangular pontoon with length C and width B: $I_x = \frac{CB^3}{12}$, $I_y = \frac{BC^3}{12}$
Step 5	Determine the metacentric height in heel ($\overline{GM}_x$) or trim ($\overline{GM}_y$) via (17.18)	$\overline{GM}_x = \frac{I_x}{V} - \overline{BG}$, $\overline{GM}_y = \frac{I_y}{V} - \overline{BG}$
Step 6	Check conditions for stability	$\overline{GM}_x > 0$, $\overline{GM}_y > 0$
Step 7	Calculate overturning moments in heel (M_x) and trim (M_y)	For wind: $M_x = F_{wind,\theta_x} h_{wind}$, $M_y = F_{wind,\theta_y} h_{wind}$ where h_{wind} is the lever arm from the resultant lateral wind force (F_{wind,θ_x} or F_{wind,θ_y}) and the CLR
Step 8	Determine the angles of heel (θ_x) and trim (θ_y) via (17.20)	$\theta_x = \frac{M_x}{mg\overline{GM}_x}$, $\theta_y = \frac{M_y}{mg\overline{GM}_y}$
Step 9	Check angles of heel and trim against stability requirements (if applicable)	E.g., Danish Maritime Authority's Technical Regulation

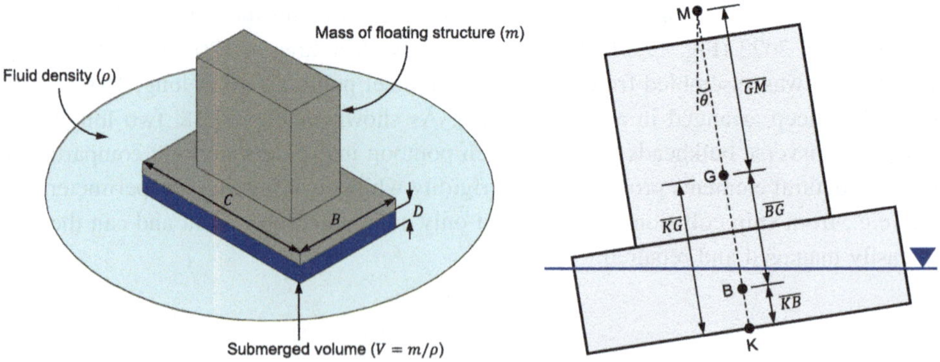

Fig. 17.16 Parameters to determine the stability of a rectangular pontoon

Fig. 17.17 Marina Bay Floating Platform in Singapore prior to its demolition (courtesy of Dietmar Rabich CC BY-SA 4.0)

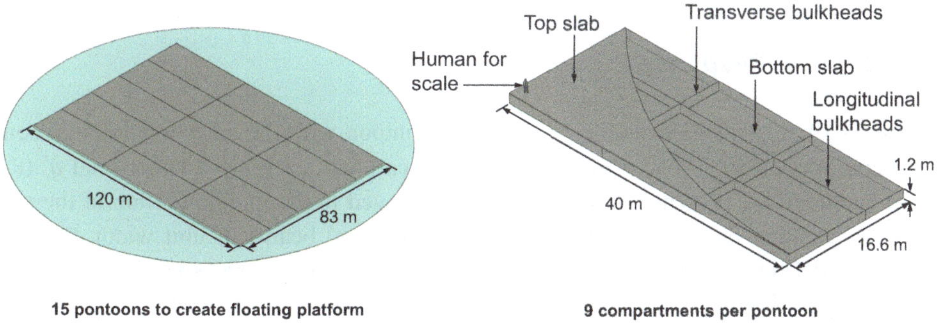

Fig. 17.18 Structural composition of the Marina Bay Floating Platform

17.9.1 Pontoon Loads

The overall area and depth of a floating platform would depend on its intended usage and the conditions necessary to achieve an acceptable level of hydrostatic stability. For structural design (and cost evaluation) purposes, it is necessary to obtain an initial assessment of the required wall thickness corresponding to the following load-resisting members (see Fig. 17.18):

- Bottom slab
- Top slab
- Longitudinal and transverse vertical bulkheads.

Following estimation of the total weight (pontoon self-weight and additional weight to be supported by the pontoon), the draft D can be determined. This draft then enables the upwards hydrostatic pressure acting uniformly on the bottom slab to be computed from $p = \rho g D$. After considering the distribution of loads supported by the pontoon via the top slab (including its self-weight), simplified moment, shear, and axial force diagrams can be produced for the aforementioned structural members via the simplified method demonstrated by Wang et al. [18]. For this purpose, it is typically more convenient to treat all area loads (with units N/m^2) as a linear load per unit width (with units N/m) as shown in Fig. 17.19. This can be done by multiplying an area load (e.g., p) by the desired unit width, say, 1 m. These loads can then be used to obtain the moment, shear, and axial force distributions. From these force diagrams, the designer can then gauge an appropriate wall thickness for the bottom and top slabs and vertical bulkheads based on their respective member capacities obtained using standard methods of structural analysis. Since floating pontoons are commonly constructed using reinforced concrete or steel, design guidelines pertaining to these materials would be adopted depending on the material selected.

17.9.2 Local Response

Figure 17.19 presents an example of a floating pontoon with the outer longitudinal compartments supporting a load q and the inner compartment carrying a larger load q' (this could be due to the presence of a superstructure located in the middle portion of the pontoon). By treating the top and bottom slabs as fixed-end beams of unit width, bending, shear, and deflections can be obtained for the top and bottom slabs spanning each compartment according to the diagrams shown in Fig. 17.20. This then allows for moment, shear, and axial forces to be determined for all remaining vertical members as shown in Fig. 17.21.

Note that this approach is inherently conservative since one-way beam behavior is assumed, while in reality, two-way slab behavior is expected. The assumption of full support fixity at each end of the "beam" is also likely to overestimate the rigidity offered by the vertical bulkheads. It is important to understand that the objective is to provide a simplified procedure that can be readily implemented by designers for initial prototyping purposes. The use of numerical methods (e.g., finite element models) could then be adopted to further refine the dimensions of all structural elements.

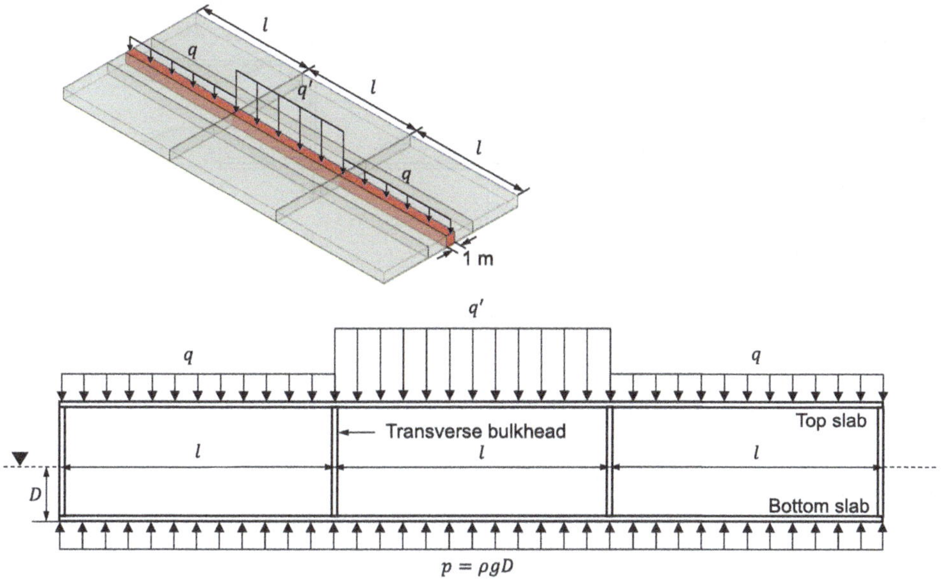

Fig. 17.19 Simplified structural analysis of a floating pontoon showing loads on top and bottom slabs

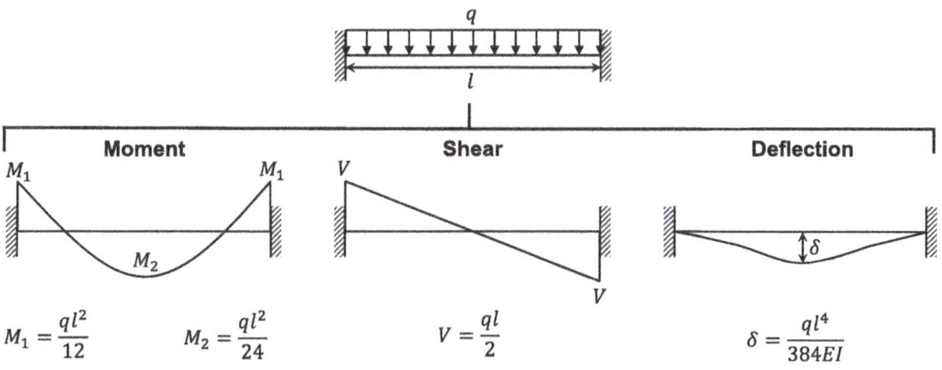

Fig. 17.20 Moment, shear, and deflection diagrams for a fixed end beam. Note that bending moments are plotted on the tension side

17.9.3 Global Response

In addition to Fig. 17.21 which shows local bending, shear, and axial behavior for individual structural elements, the global structural behavior of the entire pontoon must also be considered. This can be easily achieved by treating the pontoon as a simply supported beam with zero reaction force at the supports since the loads acting on the structure are in

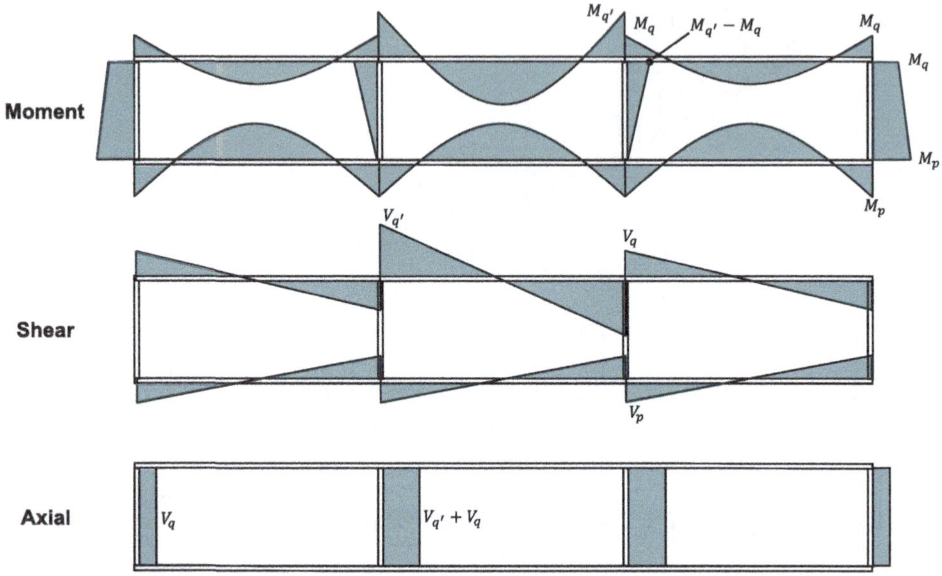

Fig. 17.21 Simplified moment (M), shear (V), and axial compression (N) diagrams (per unit width) for local behavior. Note that bending moments are plotted on the tension side

vertical equilibrium. Shear and moment diagrams for the present example are illustrated in Fig. 17.22. Here, the shear forces must be resisted by the vertical bulkheads while bending moments are carried by the top and bottom slabs.

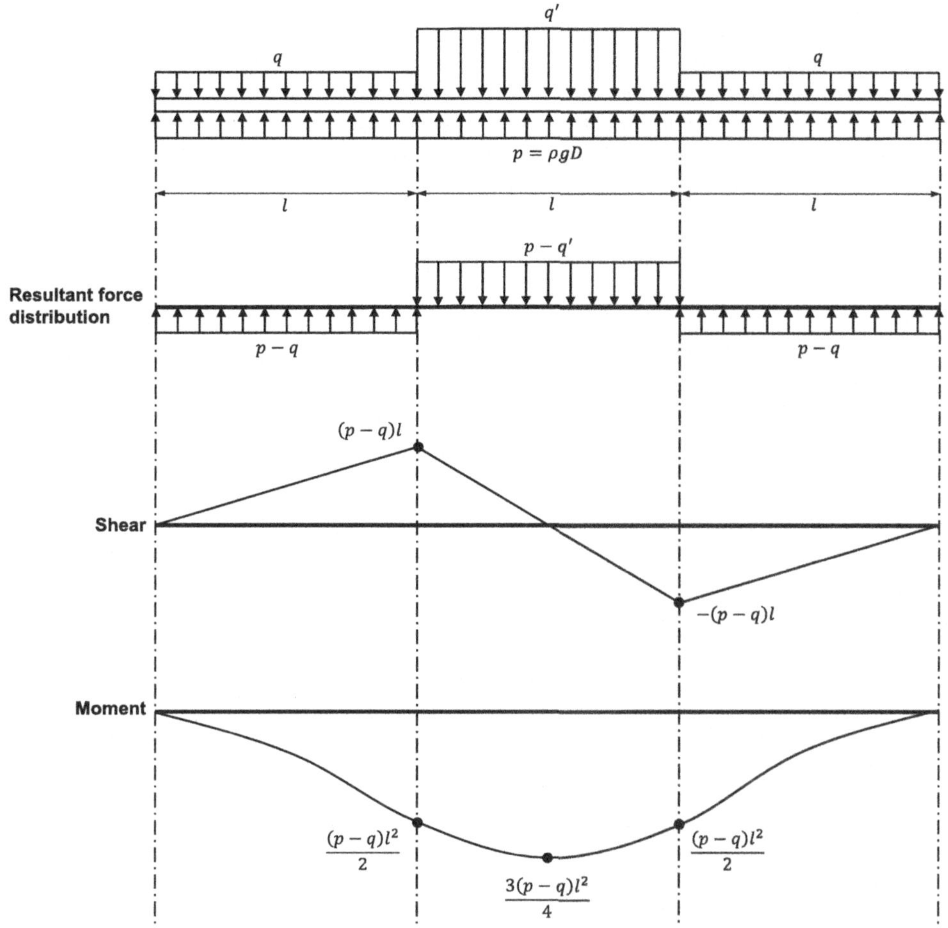

Fig. 17.22 Moment and shear diagrams (per unit width) for global behavior. Note that bending moments are plotted on the tension side

References

1. Bandrowski, T.J., 2015. Lily pad urbanism: Reimagining Seattle's floating home community (Master of Architecture). University of Washington.
2. Barltrop, N.D.P., 1998. Floating structures: a guide for design and analysis. The Centre for Marine and Petroleum Technology.
3. Danish Maritime Authority, 2007. Technical regulation on the stability and buoyancy etc. of house-boats and floating structures.
4. Hadary, T., Martinez, J.G., Sella, I., Perkol-Finkel, S., 2022. Eco-engineering for climate change - floating to the future, in: Proceedings of the Second World Conference on Floating Solutions,

Rotterdam. Presented at the WCFS2020.
5. Koh, H.S., Lim, Y.B., 2015. Floating Performance Stage at the Marina Bay, Singapore, in: Large Floating Structures: Technological Advances. Springer, Singapore, pp. 37–60.
6. Lamas-Pardo, M., Iglesias, G., Carral, L., 2015. A review of Very Large Floating Structures (VLFS) for coastal and offshore uses. Ocean Engineering 109, 677–690.
7. Leeman, W., 2021. Floating Farms: Feeding Rotterdam from Within. Architectural Design 91, 84–91.
8. Maruyama, T., Kawamura, Y., 2000. Construction of a floating swing bridge - Yumemai bridge. Osaka Technology 28–43.
9. Mohammed, A., 2019. Sustainable floating cities can offer solutions to climate change threats facing urban areas.
10. Nakajima, T., Saito, Y., Umeyama, M., 2022. A study on stability of floating architecture and its design methodology, in: Proceedings of the Second World Conference on Floating Solutions, Rotterdam. Springer, Rotterdam, pp. 273–296.
11. Neumann, B., Vafeidis, A.T., Zimmermann, J., Nicholls, R.J., 2015. Future coastal population growth and exposure to sea-level rise and coastal flooding - a global assessment. PLoS ONE 10.
12. Nguyen, T.T.T., 2022. Architectural design guideline for sustainable floating houses and floating settlements in Vietnam, in: Proceedings of the Second World Conference on Floating Solutions, Rotterdam. Presented at the WCFS2020.
13. Nicholls, R.J., Lincke, D., Hinkel, J., Brown, S., Vafeidis, A.T., Meyssignac, B., Hanson, S.E., Merkens, J.-L., Fang, J., 2021. A global analysis of subsidence, relative sea-level change and coastal flood exposure. Nature Climate Change 11, 338–342.
14. Ren, N., Zhang, C., Magee, A.R., Hellan, O., Dai, J., Ang, K.K., 2019. Hydrodynamic analysis of a modular multi-purpose floating structure system with different outermost connector types. Ocean Engineering 176, 158–168.
15. Wang, C.M., Tay, Z.Y., 2011. Very large floating structures: applications, research and development. Procedia Engineering 14, 62–72.
16. Wang, C.M., Utsunomiya, T., Wee, S.C., Choo, Y.S., 2010. Research on floating wind turbines: a literature survey. The IES Journal Part A: Civil & Structural Engineering 3, 267–277.
17. Wang, G., Drimer, N., Goldfeld, Y., 2020. Modular floating structures (MFS) for offshore dwelling a hydrodynamic analysis in the frequency domain. Ocean Engineering 216.
18. Wang, G., Goldfeld, Y., Drimer, N., 2019. Expanding coastal cities - Proof of feasibility for modular floating structures (MFS). Journal of Cleaner Production 222, 520–538.
19. Wang, S., 2025. Derivation of 3D Dynamics for Rigid Floating Structures under Directional Wave Excitation. Sustainable Marine Structures 7, 35–51.
20. Wang, S., 2022. Analytical solutions for the dynamic analysis of a modular floating structure for urban expansion. Ocean Engineering 266.
21. Watanabe, E., Wang, C.M., Utsunomiya, T., Moan, T., 2004. Very large floating structures: applications, analysis and design (No. 2004– 02). Centre for Offshore Research and Engineering, National University of Singapore.
22. Weiskel, P.K., Barlow, L.K., Smieszek, T.W., 2005. Water Resources and the Urban Environment, Lower Charles River Watershed, Massachusetts, 1630–2005. U.S. Geological Survey Circular 1280, Reston, Virginia.
23. Will, M., Berzgachina, K.V., 2022. UN-Habitat and partners unveil OCEANIX Busan, the world's first prototype floating city that adapts to sea level rise.

18 Rigid-Body Dynamics of Floating Structures

We now turn our attention to the dynamic analysis of floating structures under wave excitation. In particular, our objective is to formulate relationships that enable us to estimate the translational and rotational motions of a freely floating rigid-body as a function of wave properties pertaining to both regular and irregular waves. This will enable the dynamic analysis of structural systems supported by floating pontoons to ascertain their feasibility from a strength and serviceability perspective. We will first introduce the fundamental principles of floating body dynamics in the context of wave forcing, thereby establishing a theoretical foundation to be built upon in later chapters.

18.1 Classification of Floating Body Dynamics

The dynamic response of a floating pontoon under wave excitation can be broadly classified into hydroelastic or rigid-body regimes based on the transitional length (L_c). When approximating the structure as a beam with uniform stiffness, the transitional length suggested by Suzuki and Yoshida [6] is:

$$L_c = 2\pi \left(\frac{E_b I_b}{k_c} \right)^{1/4} \tag{18.1}$$

where $E_b I_b$ is the elastic bending stiffness and k_c denotes the heave stiffness of the hydrostatic restoring force per unit length. Large floating structures with a longitudinal length longer than L_c are governed by hydroelastic behavior (i.e., elastic deformations of the pontoon), whereas smaller pontoons shorter than L_c can be treated as a rigid body. In this text, we will assume that pontoons are sufficiently stiff such that they can be analyzed

S. Wang, *Applied Ocean Mechanics for Structural Engineering and Climate Adaptation*,
Synthesis Lectures on Ocean Systems Engineering,
https://doi.org/10.1007/978-3-032-18199-2_18

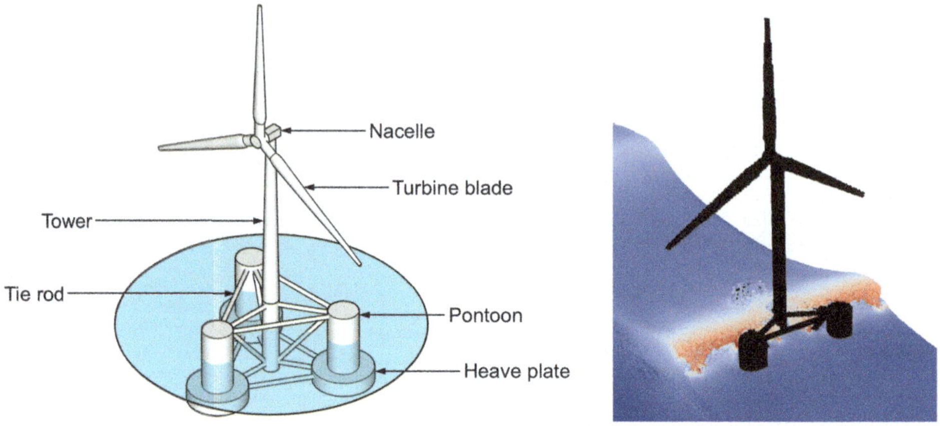

Fig. 18.1 5-MW semi-submersible OC5-DeepCwind turbine developed by NREL [4] and its numerical analysis via smoothed particle hydrodynamics by Wang and Chuang [13]

using only rigid-body dynamics in accordance with the simplified procedures given by Wang [11, 12]. Readers interested in the mechanics of hydroelastic pontoons are directed to Wang et al. [9]. In addition, we will limit our analysis to structures supported by rectangular pontoons (e.g., Fig. 17.14) since this will allow us to derive closed-form expressions that describe their motion under wave excitation. Structures that utilize complex floaters, such as the semi-submersible pontoon columns commonly found in floating wind turbines, typically require numerical methods of analysis (e.g., Fig. 18.1) and are beyond the scope of this text.

18.2 Degrees of Freedom

We begin by recalling that floating bodies can experience motion about six degrees of freedom (DoF) as illustrated in Fig. 18.2. These DoFs are:

- Surge (x)
- Sway (y)
- Heave (z)

for translations along the x, y, and z-axis, and

- Roll (θ_x)
- Pitch (θ_y)
- Yaw (θ_z)

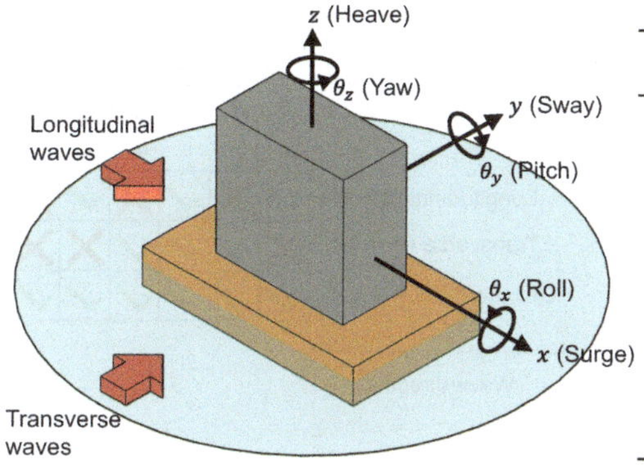

Degree of freedom		Response governed by
x	Surge	Mass
y	Sway	Mass
z	Heave	Stiffness, Mass, Damping
θ_x	Roll	Stiffness, Mass, Damping
θ_y	Pitch	Stiffness, Mass, Damping
θ_z	Yaw	Mass

Fig. 18.2 Dynamic degrees of freedom and governing response for a freely floating body

for rotations about the x, y, and z-axis, respectively.

Unlike the heel and trim rotations previously studied for statically floating structures, the abovementioned motions are time-varying. Therefore, the dynamic properties of the structure (i.e., mass, stiffness, and damping) will influence how the structure behaves in each DoF. In addition, the extent to which each of the six DoFs will be mobilized depends on the properties of incoming waves. This includes both the wave profile (e.g., larger waves will produce a more severe dynamic response) and the direction of wave propagation. For a rectangular pontoon supporting a symmetric load (e.g., Fig. 18.2), waves traveling perfectly parallel to the y-axis (transverse) will only result in sway (y), heave (z), and roll (θ_x) motions. Likewise, waves traveling parallel to the x-axis (longitudinal) will see only surge (x), heave (z), and pitch (θ_y) DoFs being activated. However, for oblique waves that are neither perfectly transverse nor longitudinal, the structure will experience motion in all six DoFs. Figure 18.3 summarizes the mobilized DoFs as a function of wave directionality, where θ denotes the angle of wave propagation relative to the pontoon's x-axis.

18.3 Principles of Dynamic Behavior

This text will only explore the rigid-body dynamics of floating bodies. Hence, we will assume that the structure is sufficiently rigid such that elastic deformations will not play a significant role in governing its response. Note that this assumption may not be appropriate for very large mat-like structures (where the length and width are significantly greater than its depth) requiring the use of hydroelasticity theory to determine its dynamic response. An example of such a structure was the 1 km long "Mega-Float" floating airport built in Tokyo Bay (Fig. 18.4). Completed in 2000 and dismantled a year later, the

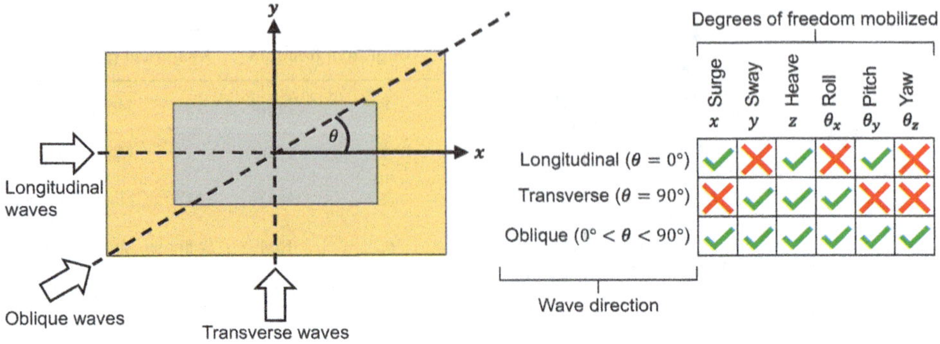

Fig. 18.3 Mobilized degrees of freedom for a symmetrically loaded rectangular pontoon

project was commissioned by the Japanese Ministry of Land, Infrastructure and Transport to study the feasibility of floating runways [3]. Despite its length and width being 1000 m and 60 m, respectively, the Mega-Float only has a depth of 3 m. Such a structure would therefore be governed by hydroelastic effects due to its slender profile and cannot be treated as a rigid body.

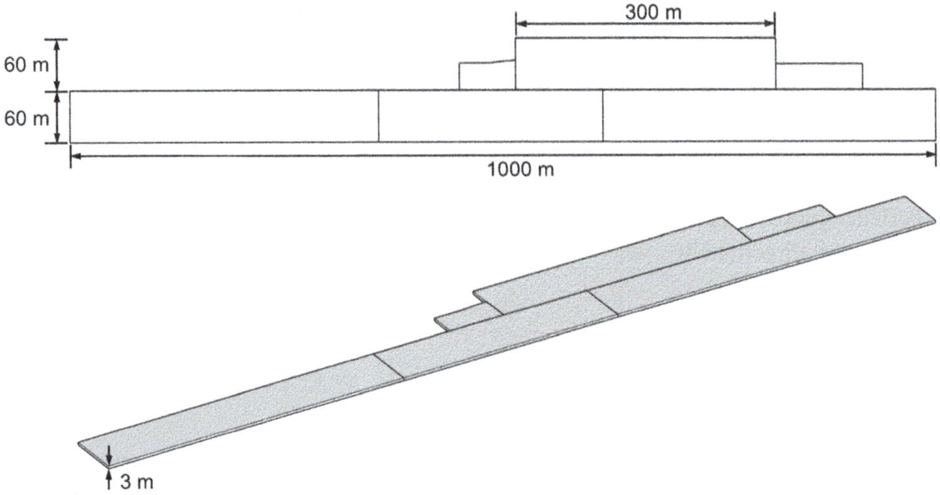

Fig. 18.4 Dimensions of the Tokyo Mega-Float

18.3.1 Equation of Motion

When considering a rigid body, the equation of motion for each of its six DoFs can be described using a standard mass-spring-damper model:

$$M\ddot{R}(t) + C\dot{R}(t) + KR(t) = F(t) \tag{18.2}$$

In this instance, R, $\dot{R}$, and $\ddot{R}$ respectively denote the time-varying translation/rotation, velocity/angular velocity, and acceleration/angular acceleration pertaining to a particular DoF and takes the following form:

$$\begin{aligned} R(t) &= R_0 \cos(\omega t - \phi) \\ \dot{R}(t) &= -R_0\omega \sin(\omega t - \phi) \\ \ddot{R}(t) &= -R_0\omega^2 \cos(\omega t - \phi) \end{aligned} \tag{18.3}$$

with R_0 being the response amplitude and ϕ is the phase shift, both of which must be solved in order to fully describe the dynamic motions of a floating body. For the sake of consistency and to avoid confusion, the harmonic response R and response amplitude R_0 will be expressed as follows when describing motion pertaining to each of the six DoFs:

$$R(t) = \begin{cases} x(t), & \text{Surge} \\ y(t), & \text{Sway} \\ z(t), & \text{Heave} \\ \theta_x(t), & \text{Roll} \\ \theta_y(t), & \text{Pitch} \\ \theta_z(t), & \text{Yaw} \end{cases}, \quad R_0 = \begin{cases} x_0, & \text{Surge} \\ y_0, & \text{Sway} \\ z_0, & \text{Heave} \\ \theta_{x0}, & \text{Roll} \\ \theta_{y0}, & \text{Pitch} \\ \theta_{z0}, & \text{Yaw} \end{cases} \tag{18.4}$$

Likewise, M, C, and K are the mass/mass moment of inertia, damping, and stiffness, and F is the harmonic wave excitation at frequency $\omega = 2\pi f$:

$$F(t) = F_0 \cos \omega t \tag{18.5}$$

with forcing amplitude F_0 (this could be a force or moment depending on the DoF under consideration). We notice that the phase ϕ features only in the response terms given by Eq. (18.3) but is not present in the harmonic forcing presented in (18.5). This property allows us to determine to what extent the response lags the wave excitation force F with $\phi = 0$.

18.3.2 Dynamic Properties

For a system with mass M and stiffness K, its natural frequency is:

$$\omega_n = 2\pi f_n = \sqrt{\frac{K}{M}} \tag{18.6}$$

which describes the body's inherent frequency of vibration in the absence of external forces. The relative importance of M, C, and K within the equation of motion given by Eq. (18.2) is dependent upon the difference in magnitude between the wave frequency ω and natural frequency ω_n:

- When the wave frequency is much lower than the natural frequency ($\omega \ll \omega_n$), the stiffness term K governs the equation of motion.
- When the wave frequency is much higher than the natural frequency ($\omega \gg \omega_n$), the mass term M governs the equation of motion.
- When the wave frequency is in close proximity to the natural frequency ($\omega \approx \omega_n$), the damping term C governs the equation of motion. This phenomenon is known as resonance and is undesirable under most circumstances due to the large motions produced. As such, the potential for resonance should be avoided for most types of floating structures.

In the context of freely floating bodies, the contribution of mass, stiffness, or damping depends on the DoF being excited (see Fig. 18.2). In this book, we will use the definition "mass dominated" to refer to surge, sway, and yaw since the body's dynamics will be largely governed by its mass M with $\omega \gg \omega_n$. On the contrary, heave, roll, and pitch will receive contributions from stiffness K, damping C, and mass M, thereby necessitating the consideration of all three terms. For simplicity, these DoFs will be referred to as "stiffness dominated" in order to emphasize the importance of the waterplane stiffness on heave, roll, and pitch motions. An intuitive explanation for this behavior is provided by the spring model in Fig. 18.5 which depicts a freely floating body (recall from Chap. 17 on hydrostatics). We see that the vertical or rotational springs are only mobilized under roll, pitch, or heave motions. Hence, the remaining surge, sway, and yaw DoFs are governed mainly by the body's inertia when assuming the lack of very stiff lateral restraints. We must now solve Eq. (18.2) to obtain solutions that are applicable towards both stiffness and mass dominated DoFs.

18.3.3 General Solution for "Stiffness Dominated" Degrees of Freedom

In dynamic analysis, it is often useful to express the dynamic response relative to its quasi-static behavior (i.e., F_0/K) by means of a dynamic amplification factor. To derive an expression for this quantity, we first substitute (18.3) into (18.2) in order to obtain the full equation of motion:

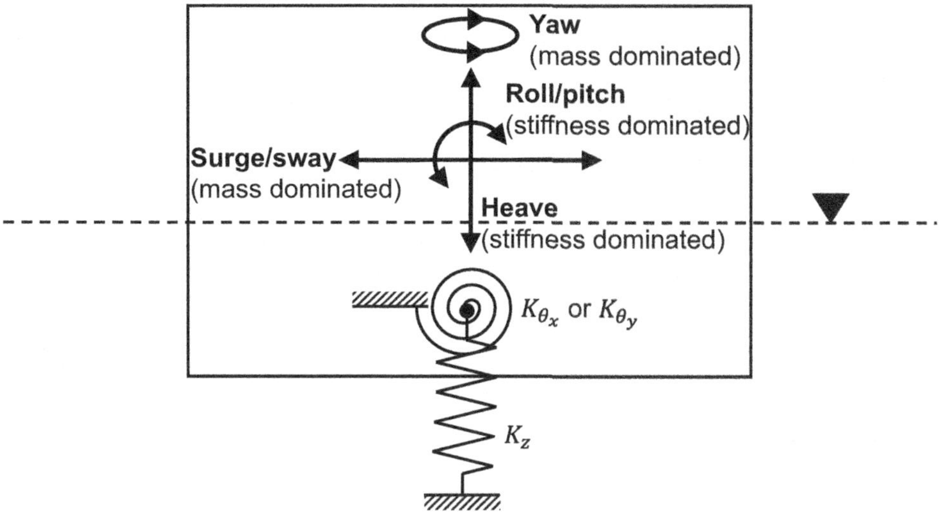

Fig. 18.5 Idealized spring model of a freely floating body

$$-MR_0\omega^2\cos(\omega t-\phi)-CR_0\omega\sin(\omega t-\phi)+KR_0\cos(\omega t-\phi)=F_0\cos\omega t \tag{18.7}$$

which can be rearranged into:

$$R_0\left[\left(K-M\omega^2\right)\cos(\omega t-\phi)-C\omega\sin(\omega t-\phi)\right]=F_0\cos\omega t \tag{18.8}$$

Subsequently dividing both sides of (18.8) by K gives:

$$R_0\left[\left(1-\frac{M}{K}\omega^2\right)\cos(\omega t-\phi)-C\frac{\omega}{K}\sin(\omega t-\phi)\right]=\frac{F_0}{K}\cos\omega t \tag{18.9}$$

We know that the natural frequency of vibration (in rad s^{-1}) is:

$$\omega_n=\sqrt{\frac{K}{M}} \tag{18.10}$$

In addition, the damping C is commonly expressed in terms of the damping ratio:

$$\xi=\frac{C}{C_{crit}}=\frac{C}{2\sqrt{MK}} \tag{18.11}$$

where $C_{crit}=2\sqrt{MK}$ is the critical damping. Substituting (18.10) and (18.11) into (18.9) gives:

$$R_0\left[\left(1-\frac{\omega^2}{\omega_n^2}\right)\cos(\omega t-\phi)-2\xi\frac{\omega}{\omega_n}\sin(\omega t-\phi)\right]=\frac{F_0}{K}\cos\omega t \tag{18.12}$$

From Ptolemy's theorem, the sine and cosine terms can be expanded into the following:

$$\begin{aligned}\cos(\omega t - \phi) &= \cos(\omega t)\cos\phi + \sin(\omega t)\sin\phi \\ \sin(\omega t - \phi) &= \sin(\omega t)\cos\phi - \cos(\omega t)\sin\phi\end{aligned} \quad (18.13)$$

which when substituted into (18.12) yields two separate equations after collecting the sine and cosine terms:

$$R_0\left[\left(1 - \frac{\omega^2}{\omega_n^2}\right)\cos(\phi) + 2\xi\frac{\omega}{\omega_n}\sin(\phi)\right]\cos(\omega t) = \frac{F_0}{K}\cos\omega t \quad (18.14)$$

$$R_0\left[\left(1 - \frac{\omega^2}{\omega_n^2}\right)\sin(\phi) - 2\xi\frac{\omega}{\omega_n}\cos(\phi)\right]\sin(\omega t) = 0 \quad (18.15)$$

Since we have two equations, we can now solve for both R_0 and ϕ:

$$R_0 = \frac{F_0/K}{\sqrt{\left(1 - \frac{\omega^2}{\omega_n^2}\right)^2 + \left(2\xi\frac{\omega}{\omega_n}\right)^2}} = \frac{F_0/K}{\sqrt{\left(1 - \frac{f^2}{f_n^2}\right)^2 + \left(2\xi\frac{f}{f_n}\right)^2}} \quad (18.16)$$

$$\phi = \arctan\left(\frac{2\xi\frac{\omega}{\omega_n}}{1 - \frac{\omega^2}{\omega_n^2}}\right) = \arctan\left(\frac{2\xi\frac{f}{f_n}}{1 - \frac{f^2}{f_n^2}}\right) \quad (18.17)$$

where f_n is the natural frequency (in s^{-1} or Hz):

$$f_n = \frac{\omega_n}{2\pi} = \frac{1}{2\pi}\sqrt{\frac{K}{M}} \quad (18.18)$$

Note that extra care must be taken when computing ϕ using Eq. (18.17) due to the arctangent problem (Fig. 18.6). It is clear that the numerator within the parenthesis, $2\xi f/f_n$, must be greater than or equal to zero. Therefore, ϕ can only exist in quadrants I or II of the arctangent graph and the condition $\phi \geq 0$ must hold. This is not an issue when $f < f_n$ since both the numerator and denominator will be positive (quadrant I). However, if $f > f_n$, ϕ may be incorrectly calculated by calculators or computer programs since it becomes unclear whether the negative sign exists in the numerator or denominator. If it is incorrectly assumed to exist in the numerator (as is the case with most basic calculators), then ϕ will be incorrectly placed within quadrant IV and take a negative value. To correct for this, we must add π to the incorrectly calculated phase in order to "transfer" ϕ back into quadrant II. The phase ϕ may therefore be more explicitly determined as follows:

$$\phi(f) = \begin{cases} \arctan\left(\dfrac{2\xi \frac{f}{f_n}}{1-\frac{f^2}{f_n^2}}\right), & f < f_n \\ \arctan\left(\dfrac{-2\xi \frac{f}{f_n}}{\left|1-\frac{f^2}{f_n^2}\right|}\right) + \pi, & f > f_n \\ \frac{\pi}{2}, & f = f_n \end{cases} \tag{18.19}$$

with f_n calculated from Eq. (18.18).

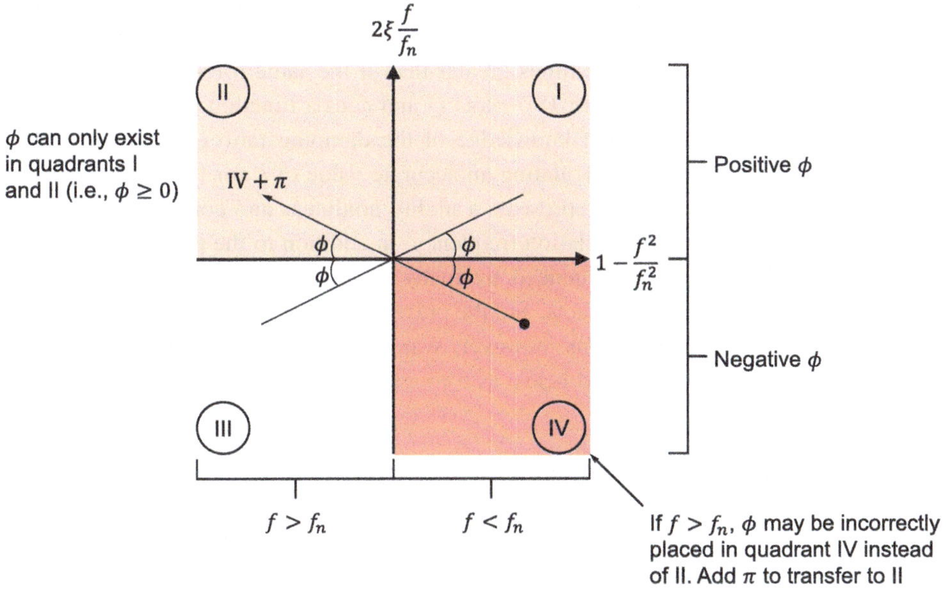

Fig. 18.6 The arctangent problem for ϕ

The response amplitude from Eq. (18.16) is often written in the form of:

$$R_0 = \frac{QF_0}{K} \tag{18.20}$$

where Q is the dynamic amplification factor:

$$Q(f) = \frac{R_0}{R_{static}} = \frac{1}{\sqrt{\left(1 - \frac{f^2}{f_n^2}\right)^2 + \left(2\xi\frac{f}{f_n}\right)^2}} \tag{18.21}$$

The dynamic amplification factor Q therefore expresses the ratio between the dynamic amplitude R_0 and the quasi-static response $R_{static} = F_0/K$. As an example, $Q = 5$ signifies that the dynamic motions are five times greater than if the same force F_0 was applied as a static load (i.e., if $\omega \approx 0$). Figure 18.7 plots Q and ϕ as a function of f/f_n and ξ.

From Eq. (18.21), we see that knowledge of the damping ratio ξ is required before determining Q. Unfortunately, calculating an accurate value of ξ for floating structures is not a trivial task since damping represents a highly nonlinear and complex phenomenon that is also dependent on the oscillation frequency. In addition to the pontoon's geometry, the damping ratio is further influenced by many factors including skin friction, eddy shedding, wave conditions, lift effects, and the presence of motion stabilizers such as bilge keels [2]. For rectangular pontoons, however, Wang [10, 12] demonstrated that $\xi \approx 0.1$ may be applied for roll, pitch, and heave as a rough approximation.

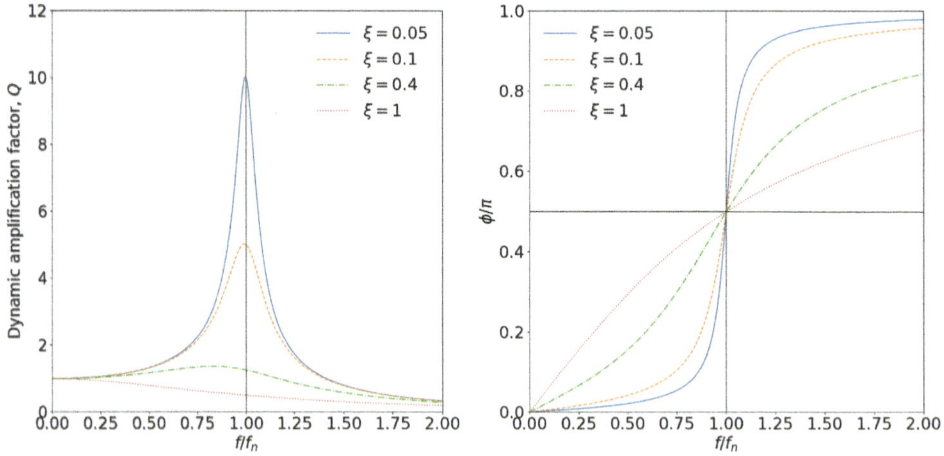

Fig. 18.7 Dynamic amplification factor (left) and phase (right) for various damping and f/f_n ratios

18.3.4 Solution for "Mass Dominated" Degrees of Freedom

Equation (18.20) is relevant for heave, roll, and pitch DoFs that feature stiffness K, damping C, and mass M contributions within the equation of motion and are referred to as "stiffness dominated". For "mass dominated" DoFs (surge, sway, and yaw), the waterplane stiffness is neglected and Eq. (18.2) reduces to:

$$-MR_0\omega^2\cos(\omega t - \phi) = F_0\cos(\omega t) \tag{18.22}$$

which expands to:

$$-MR_0\omega^2[\cos(\omega t)\cos\phi + \sin(\omega t)\sin\phi] = F_0\cos(\omega t) \tag{18.23}$$

For mass dominated systems, $f \gg f_n$ and the damping is negligible. Therefore, $\phi \approx \pi$ (see Fig. 18.7) and (18.23) becomes:

$$MR_0\omega^2\cos(\omega t) = F_0\cos(\omega t) \tag{18.24}$$

Rearranging for R_0 then gives the response amplitude for mass dominated systems:

$$R_0 = \frac{F_0}{M\omega^2} \tag{18.25}$$

18.3.5 Response Amplitude Operator

A summary of the dynamic response for mass or stiffness dominated DoFs is presented in Table 18.1. When applying Eqs. (18.20) and (18.25), R_0 is typically expressed as the response amplitude per unit wave amplitude (R_0/A). This ratio is known as the "response amplitude operator" (RAO) and is commonly referenced in dynamic analysis:

$$\text{RAO}_R = \frac{R_0}{A} \tag{18.26}$$

where the subscript R denotes the particular DoF expressed in (18.4). The RAO is sometimes written as $|H(f)|$ or $|H(\omega)|$ and can be interpreted as the rate at which the maximum structural response increases with increasing wave amplitude. Plotting the RAO as a function of the wave frequency f thus elucidates the maximum response over a range of frequencies pertaining to regular waves.

Finally, from (18.20) and (18.25), we discover that in order to calculate the response amplitude, we must first determine the mass (M), stiffness (K), and forcing amplitude (F_0) associated with the DoF under consideration.

Table 18.1 Dynamic response for mass and stiffness dominated DoFs of a freely floating structure

	Mass dominated (surge, sway, yaw)	Stiffness dominated (heave, roll, pitch)
Translation or rotation	Response: $R(t) = R_0 \cos(\omega t - \pi)$ Amplitude: $R_0 = A \times \text{RAO}_R$ RAO: $\text{RAO}_R = \frac{F_0\|_{A=1}}{M\omega^2}$	Response: $R(t) = R_0 \cos(\omega t - \phi)$ Amplitude: $R_0 = A \times \text{RAO}_R$ RAO: $\text{RAO}_R = \frac{Q F_0\|_{A=1}}{K}$
Velocity	$\dot{R}(t) = \frac{dR}{dt} = -\omega R_0 \sin(\omega t - \pi)$	$\dot{R}(t) = \frac{dR}{dt} = -\omega R_0 \sin(\omega t - \phi)$
Acceleration	$\ddot{R}(t) = \frac{d^2R}{dt^2} = -\omega^2 R_0 \cos(\omega t - \pi)$	$\ddot{R}(t) = \frac{d^2R}{dt^2} = -\omega^2 R_0 \cos(\omega t - \phi)$

Note The phase ϕ is obtained using (18.19) and A is the wave amplitude. When determining the RAO, the wave forcing amplitude F_0 is computed assuming $A = 1$

18.4 Waterplane Stiffness

The waterplane stiffness K is straightforward since from Chap. 17 on hydrostatics, we know that the rotational and heave stiffness of a floating body displacing a volume V with waterplane area A_w is:

$$K = \begin{cases} K_{\theta_x} = \rho g \left(I_x - \overline{BG} V\right), & \text{Roll} \\ K_{\theta_y} = \rho g \left(I_y - \overline{BG} V\right), & \text{Pitch} \\ K_z = \rho g A_w, & \text{Heave} \end{cases} \tag{18.27}$$

where I_x and I_y are the waterplane second moments of area about the x and y-axis, respectively, and $\overline{BG}$ is the distance from the center of buoyancy to the body's center of gravity.

Note that the presence of moorings used to anchor floating structures to the seabed may also contribute to the overall stiffness for each DoF. However, if compliant mooring systems (e.g., slack mooring lines) are adopted, their properties might not have a major effect on the dynamics of floating bodies [7], except when the structure is specifically designed to resonate with waves, e.g., when used as wave energy converters. For simplicity, we will decouple pontoon movements from potential mooring forces by assuming that such forces are much smaller than the wave-induced loads featured within the pontoon's equation of motion [5].

18.5 Forcing Amplitude

We now turn our attention to the determination of the forcing amplitude F_0. For the propagation of linear waves, recall that the dynamic pressure is (see Chap. 3 on linear wave theory):

$$p(t) = \begin{cases} \rho g A\left(\frac{\cosh[k(h+z)]}{\cosh(kh)}\right)\cos(kx-\omega t), & \text{Longitudinal waves} \\ \rho g A\left(\frac{\cosh[k(h+z)]}{\cosh(kh)}\right)\cos(ky-\omega t), & \text{Transverse waves} \end{cases} \tag{18.28}$$

where longitudinal and transverse waves refer to waves propagating in the direction of positive x and y, respectively. We can obtain the total wave force acting on a floating body via the integration of p over the surface in contact with the moving fluid. This dynamic force is commonly referred to as the Froude-Krylov force, expressed as:

$$F_{FK}(t) = \iint_S p(t)dS \tag{18.29}$$

where S is the wetted area of the body. However, (18.28) does not account for the presence of a body that acts to interfere and distort the flow field (since waves will be diffracting around the body). In order to adjust for this effect, we can introduce an additional mass m_a, to which its acceleration produces an additional force that acts in conjunction with the Froude-Krylov force. m_a is known as the added mass and can be thought of as an additional quantity of fluid being "trapped" by the body as it moves. The total force is therefore:

$$F(t) = F_{FK}(t) + F_{AM}(t) \tag{18.30}$$

where $F_{FK}(t)$ is the Froude-Krylov force per Eq. (18.29) and:

$$F_{AM}(t) = m_a a(t) \tag{18.31}$$

is the added mass force with $a(t)$ being the horizontal (a_x and a_y) or vertical (a_z) fluid acceleration at the location of the added mass (when treating it as a lumped mass):

$$a_x(t) = A\omega^2\left(\frac{\cosh[k(h+z)]}{\sinh(kh)}\right)\sin(kx-\omega t) \tag{18.32}$$

$$a_y(t) = A\omega^2\left(\frac{\cosh[k(h+z)]}{\sinh(kh)}\right)\sin(ky-\omega t) \tag{18.33}$$

$$a_z(t) = \begin{cases} -A\omega^2\left(\frac{\sinh[k(h+z)]}{\sinh(kh)}\right)\cos(kx-\omega t), & \text{Longitudinal waves} \\ -A\omega^2\left(\frac{\sinh[k(h+z)]}{\sinh(kh)}\right)\cos(ky-\omega t), & \text{Transverse waves} \end{cases} \tag{18.34}$$

Note that $F(t)$ in Eq. (18.30) denotes the time-varying total force since both the pressure $p(t)$ and accelerations $a(t)$ change with time t. However, we are more interested in determining the forcing amplitude F_0 which dictates the largest magnitude of the harmonic response for substitution into Eqs. (18.20) or (18.25). To do so, we must set the time t to a constant t_0 that results in a surface profile which yields the maximum response for the desired DoF. After substituting t_0 into (18.28) or (18.32) to (18.34), Eq. (18.30) then becomes:

$$F_0 = |F(t_0)| = |F_{FK}(t_0) + F_{AM}(t_0)| \tag{18.35}$$

For rotational DoFs, Eq. (18.35) instead depicts the moment amplitude produced by multiplying the force by its lever arm to the axis of rotation. Naturally, t_0 depends on the excited DoF. As an example, $t_0 = 0$ for the heave response since this corresponds to a wave crest directly below the axis origin (passing though the center of gravity) and therefore will produce the greatest vertical deviation from the still water level. The selection of t_0 for all six DoFs is:

$$t_0 = \begin{cases} 0, & \text{Heave, Yaw} \\ -\frac{T}{4}, & \text{Surge, Sway, Pitch, Roll} \end{cases} \tag{18.36}$$

where $T = 1/f = 2\pi/\omega$ is the wave period. Equation (18.36) is visualized in Fig. 18.8.

Evidently, Eqs. (18.28) and (18.32) to (18.34) pertain only to waves traveling along the longitudinal ($\theta = 0°$) or transverse ($\theta = 90°$) directions. For waves propagating in any arbitrary direction θ along an axis x_θ, the Airy wave pressure per unit amplitude ($A = 1$) at t_0 becomes:

$$p_0(x_\theta) = \begin{cases} \rho g \left(\frac{\cosh[k(h+z)]}{\cosh(kh)}\right) \cos(kx_\theta), & t_0 = 0 \\ -\rho g \left(\frac{\cosh[k(h+z)]}{\cosh(kh)}\right) \sin(kx_\theta), & t_0 = -T/4 \end{cases} \tag{18.37}$$

while the horizontal (a_{0x}) or vertical (a_{0z}) Airy wave acceleration per unit amplitude ($A = 1$) at t_0 are:

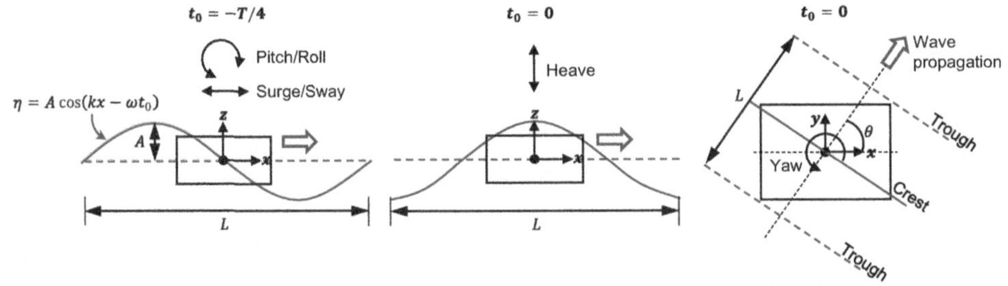

Fig. 18.8 Time t_0 yielding the maximum motion response pertaining to various DoFs

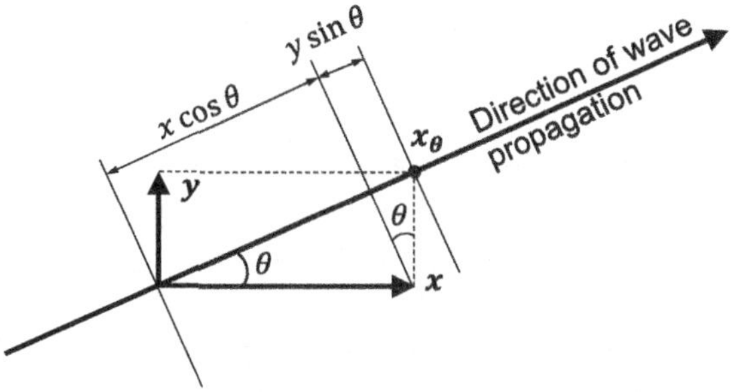

Fig. 18.9 Relationship between the xy-axis and the direction of wave propagation x_θ

$$a_{0x}(x_\theta) = \begin{cases} \omega^2\left(\frac{\cosh[k(h+z)]}{\sinh(kh)}\right)\sin(kx_\theta), & t_0 = 0 \\ \omega^2\left(\frac{\cosh[k(h+z)]}{\sinh(kh)}\right)\cos(kx_\theta), & t_0 = -T/4 \end{cases} \tag{18.38}$$

$$a_{0z}(x_\theta) = \begin{cases} -\omega^2\left(\frac{\sinh[k(h+z)]}{\sinh(kh)}\right)\cos(kx_\theta), & t_0 = 0 \\ \omega^2\left(\frac{\sinh[k(h+z)]}{\sinh(kh)}\right)\sin(kx_\theta), & t_0 = -T/4 \end{cases} \tag{18.39}$$

Note that x_θ denotes the direction of wave propagation at an angle θ counterclockwise from the x (surge) axis and $k = 2\pi/L$ is the wavenumber (where L is the wavelength) determined via the dispersion relation for a given wave frequency $\omega = 2\pi f$:

$$\omega = \sqrt{gk\tanh(kh)} \tag{18.40}$$

The axis x_θ can also be expressed in terms of x and y via:

$$x_\theta = x\cos\theta + y\sin\theta \tag{18.41}$$

and is illustrated in Fig. 18.9.

18.6 Added Mass

As previously discussed, Froude-Krylov pressures cannot fully account for the wave forces exerted upon a floating body since waves will diffract around an object in its path. In addition to wave diffraction, the motion of a floating body triggered by waves will also produce outgoing waves of its own, referred to as radiation. To account for such effects, an additional source of forcing associated with the movement of an "added mass" of water surrounding the floating body is commonly adopted. Technically speaking, the added mass m_a is influenced by the body's geometry, oscillation frequency, and direction of motion. Hence, the determination of m_a is not trivial and requires the use of diffraction theory which is beyond the scope of this text.

In this book, we will utilize frequency-independent approximations for the treatment of added mass applicable towards rectangular pontoons (with length C_p, width B_p, and draft D) as illustrated in Fig. 18.10. When using this approach, the added mass is treated as prismatic volume of water moving in sync with the floating body [8]. The shape of this volume is dependent on the DoF under consideration. Note that for translational DoFs (surge, sway, and heave), the added mass is simply the total mass of the water volume depicted in Fig. 18.10. For rotational DoFs (roll, pitch, and yaw), however, m_a refers to the mass moment of inertia associated with the water volume when taken about the center of rotation (CoR).

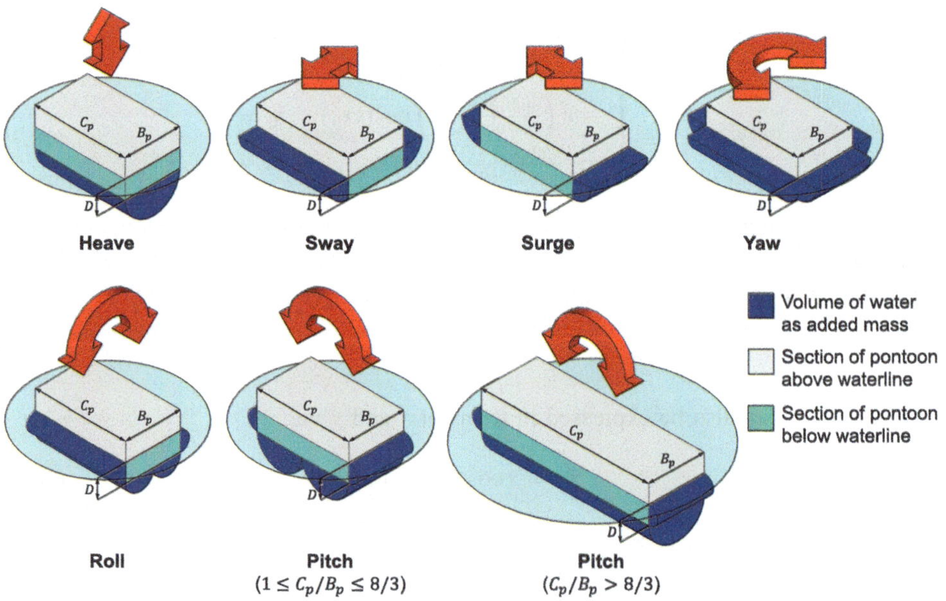

Fig. 18.10 Added mass for rectangular pontoons represented as a prismatic volume of water for various DoFs

Across all six DoFs, the added masses are:

$$m_a = \begin{cases} m_{ax} = \frac{1}{2}\rho\pi D^2 B_p, & \text{Surge} \\ m_{ay} = \frac{1}{2}\rho\pi D^2 C_p, & \text{Sway} \\ m_{az} = \frac{1}{8}\rho\pi B_p^2 C_p C_{zh} C_{zb}, & \text{Heave} \\ m_{a\theta_x} = \rho\pi\left[\left(\frac{B_p}{4}\right)^4 + \left(\frac{D}{2}\right)^2\left(\overline{KR}_{\theta_x} - \frac{D}{2}\right)^2\right]C_p, & \text{Roll} \\ m_{a\theta_y} = \rho\pi\left[\left(\frac{C_p}{4}\right)^4 + \left(\frac{D}{2}\right)^2\left(\overline{KR}_{\theta_y} - \frac{D}{2}\right)^2\right]B_p, & \text{Pitch } (C_p/B_p \le 8/3) \\ m_{a\theta_y} = \rho\pi\left[\frac{B_p}{12}\left(\frac{C_p}{2}\right)^3 + \left(\frac{D}{2}\right)^2\left(\overline{KR}_{\theta_y} - \frac{D}{2}\right)^2\right]B_p, & \text{Pitch } (C_p/B_p > 8/3) \\ m_{a\theta_z} = \frac{\rho\pi D^2}{2}\left(\frac{C_p^3}{12} + \frac{B_p^3}{12}\right), & \text{Yaw} \end{cases} \quad (18.42)$$

with $\overline{KR}_{\theta_x}$ and $\overline{KR}_{\theta_y}$ denoting the distance from the keel to the center of rotation (CoR) for roll and pitch, respectively, which will be discussed in the next section. The added mass in heave (m_{az}) is influenced by relationships between the draft (D), pontoon width (B_p), and water depth (h). Through applying a power law to the experimental results obtained by Brennen [1], the following simplified heave added mass modification factors C_{zh} and C_{zb} may be adopted:

$$C_{zh} = \begin{cases} 1.518\left(\frac{h-D}{D}\right)^{-0.365}, & \frac{h-D}{D} < 3 \\ 1, & \frac{h-D}{D} \ge 3 \end{cases}$$

$$C_{zb} = 1.073\left(\frac{B_p}{D}\right)^{-0.109} \quad (18.43)$$

In addition, the added mass moment of inertia for pitch ($m_{a\theta_y}$) is dependent on the ratio between the length (C_p) and width (B_p) of the pontoon. For stockier pontoons that are more square-like, the mobilized volume of fluid interacting with the keel can be approximated as two semicircular sections oriented along the transverse direction parallel to the axis of pitch rotation (y-axis). This treatment is consistent with the approach for roll (see Fig. 18.10). However, for slender pontoons where its length is much greater than its width, the keel added mass in pitch takes the form of a single semicircular prismatic volume oriented longitudinally along the keel.

Figure 18.11 illustrates the calculation of $m_{a\theta_y}$ by treating the added masses distributed about the wetted perimeter of the pontoon as a series of lumped masses attached to the keel and side walls with the direction of positive acceleration indicated. The mass moment of inertia is determined by multiplying each lumped mass by the squared distance to the center of rotation taken perpendicular to the direction of acceleration. When the pitch added mass along the keel is treated as a single semicircular prismatic volume (for more slender pontoon geometries), this total volume is discretized into individual slices of unit thickness with each slice representing a single lumped mass. For stocky pontoons, the mass moment of inertia is approximated as:

$$m_{a\theta_y} = \sum_{i=1}^{4} m_i r_i^2 \tag{18.44}$$

where m_i, r_i, and m' are given in Fig. 18.11. For slender pontoons with length (C_p) much longer than the width (B_p), the mass moment of inertia becomes:

$$m_{a\theta_y} = \sum_{i=1}^{2} m_i r_i^2 + \int_{-C_p/2}^{C_p/2} m' x^2 dx \tag{18.45}$$

Equating these two expressions will produce:

$$\sum_{i=1}^{4} m_i r_i^2 = \sum_{i=1}^{2} m_i r_i^2 + \int_{-C_p/2}^{C_p/2} m' x^2 dx \tag{18.46}$$

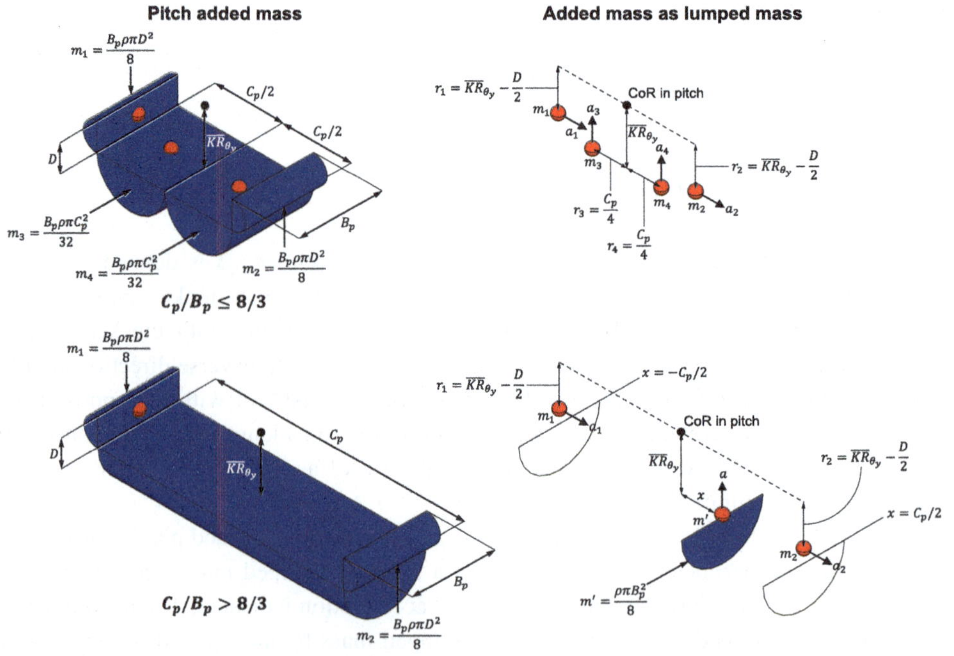

Fig. 18.11 Treatment of added mass via lumped mass approach for determination of added mass moment of inertia in pitch for stocky (top) and slender (bottom) pontoons

Equation (18.46) is satisfied when $C_p/B_p = 8/3$. A C_p/B_p ratio of 8/3 ($\approx$ 2.67) is therefore suggested as a rough differentiator to separate how added mass should be treated for stocky and slender pontoons. Note that the added mass expressions in Eq. (18.42) for the other DoFs were obtained via the same lumped mass approach exemplified in Fig. 18.11.

The total mass M used for dynamic analysis is then the sum of the total structural mass (m) and the added mass (m_a) associated with the DoF of motion:

$$M = m + m_a \tag{18.47}$$

The total mass (M) associated with each DoF is thus:

$$M = \begin{cases} M_x = m + m_{ax}, & \text{Surge} \\ M_y = m + m_{ay}, & \text{Sway} \\ M_z = m + m_{az}, & \text{Heave} \\ M_{\theta_x} = m_{\theta_x} + m\left(\overline{KG} - \overline{KR}_{\theta_x}\right)^2 + m_{a\theta_x}, & \text{Roll} \\ M_{\theta_y} = m_{\theta_y} + m\left(\overline{KG} - \overline{KR}_{\theta_y}\right)^2 + m_{a\theta_y}, & \text{Pitch} \\ M_{\theta_z} = m_{\theta_z} + m_{a\theta_z}, & \text{Yaw} \end{cases} \tag{18.48}$$

where m_{θ_x}, m_{θ_y}, and m_{θ_z} are the mass moments of inertia of the floating structure taken about the x, y, and z axes for roll, pitch, and yaw, respectively. Note that m_{θ_z} assumes that the z-axis passes through the CoG of the added mass in yaw as shown in Fig. 18.12.

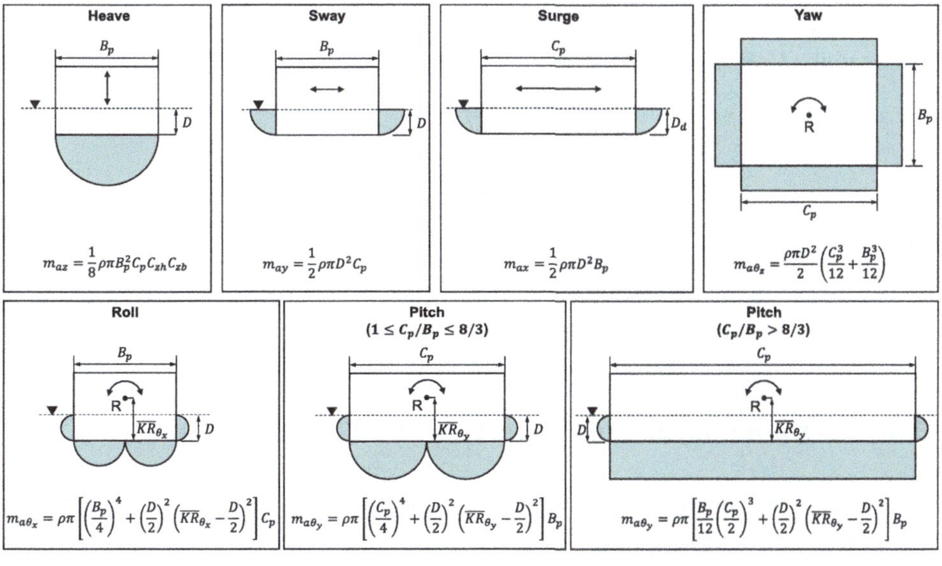

Fig. 18.12 Added mass equations pertaining to all six DoFs. "R" denotes center of rotation

For a rectangular object of uniform density with length C, width B, height D, and mass m, the mass moments of inertia in roll (m_{θ_x}), pitch (m_{θ_y}), and yaw (m_{θ_z}) are respectively:

$$m_{\theta_x} = \frac{m}{12}\left(B^2 + D^2\right) \tag{18.49}$$

$$m_{\theta_y} = \frac{m}{12}\left(C^2 + D^2\right) \tag{18.50}$$

$$m_{\theta_z} = \frac{m}{12}\left(C^2 + B^2\right) \tag{18.51}$$

If a structure is composed of multiple rectangular objects (e.g., pontoon + building), the parallel axis theorem can then be used to find the overall mass moments of inertia.

18.7 Center of Rotation

In this text, sway-roll and surge-pitch DoFs are decoupled and solved independently in order to avoid the use of mass, stiffness, and damping matrices. This operation is necessary because surge-pitch and sway-roll motions are inherently coupled as horizontal motions along x and y would naturally induce rotations about the y and x axes, respectively. However, the translational (surge/sway) and rotational (roll/pitch) response may be approximately decoupled and evaluated separately if the x and y axes of rotation are selected to coincide with the CoG associated with the structural mass (m) and added mass pertaining to translational motion m_{ax} or m_{ay} (i.e., surge added mass for pitch and sway added mass for roll). This is referred to as the center of rotation (CoR). The distance from the keel to the CoR in roll ($\overline{KR}_{\theta_x}$) and pitch ($\overline{KR}_{\theta_y}$) are respectively given as:

$$\begin{aligned} \overline{KR}_{\theta_x} &= \frac{m\overline{KG} + m_{ay}\left(\frac{D}{2}\right)}{m + m_{ay}} \\ \overline{KR}_{\theta_y} &= \frac{m\overline{KG} + m_{ax}\left(\frac{D}{2}\right)}{m + m_{ax}} \end{aligned} \tag{18.52}$$

and illustrated via Fig. 18.13. Note that the center of rotation in yaw is taken as the centroid of the pontoon (Fig. 18.12).

18.8 Example: Dynamic Properties of a Floating House

We will adopt the floating house showcased in Sect. 17.7 to illustrate the procedure by which some of the properties used for dynamic analysis can be obtained. For this example, we will assume that the structure is placed in water with a depth of 10 m. Relevant

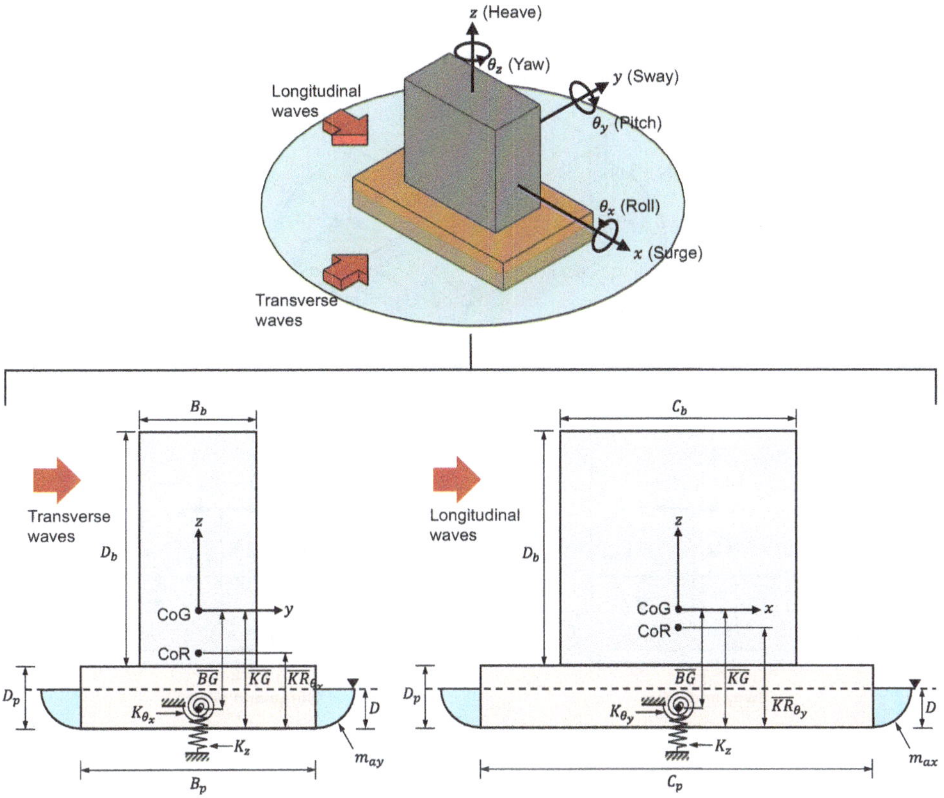

Fig. 18.13 Specification of CoR for (left) roll and (right) pitch DoFs

geometric parameters for this analysis are shown in Fig. 18.14 and are summarized in Table 18.2, of which many were previously obtained in Sect. 17.7.

The aim of this example will be to use the parameters in Table 18.2 to determine the following dynamic properties:

- Waterplane stiffness in heave, roll, and pitch
- Added mass (translational DoFs) and mass moment of inertia (rotational DoFs)
- Total mass (translational DoFs) and mass moment of inertia (rotational DoFs)
- Natural frequency of vibration for stiffness-dominated DoFs.

Note, however, that the wave forcing amplitude pertaining to each DoF is beyond the scope of this exercise and will be discussed in detail in Chap. 19.

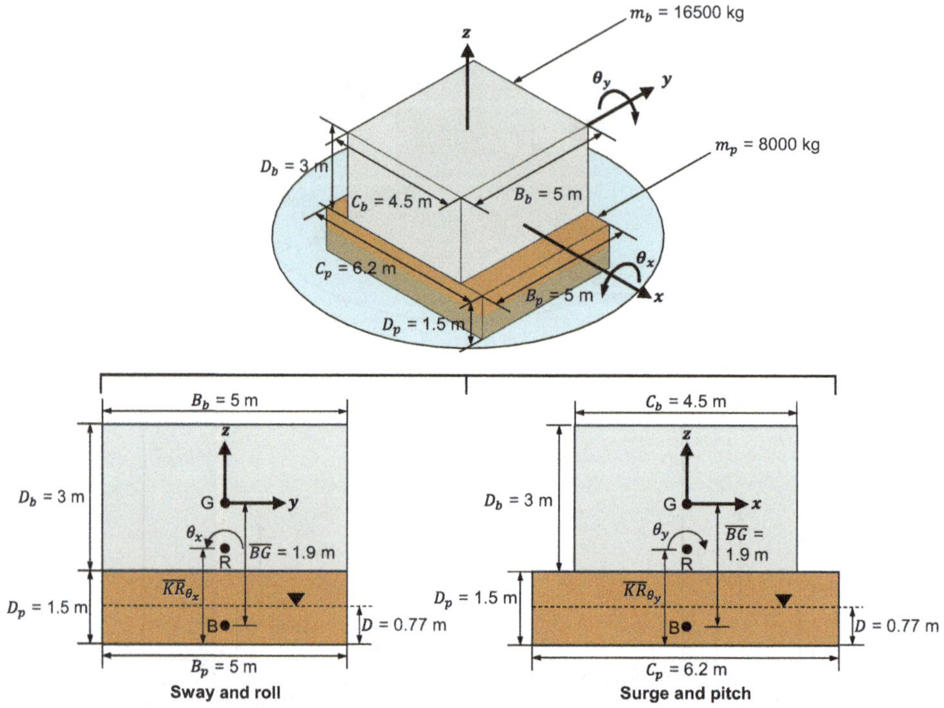

Fig. 18.14 Geometry of floating house showing the parameters used for dynamic analysis

18.8.1 Waterplane Stiffness

The waterplane stiffness can be directly obtained from Eq. (18.27). These dynamic properties will only be relevant for the stiffness-dominated roll, pitch, and heave DoFs, being:

$$K_{\theta_x} = \rho g\left(I_x - \overline{BG}V\right) = 1025 \times 9.81 \times (64.6 - 1.9 \times 23.9) = 192{,}960\,\text{N m/rad}$$
$$K_{\theta_y} = \rho g\left(I_y - \overline{BG}V\right) = 1025 \times 9.81 \times (99.3 - 1.9 \times 23.9) = 541{,}880\,\text{N m/rad}$$
$$K_z = \rho g A_w = 1025 \times 9.81 \times 31 = 311{,}710\,\text{N/m}$$

18.8.2 Added Mass

We can next determine the added mass (or mass moment of inertia in the case of rotational motion) about all six DoFs via Eq. (18.42). The added mass in surge and sway are respectively:

Table 18.2 Parameters of floating building shown in Fig. 18.14

Parameter	Variable	Value
Length of pontoon	C_p	6.2 m
Width of pontoon	B_p	5 m
Height of pontoon	D_p	1.5 m
Length of building	C_b	4.5 m
Width of building	B_b	5 m
Height of building	D_b	3 m
Mass of pontoon	m_p	8000 kg
Mass of building	m_b	16,500 kg
Total mass of floating structure	m	24,500 kg
Distance from CoB to CoG	$\overline{BG}$	1.9 m
Distance from keel to CoG	$\overline{KG}$	2.27 m
Water depth	h	10 m
Water density	ρ	1025 kg m^{-3}
Draft	D	0.77 m
Displaced volume	V	23.9 m^3
Waterplane area	A_w	31 m^2
Waterplane second moment of area in roll	I_x	64.6 m^4
Waterplane second moment of area in pitch	I_y	99.3 m^4

$$m_{ax} = \frac{1}{2}\rho\pi D^2 B_p = \frac{1}{2} \times 1025 \times \pi \times 0.77^2 \times 5 = 4773\,\text{kg}$$

$$m_{ay} = \frac{1}{2}\rho\pi D^2 C_p = \frac{1}{2} \times 1025 \times \pi \times 0.77^2 \times 6.2 = 5919\,\text{kg}$$

For heave, the added mass modification factors are first determined from Eq. (18.43):

$$\frac{h - D}{D} = \frac{10 - 0.77}{0.77} = 12 > 3$$

Hence:

$$C_{zh} = 1$$

and:

$$C_{zb} = 1.073\left(\frac{B_p}{D}\right)^{-0.109} = 1.073\left(\frac{5}{0.77}\right)^{-0.109} = 0.875$$

The heave added mass is then:

$$m_{az} = \frac{1}{8}\rho\pi B_p^2 C_p C_{zh} C_{zb} = \frac{1}{8} \times 1025 \times \pi \times 5^2 \times 6.2 \times 1 \times 0.875 = 54{,}590\,\text{kg}$$

To determine the added mass moments of inertia in roll or pitch, we need to first find the elevation of the centers of rotation (CoR) from the keel via Eq. (18.52):

$$\overline{KR}_{\theta_x} = \frac{m\overline{KG} + m_{ay}\left(\frac{D}{2}\right)}{m + m_{ay}} = \frac{24{,}500 \times 2.27 + 5919\left(\frac{0.77}{2}\right)}{24{,}500 + 5919} = 1.90\,\text{m}$$

$$\overline{KR}_{\theta_y} = \frac{m\overline{KG} + m_{ax}\left(\frac{D}{2}\right)}{m + m_{ax}} = \frac{24{,}500 \times 2.27 + 4773\left(\frac{0.77}{2}\right)}{24{,}500 + 4773} = 1.96\,\text{m}$$

We notice that the CoR in pitch is slightly higher than that of roll. This makes intuitive sense since the increased sway added mass (due to the longer side length C_p) mobilized by transverse waves will result in the overall center of gravity being weighted more towards the keel when compared to surge. Next, we can obtain the roll added mass moment of inertia from the fourth equation in (18.42):

$$\begin{aligned} m_{a\theta_x} &= \rho\pi\left[\left(\frac{B_p}{4}\right)^4 + \left(\frac{D}{2}\right)^2\left(\overline{KR}_{\theta_x} - \frac{D}{2}\right)^2\right]C_p \\ &= 1025 \times \pi\left[\left(\frac{5}{4}\right)^4 + \left(\frac{0.77}{2}\right)^2\left(1.9 - \frac{0.77}{2}\right)^2\right] \times 6.2 = 55{,}534\,\text{kg m}^2 \end{aligned}$$

For the pitch added mass moment of inertia, we first need to determine the slenderness of the pontoon as expressed by its C_p/B_p ratio:

$$\frac{C_p}{B_p} = \frac{6.2}{5} = 1.24 < \frac{8}{3}$$

which leads to:

$$\begin{aligned} m_{a\theta_y} &= \rho\pi\left[\left(\frac{C_p}{4}\right)^4 + \left(\frac{D}{2}\right)^2\left(\overline{KR}_{\theta_y} - \frac{D}{2}\right)^2\right]B_p \\ &= 1025 \times \pi\left[\left(\frac{6.2}{4}\right)^4 + \left(\frac{0.77}{2}\right)^2\left(1.96 - \frac{0.77}{2}\right)^2\right] \times 5 = 98{,}850\,\text{kg m}^2 \end{aligned}$$

Finally, the yaw added mass moment of inertia can be obtained from the last equation in (18.42):

$$m_{a\theta_z} = \frac{\rho\pi D^2}{2}\left(\frac{C_p^3}{12} + \frac{B_p^3}{12}\right) = \frac{1025 \times \pi \times 0.77^2}{2}\left(\frac{6.2^3}{12} + \frac{5^3}{12}\right) = 28{,}900\,\text{kg m}^2$$

18.8.3 Total Mass

The total mass (or mass moment of inertia in the case of rotational motion), accounting for both the structural mass (m) and added mass (m_a), that will be mobilized about all six DoFs can now be found from Eq. (18.48). In surge, sway, and heave, they are:

$$M_x = m + m_{ax} = 24{,}500 + 4773 = 29{,}273\,\text{kg}$$
$$M_y = m + m_{ay} = 24{,}500 + 5919 = 30{,}419\,\text{kg}$$
$$M_z = m + m_{az} = 24{,}500 + 54{,}590 = 79{,}090\,\text{kg}$$

To find the total mass moments of inertia, we must first calculate the mass moments of inertia of the floating structure in roll, pitch, and yaw. Since this structure consists of a rectangular pontoon supporting a rectangular building, the mass moments of inertia for each component must first be determined before combining them using the parallel axis theorem. For the pontoon, we have:

$$m_{p\theta_x} = \frac{m_p}{12}\left(B_p^2 + D_p^2\right) = \frac{8000}{12}\left(5^2 + 1.5^2\right) = 18{,}170\,\text{kg m}^2$$
$$m_{p\theta_y} = \frac{m_p}{12}\left(C_p^2 + D_p^2\right) = \frac{8000}{12}\left(6.2^2 + 1.5^2\right) = 27{,}130\,\text{kg m}^2$$
$$m_{p\theta_z} = \frac{m_p}{12}\left(C_p^2 + B_p^2\right) = \frac{8000}{12}\left(6.2^2 + 5^2\right) = 42{,}290\,\text{kg m}^2$$

while the building moments of inertia are:

$$m_{b\theta_x} = \frac{m_b}{12}(B_b^2 + D_b^2) = \frac{16{,}500}{12}(5^2 + 3^2) = 46{,}750\,\text{kg m}^2$$
$$m_{b\theta_y} = \frac{m_b}{12}(C_b^2 + D_b^2) = \frac{16{,}500}{12}(4.5^2 + 3^2) = 40{,}220\,\text{kg m}^2$$
$$m_{b\theta_z} = \frac{m_b}{12}(C_b^2 + B_b^2) = \frac{16{,}500}{12}(4.5^2 + 5^2) = 62{,}220\,\text{kg m}^2$$

The combined mass moments of inertia in roll, pitch, and yaw of the floating structure are:

$$\begin{aligned} m_{\theta_x} &= m_{p\theta_x} + m_p\left(\overline{KG} - \overline{KG}_p\right)^2 + m_{b\theta_x} + m_b\left(\overline{KG} - \overline{KG}_b\right)^2 \\ &= 18{,}170 + 8000(2.27 - 0.75)^2 + 46{,}750 + 16{,}500(2.27 - 3)^2 = 92{,}200\,\text{kg m}^2 \end{aligned}$$
$$\begin{aligned} m_{\theta_y} &= m_{p\theta_y} + m_p\left(\overline{KG} - \overline{KG}_p\right)^2 + m_{b\theta_y} + m_b\left(\overline{KG} - \overline{KG}_b\right)^2 \\ &= 27{,}130 + 8000(2.27 - 0.75)^2 + 40{,}220 + 16{,}500(2.27 - 3)^2 = 94{,}630\,\text{kg m}^2 \end{aligned}$$
$$m_{\theta_z} = m_{p\theta_z} + m_{b\theta_z} = 42{,}290 + 62{,}220 = 104{,}510\,\text{kg m}^2$$

where $\overline{KG}_p$ and $\overline{KG}_b$ denote elevation of the CoG of the pontoon and building above the keel, respectively. The total mass moments of inertia in roll, pitch, and yaw when accounting for both the structure and added mass about the center of rotation then becomes:

$$M_{\theta_x} = m_{\theta_x} + m\left(\overline{KG} - \overline{KR}_{\theta_x}\right)^2 + m_{a\theta_x} = 92{,}200 + 24{,}500(2.27 - 1.9)^2 + 55{,}534$$
$$= 151{,}090\,\text{kg m}^2$$

$$M_{\theta_y} = m_{\theta_y} + m\left(\overline{KG} - \overline{KR}_{\theta_y}\right)^2 + m_{a\theta_y} = 94{,}630 + 24{,}500(2.27 - 1.96)^2 + 98{,}850$$
$$= 195{,}830\,\text{kg m}^2$$

$$M_{\theta_z} = m_{\theta_z} + m_{a\theta_z} = 104{,}510 + 28{,}900 = 133{,}410\,\text{kg m}^2$$

18.8.4 Natural Frequency

With knowledge of the total hydrodynamic mass and the waterplane stiffness, we can now determine the natural frequency of vibration for the three stiffness-dominated DoFs of heave, roll, and pitch from Eq. (18.18):

$$f_{n,z} = \frac{1}{2\pi}\sqrt{\frac{K_z}{M_z}} = \frac{1}{2\pi}\sqrt{\frac{311{,}710}{79{,}090}} = 0.316\,\text{Hz}$$

$$f_{n,\theta_x} = \frac{1}{2\pi}\sqrt{\frac{K_{\theta_x}}{M_{\theta_x}}} = \frac{1}{2\pi}\sqrt{\frac{192{,}960}{151{,}090}} = 0.180\,\text{Hz}$$

$$f_{n,\theta_y} = \frac{1}{2\pi}\sqrt{\frac{K_{\theta_y}}{M_{\theta_y}}} = \frac{1}{2\pi}\sqrt{\frac{541{,}880}{195{,}830}} = 0.265\,\text{Hz}$$

Waves traveling with a frequency f in close proximity to any of these natural frequencies would therefore be expected to induce large motions of the floating structure due to resonance.

This example will only illustrate the determination of the waterplane stiffness, hydrodynamic mass, and natural frequency. In the next chapter, we will discuss how the forcing amplitude pertaining to each DoF can be obtained under regular waves to ultimately formulate their response amplitude operators (RAO).

References

1. Brennen, C.E., 1982. A review of added mass and fluid inertial forces. Naval Civil Engineering Laboratory, Port Hueneme, California.
2. Chakrabarti, S., 2001. Empirical calculation of roll damping for ships and barges. Ocean Engineering 28, 915–932.

3. Fujikubo, M., Suzuki, H., 2015. Mega-Float, in: Large Floating Structures: Technological Advances. Springer, Singapore, pp. 197–220.
4. Robertson, A.N., Wendt, F., Jonkman, J.M., Popko, W., Dagher, H., Geuydon, S., Qvist, J., Vittori, F., Azcona, J., Uzunoglu, E., Soares, C.G., Harries, R., Yde, A., Galinos, C., Hermans, K., de Vaal, J.B., Bozonnet, P., Bouy, L., Bayati, I., Bergua, R., Galvan, J., Mendikoa, I., Sanchez, C.B., Shin, H., Oh, S., Molins, C., Debruyne, Y., 2017. OC5 project phase II: Validation of global loads of the DeepCwind floating semisubmersible wind turbine. Energy Procedia 137, 38–57.
5. Ruzzo, C., Muggiasca, S., Malara, G., Taruffi, F., Belloli, M., Collu, M., Li, L., Brizzi, G., Arena, F., 2021. Scaling strategies for multi-purpose floating structures physical modeling: state of art and new perspectives. Applied Ocean Research 108, 102487.
6. Suzuki, H., Yoshida, K., 1996. Design flow and strategy for safety of very large floating structure, in: Proceedings of the International Workshop on Very Large Floating Structures. Hayama, Japan, pp. 21–27.
7. Trubat, P., Molins, C., Gironella, X., 2020. Wave hydrodynamic forces over mooring lines on floating offshore wind turbines. Ocean Engineering 195, 106730.
8. Vugts, J.H., 1968. The hydrodynamic coefficients for swaying, heaving, and rolling cylinders in a free surface (No. 194). Delft University of Technology, Delft, Netherlands.
9. Wang, C.M., Watanabe, E., Utsunomiya, T., 2008. Very large floating structures. Taylor and Francis.
10. Wang, S., 2025. Derivation of 3D Dynamics for Rigid Floating Structures under Directional Wave Excitation. Sustainable Marine Structures 7, 35–51.
11. Wang, S., 2023. Simplified analytical solutions to the yaw dynamics of modular floating structures. Ocean Engineering 276.
12. Wang, S., 2022. Analytical solutions for the dynamic analysis of a modular floating structure for urban expansion. Ocean Engineering 266.
13. Wang, S., Chuang, W.-L., 2025. Dynamic analysis of breaking wave impact on a floating offshore wind turbine via smoothed particle hydrodynamics. Marine Structures 100.

Dynamics of Floating Structures Under Regular Waves

19

Chapter 18 introduced three components that are necessary to predict the motion of floating structures:

- Waterplane stiffness K
- Total mass (structural mass + added mass) M
- Wave forcing amplitude F_0

In this chapter, we will reveal how these components fit together to ultimately produce closed-form equations that describe the dynamics of floating bodies across all six DoFs for waves approaching from any arbitrary direction. For simplicity, we will limit our analysis to structures supported by rectangular pontoons.

19.1 The Response Amplitude Operator

Under regular waves (i.e., waves with a single amplitude and frequency), the dynamic motions of a floating structure for each of its six degrees of freedom will follow a sinusoidal profile with time, being:

$$R(t) = R_0 \cos(\omega t - \phi) \tag{19.1}$$

The original version of the chapter has been revised. A correction to this chapter can be found at https://doi.org/10.1007/978-3-032-18199-2_22

S. Wang, *Applied Ocean Mechanics for Structural Engineering and Climate Adaptation*, Synthesis Lectures on Ocean Systems Engineering,
https://doi.org/10.1007/978-3-032-18199-2_19

Since regular waves have only a single frequency ω, the frequency of motion associated with each of its DoFs will match the frequency of wave propagation. Therefore, to fully capture the dynamic response, only the amplitude of motion (R_0) and its phase (ϕ) need to be determined. When adopting linear wave theory, R_0 is linearly correlated with the wave amplitude (A) via the response amplitude operator (RAO) as shown in Fig. 19.1.

The response amplitude of a floating structure excited by regular waves for mass and stiffness dominated DoFs are restated as follows:

$$R_0 = A \times \text{RAO}_R = \begin{cases} A \times \frac{Q_R F_{0R}}{K_R}, & \text{Stiffness dominated (heave, roll, pitch)} \\ A \times \frac{F_{0R}}{M_R \omega^2}, & \text{Mass dominated (sway, surge, yaw)} \end{cases} \tag{19.2}$$

where the subscript R can be replaced by x, y, z, θ_x, θ_y, or θ_z to represent motion about its respective DoF (e.g., RAO_x = surge RAO). Note that F_{0R} refers to the forcing amplitude produced by waves with amplitude $A = 1$. For stiffness-dominated DoFs, the dynamic amplification factor Q_R is:

$$Q_R = \frac{1}{\sqrt{\left(1-\frac{f^2}{f_{nR}^2}\right)^2 + \left(2\xi_R \frac{f}{f_{nR}}\right)^2}}, \quad R = \begin{cases} z, & \text{Heave} \\ \theta_x, & \text{Roll} \\ \theta_y, & \text{Pitch} \end{cases} \tag{19.3}$$

with ξ_R denoting the damping ratio and f_{nR} is the natural frequency of DoF R:

$$f_{nR} = \frac{1}{2\pi}\sqrt{\frac{K_R}{M_R}}, \quad R = \begin{cases} z, & \text{Heave} \\ \theta_x, & \text{Roll} \\ \theta_y, & \text{Pitch} \end{cases} \tag{19.4}$$

Here, the waterplane stiffness K_R and total mass M_R can be obtained from Eqs. (18.27) and (18.48), respectively. Determining the RAO therefore gives the harmonic motion time

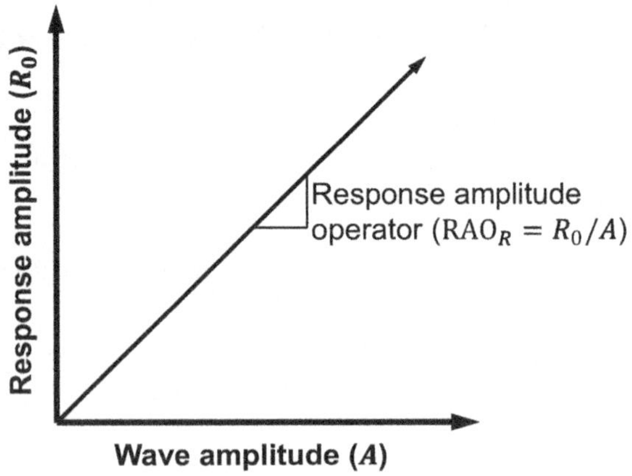

Fig. 19.1 Relationship between R_0 and A

history (pertaining to DoF R) of a floating structure under regular waves with amplitude A and frequency $\omega = 2\pi/T$:

$$R(t) = R_0 \cos(\omega t - \phi) = A \times \mathrm{RAO}_R \cos(\omega t - \phi) \tag{19.5}$$

where ϕ is the phase lag given by:

$$\phi = \begin{cases} \arctan\left(\dfrac{2\xi_R \frac{f}{f_{nR}}}{1-\frac{f^2}{f_{nR}^2}}\right), & R = \{z, \theta_x, \theta_y\} \\ \pi, \quad R = \{x, y, \theta_z\} \end{cases} \tag{19.6}$$

To determine the response amplitude operator, we must obtain the forcing amplitude F_{0R} for each DoF. From Eq. (18.35), we see that this amplitude is dependent upon the Froude-Krylov (F_{FKR}) and added mass (F_{AMR}) forces:

$$F_{0R} = |F_{FKR}(t_0) + F_{AMR}(t_0)|\,, \quad \begin{cases} t_0 = 0, & R = \{z, \theta_z\} \\ t_0 = -\frac{T}{4}, & R = \{x, y, \theta_x, \theta_y\} \end{cases} \tag{19.7}$$

where $T = 2\pi/\omega$ is the wave period. These forces are also highly dependent upon the direction of wave propagation relative to the pontoon axes. Figure 19.2 illustrates the convention adopted in this text where the angle of wave propagation (θ) is taken positive counterclockwise from the x (surge) axis. Next, it is necessary to define a relationship between the xy-axis mapping the coordinates of the pontoon and the x_θ-axis coinciding with the direction of wave propagation. This transformation is given by:

$$x_\theta = x\cos\theta + y\sin\theta \tag{19.8}$$

For convenience, transformations for the two longitudinal (x_1 and x_2) and transverse (y_1 and y_2) edges delineating the perimeter of the pontoon are further listed as:

$$\begin{aligned} x_1 &= x\cos\theta - \tfrac{B_p}{2}\sin\theta, & x_2 &= x\cos\theta + \tfrac{B_p}{2}\sin\theta \\ y_1 &= y\sin\theta - \tfrac{C_p}{2}\cos\theta, & y_2 &= y\sin\theta + \tfrac{C_p}{2}\cos\theta \end{aligned} \tag{19.9}$$

To further simply the analytical expressions presented in this chapter, the following abbreviations are introduced:

$$\begin{aligned} s_B &= \sin\left(\tfrac{B_p k}{2}\sin\theta\right) & s_C &= \sin\left(\tfrac{C_p k}{2}\cos\theta\right) \\ c_B &= \cos\left(\tfrac{B_p k}{2}\sin\theta\right) & c_C &= \cos\left(\tfrac{C_p k}{2}\cos\theta\right) \end{aligned} \tag{19.10}$$

where B_p and C_p respectively denote the width and length of the pontoon, and k is the wavenumber. Due to the double symmetry of rectangular pontoons, θ need only be taken between 0 and 90°. However, to prevent the occurrence of singularities (i.e., division by zero) when θ is exactly equal to 0 or 90°, it is recommended that the following

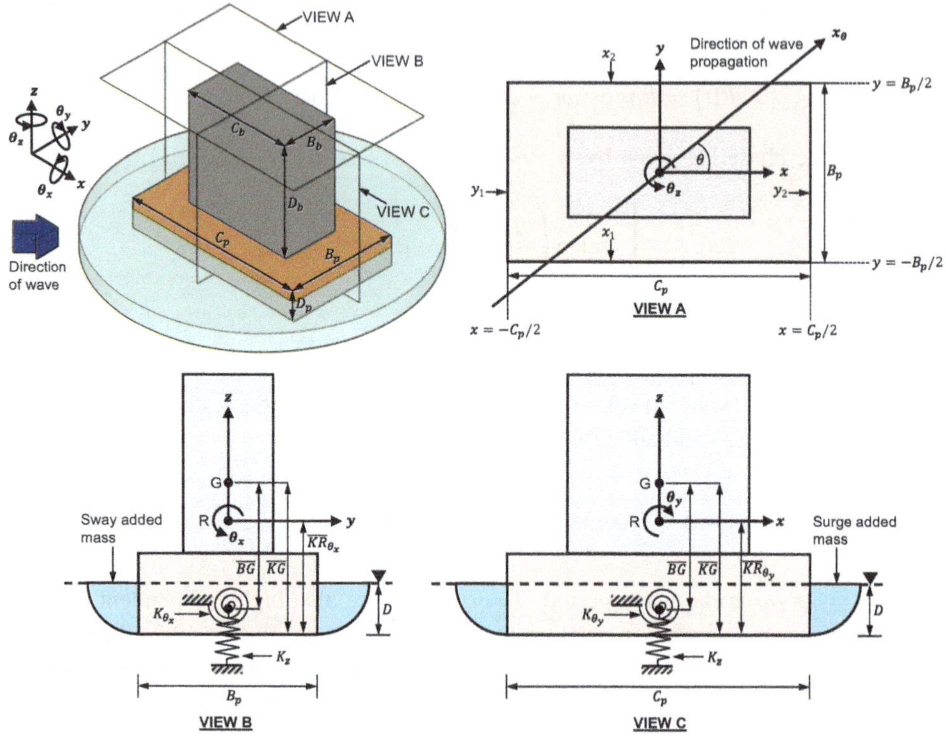

Fig. 19.2 Geometric properties of a rectangular floating pontoon supporting a rectangular superstructure

adjustments be applied for the treatment of θ:

$$\theta = \begin{cases} \varepsilon, & \theta = 0^\circ \\ 90^\circ - \varepsilon, & \theta = 90^\circ \\ \theta, & 0^\circ < \theta < 90^\circ \end{cases} \tag{19.11}$$

where ε represents an arbitrarily small positive quantity (e.g., $\varepsilon = 0.1°$).

19.2 Geometric Definitions of a Floating Structure

In this chapter, we will illustrate the application of Eq. (19.2) to obtain expressions for the dynamic response of a hypothetical floating structure shown in Fig. 19.3 subject to regular waves propagating with amplitude A and frequency $\omega = 2\pi f$ in water of depth $h = 30\,\text{m}$. In particular, we are interested in determining the response amplitude operator (RAO) across all six DoFs which we can then multiply with the wave amplitude to obtain R_0

(i.e., $R_0 = A \times \text{RAO}_R$). As such, the wave amplitude A featured in Eqs. (18.28) and (18.32) to (18.34) used to respectively determine the Froude-Krylov and added mass forces will be taken as unity ($A = 1$) since by definition, the RAO is the structural response per unit wave amplitude.

The pontoon has mass $m_p = 5000\,\text{t}$ (where 1 t = 1000 kg) with width $B_p = 30\,\text{m}$, length $C_p = 50\,\text{m}$, and depth $D_p = 8\,\text{m}$, and supports a building (centered on the pontoon) of mass $m_b = 2700\,\text{t}$ with dimensions $B_b = 15\,\text{m}$, $C_b = 30\,\text{m}$, and $D_b = 30\,\text{m}$. We will assume that the mass of the pontoon and building are uniformly distributed. The total mass is therefore $m = m_p + m_s = 7700\,\text{t}$ and the draft $D = m/(\rho B_p C_p) = 5\,\text{m}$ can be easily determined. Per standard convention, the z-axis travels through the center of mass and the x-axis is oriented parallel to the pontoon's longer dimension C_p as shown in Fig. 19.3. All structural properties and their associated values pertaining to the pontoon, building, and complete structure are summarized in Tables 19.1, 19.2, and 19.3, respectively.

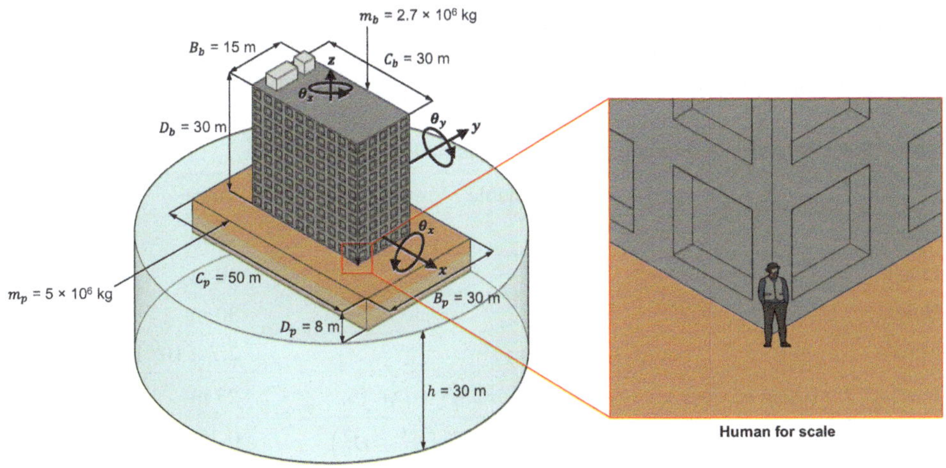

Fig. 19.3 Geometric properties of the floating structure to be analyzed under regular waves

Table 19.1 Summary of pontoon properties

Property	Variable	Value
Width	B_p	30 m
Length	C_p	50 m
Depth	D_p	8 m
Mass	m_p	5×10^6 kg
Distance from keel to CoG	$\overline{KG}_p = \frac{D_p}{2}$	4 m
Mass moment of inertia about x	$m_{p\theta_x} = \frac{m_p}{12}\left(B_p^2 + D_p^2\right)$	0.4×10^9 kg m^2
Mass moment of inertia about y	$m_{p\theta_y} = \frac{m_p}{12}\left(C_p^2 + D_p^2\right)$	1.07×10^9 kg m^2
Mass moment of inertia about z	$m_{p\theta_z} = \frac{m_p}{12}\left(C_p^2 + B_p^2\right)$	1.42×10^9 kg m^2
Waterplane second moment of area about x	$I_x = \frac{C_p B_p^3}{12}$	112,500 m^4
Waterplane second moment of area about y	$I_y = \frac{B_p C_p^3}{12}$	312,500 m^4

Table 19.2 Summary of building properties supported by pontoon

Property	Variable	Value
Width	B_b	15 m
Length	C_b	30 m
Height	D_b	30 m
Mass	m_b	2.7×10^6 kg
Distance from keel to CoG	$\overline{KG}_b = \frac{D_b}{2} + D_p$	23 m
Mass moment of inertia about x	$m_{b\theta_x} = \frac{m_b}{12}\left(B_b^2 + D_b^2\right)$	0.25×10^9 kg m^2
Mass moment of inertia about y	$m_{b\theta_y} = \frac{m_b}{12}\left(C_b^2 + D_b^2\right)$	0.41×10^9 kg m^2
Mass moment of inertia about z	$m_{b\theta_z} = \frac{m_b}{12}\left(C_b^2 + B_b^2\right)$	0.25×10^9 kg m^2

19.3 Dynamic Response in Surge

19.3.1 Froude-Krylov Force

Surge motions are induced by the component of wave forcing along the x-axis with its dynamic behavior being mass dominated. From Fig. 18.8, we also see that surge motions are at a maximum when $t_0 = -T/4$, where T is the wave period. The Froude-Krylov force per unit wave amplitude ($A = 1$) for surge $F_{FKx}(t_0)$ is obtained by integrating the

Table 19.3 Summary of complete floating structure (pontoon + building) properties

Property	Variable	Value
Mass	$m = m_p + m_b$	7.7×10^6 kg
Draft	$D = \frac{m}{\rho B_p C_p}$	5 m
Distance from keel to CoG	$\overline{KG} = \frac{m_p \overline{KG}_p + m_b \overline{KG}_b}{m_p + m_b}$	10.7 m
Distance from CoB to CoG	$\overline{BG} = \overline{KG} - \frac{D}{2}$	8.2 m
Metacentric height in heel	$\overline{GM}_x = \frac{I_x}{V} - \overline{BG}$	6.2 m
Metacentric height in trim	$\overline{GM}_y = \frac{I_y}{V} - \overline{BG}$	33.4 m
Mass moment of inertia about x	$m_{\theta_x} = m_{p\theta_x} + m_p(\overline{KG} - \overline{KG}_p)^2 + m_{b\theta_x} + m_b(\overline{KG} - \overline{KG}_b)^2$	1.29×10^9 kg m^2
Mass moment of inertia about y	$m_{\theta_y} = m_{p\theta_y} + m_p(\overline{KG} - \overline{KG}_p)^2 + m_{b\theta_y} + m_b(\overline{KG} - \overline{KG}_b)^2$	2.11×10^9 kg m^2
Mass moment of inertia about z	$m_{\theta_z} = m_{p\theta_z} + m_{b\theta_z}$	1.67×10^9 kg m^2

dynamic pressure $p(t_0)$ up the sidewalls from $z = -D$ to $z = 0$ along the transverse side walls (y_1 and y_2) of the pontoon as shown in Fig. 19.4. This force is mathematically expressed as:

$$F_{FKx}(t_0) = \int_{-D}^{0} \int_{-B_p/2}^{B_p/2} [p_0(y_1) - p_0(y_2)] dydz \tag{19.12}$$

where p_0 is from Eq. (18.37). Evaluating the integrals then gives:

$$F_{FKx}(t_0) = \rho g \zeta_{FKxy} \gamma_{FKx} \tag{19.13}$$

where:

$$\zeta_{FKxy} = \int_{-D}^{0} \frac{\cosh[k(h + z)]}{\cosh(kh)} dz \tag{19.14}$$

While Eq. (19.14) is technically applicable to all water depths, care should be taken when h is large since the $\cosh(kh)$ term will rapidly tend to infinity. Hence, in deep water ($kh \gg 1$), an alternative form is adopted that is independent of h:

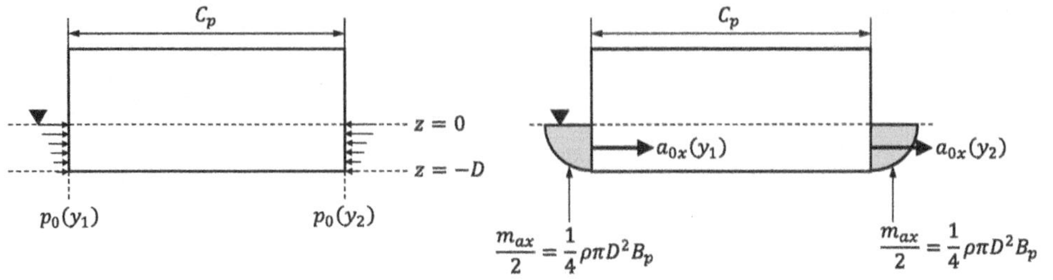

Fig. 19.4 Froude-Krylov (left) and added mass (right) components for surge

$$\zeta_{FKxy} = \int_{-D}^{0} \exp(kz)dz, \quad kh \gg 1 \tag{19.15}$$

After resolving the integrals in (19.14) and (19.15), the ζ_{FKxy} term is expressed as:

$$\zeta_{FKxy} = \begin{cases} \frac{\operatorname{sech}(kh)\{\sinh[k(D-h)]+\sinh(kh)\}}{k}, & \text{All depths} \\ \frac{1-\exp(-kD)}{k}, & \text{Deep water } (kh \gg 1) \end{cases} \tag{19.16}$$

and:

$$\gamma_{FKx} = \frac{4s_B s_C}{k \sin\theta} \tag{19.17}$$

with s_B and s_C given in Eq. (19.10).

19.3.2 Added Mass Force

The added mass force per unit wave amplitude ($A = 1$) for surge $F_{AMx}(t_0)$ is the horizontal fluid acceleration $a_x(t_0)$ acting on the transverse sidewalls at y_1 and y_2 multiplied by the added mass on each side (see Fig. 19.4). For simplicity, the acceleration is taken to act at $z = -D/2$ below the waterplane to represent the approximate centroid of the added mass in surge. This yields:

$$F_{AMx}(t_0) = \frac{\rho\pi D^2}{4} \int_{-B_p/2}^{B_p/2} \left[a_{0x}(y_1) + a_{0x}(y_2)\right] \cos\theta dy \tag{19.18}$$

where a_{0x} is per Eq. (18.38). Evaluating the integral gives:

$$F_{AMx}(t_0) = \frac{\rho\pi D^2}{4}\omega^2 \zeta_{AMxy}\gamma_{AMx} \cos\theta \tag{19.19}$$

with:

$$\zeta_{AMxy} = \begin{cases} \frac{\cosh\left[k\left(h-\frac{D}{2}\right)\right]}{\sinh(kh)}, & \text{All depths} \\ \exp\left(-\frac{kD}{2}\right), & \text{Deep water } (kh \gg 1) \end{cases} \tag{19.20}$$

and:

$$\gamma_{AMx} = \frac{4 s_{BC} c_C}{k \sin\theta} \tag{19.21}$$

with and given in Equation (19.10).

19.3.3 Response Amplitude Operator

Knowledge of $F_{FKx}(t_0)$ and $F_{AMx}(t_0)$ allows us to calculate the RAO for surge:

$$\text{RAO}_x = \frac{x_0}{A} = \frac{|F_{FKx}(t_0) + F_{AMx}(t_0)|}{M_x \omega^2} \tag{19.22}$$

where x_0 denotes the harmonic surge amplitude and M_x can be obtained using Eq. (18.48). Figure 19.5 illustrates the surge RAO as a function of the wave period from 5 to 20 s for the floating structure described in Sect. 19.2 in response to waves traveling along the longitudinal direction ($\theta \approx 0°$). Also shown is its motion as a function of time along the x-axis for regular waves with period $T = 10$ s and an amplitude of $A = 1$ m. The amplitude of this motion corresponds to the RAO at $T = 10$ s multiplied by the wave amplitude A.

19.4 Dynamic Response in Sway

19.4.1 Froude-Krylov Force

Similar to that of surge, the Froude-Krylov force amplitude in sway $F_{FKy}(t_0)$ can be determined via the integration of p_0 corresponding to $t_0 = -T/4$ along the longitudinal side walls (x_1 and x_2) of the pontoon as shown in Fig. 19.6:

$$F_{FKy}(t_0) = \int_{-D}^{0} \int_{-C_p/2}^{C_p/2} \left[p_0(x_1) - p_0(x_2)\right] dx dz \tag{19.23}$$

yielding:

$$F_{FKy}(t_0) = \rho g \zeta_{FKxy} \gamma_{FKy} \tag{19.24}$$

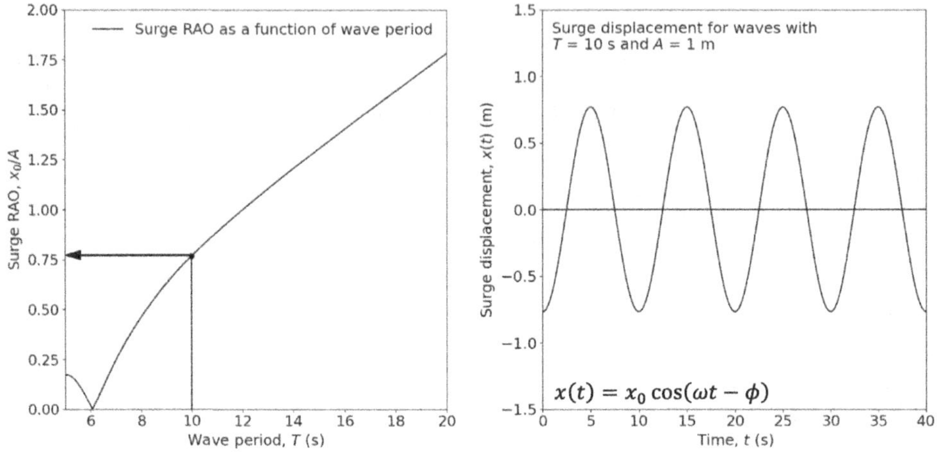

Fig. 19.5 Left: Surge RAO as a function of wave period. Right: Surge motion under waves with $T = 10\,\text{s}$ and $A = 1\,\text{m}$ approaching from the longitudinal direction ($\theta \approx 0°$)

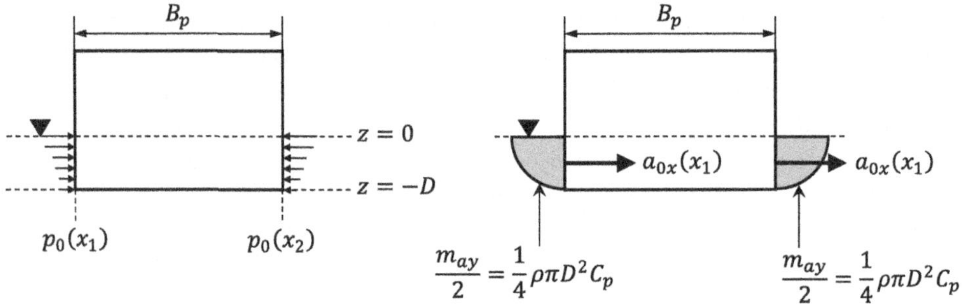

Fig. 19.6 Froude-Krylov (left) and added mass (right) components for sway

where ζ_{FKxy} is from Eq. (19.16) and:

$$\gamma_{FKy} = \frac{4 s_B s_C}{k \cos\theta} \tag{19.25}$$

with s_B and s_C given in Eq. (19.10).

19.4.2 Added Mass Force

The added mass force amplitude $F_{AMy}(t_0)$ arises from the sway added mass m_{ay} experiencing an acceleration a_{0x} oriented towards the y direction (Fig. 19.6). This acceleration must be integrated along the longitudinal side walls x_1 and x_2 at $z = -D/2$ below the waterplane to represent the approximate centroid of the added mass in sway, resulting in:

$$F_{AMy}(t_0) = \frac{\rho\pi D^2}{4} \int_{-C_p/2}^{C_p/2} [a_{0x}(x_1) + a_{0x}(x_2)] \sin\theta dx \tag{19.26}$$

Evaluating the integral gives:

$$F_{AMy}(t_0) = \frac{\rho\pi D^2}{4} \omega^2 \zeta_{AMxy} \gamma_{AMy} \sin\theta \tag{19.27}$$

where ζ_{AMxy} is per Eq. (19.20) and:

$$\gamma_{AMy} = \frac{4 c_B s_C}{k \cos\theta} \tag{19.28}$$

with c_B and s_C given in Eq. (19.10).

19.4.3 Response Amplitude Operator

Knowledge of $F_{FKy}(t_0)$ and $F_{AMy}(t_0)$ allows us to calculate the RAO for sway:

$$\text{RAO}_y = \frac{y_0}{A} = \frac{|F_{FKy}(t_0) + F_{AMy}(t_0)|}{M_y\omega^2} \tag{19.29}$$

where M_y is from Eq. (18.48). Figure 19.7 illustrates the sway RAO as a function of the wave period from 5 to 20 s for the floating structure described in Sect. 19.2 in response to waves traveling along the transverse direction ($\theta \approx 90°$). Also shown is its motion as a function of time along the y-axis for regular waves with period $T = 10$ s and an amplitude of $A = 1$ m. The amplitude of this motion corresponds to the RAO at $T = 10$ s multiplied by the wave amplitude A.

19.5 Dynamic Response in Heave

19.5.1 Froude-Krylov Force

The heave response is at a maximum under a wave crest (i.e., $t_0 = 0$). As such, the Froude-Krylov amplitude in heave $F_{FKz}(t_0)$ is determined via the integration of p_0 corresponding to $t_0 = 0$ over the keel of the pontoon located at $z = -D$:

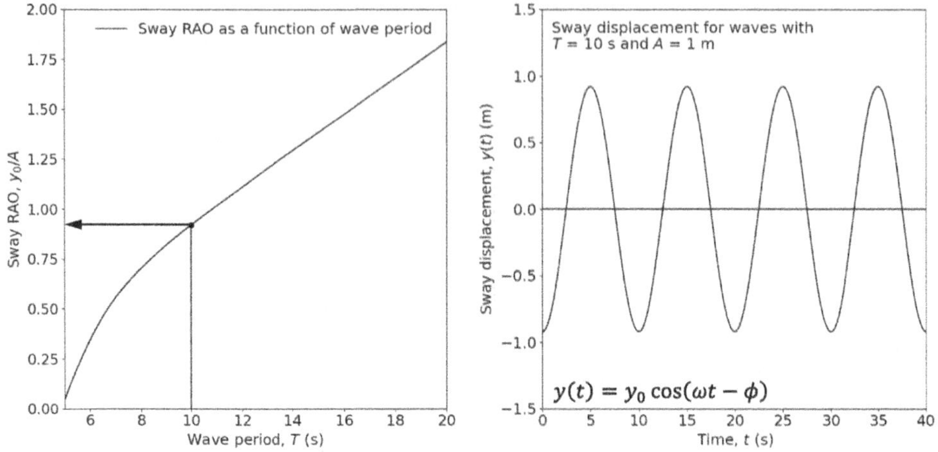

Fig. 19.7 Left: Sway RAO as a function of wave period. Right: Sway motion under waves with $T = 10$ s and $A = 1$ m approaching from the transverse direction ($\theta \approx 90°$)

$$F_{FKz}(t_0) = \int_{-B_p/2}^{B_p/2} \int_{-C_p/2}^{C_p/2} p_0(x_\theta)dxdy \tag{19.30}$$

where x_θ is given by Eq. (19.8). Resolving the integrals produce:

$$F_{FKz}(t_0) = \rho g \zeta_{FKz} \gamma_z \tag{19.31}$$

with:

$$\zeta_{FKz} = \begin{cases} \frac{\cosh[k(h-D)]}{\cosh(kh)}, & \text{All depths} \\ \exp(-kD), & \text{Deep water } (kh \gg 1) \end{cases} \tag{19.32}$$

and:

$$\gamma_z = \frac{4 s_B s_C}{k^2 \sin\theta \cos\theta} \tag{19.33}$$

where s_B and s_C are given in Eq. (19.10).

19.5.2 Added Mass Force

The added mass forcing amplitude $F_{AMz}(t_0)$ arises from the integration of a_{0z} (vertical acceleration) over the keel of the pontoon ($z = -D$) multiplied by the heave added mass per unit area:

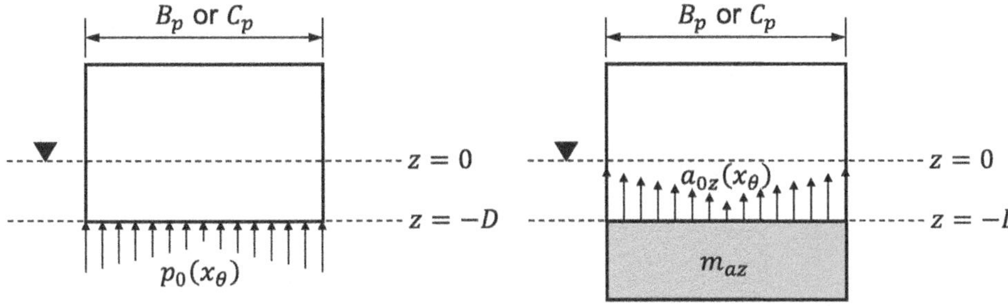

Fig. 19.8 Froude-Krylov (left) and added mass (right) components for heave

$$F_{AMz}(t_0) = \frac{m_{az}}{B_p C_p} \int_{-B_p/2}^{B_p/2} \int_{-C_p/2}^{C_p/2} a_{0z}(x_\theta) dx dy \tag{19.34}$$

where m_{az} denotes the total heave added mass given in Eq. (18.42). Here, the heave added mass is approximated as a rectangular block as shown in Fig. 19.8. Evaluating the integrals results in:

$$F_{AMz}(t_0) = -\frac{m_{az}}{B_p C_p} \omega^2 \zeta_{AMz} \gamma_z \tag{19.35}$$

with:

$$\zeta_{AMz} = \begin{cases} \frac{\sinh[k(h-D)]}{\sinh(kh)}, & \text{All depths} \\ \exp(-kD), & \text{Deep water } (kh \gg 1) \end{cases} \tag{19.36}$$

and γ_z is from Eq. (19.33).

19.5.3 Response Amplitude Operator

Knowledge of $F_{FKz}(t_0)$ and $F_{AMz}(t_0)$ allows us to calculate the heave RAO:

$$\text{RAO}_z = \frac{z_0}{A} = \frac{Q_z |F_{FKz}(t_0) + F_{AMz}(t_0)|}{K_z} \tag{19.37}$$

where z_0 denotes the harmonic heave amplitude. The response amplitude operator Q_z and waterplane stiffness in heave K_z are from Eqs. (19.3) and (18.27), respectively. Figure 19.9 illustrates the heave RAO as a function of the wave period from 5 to 20 s for the floating structure described in Sect. 19.2 in response to waves traveling along the longitudinal ($\theta \approx 0°$) or transverse ($\theta \approx 90°$) direction. Note that a heave damping ratio of $\xi_z = 0.1$ is assumed per Wang [2]. Also shown is its motion as a function of time along the z-axis for

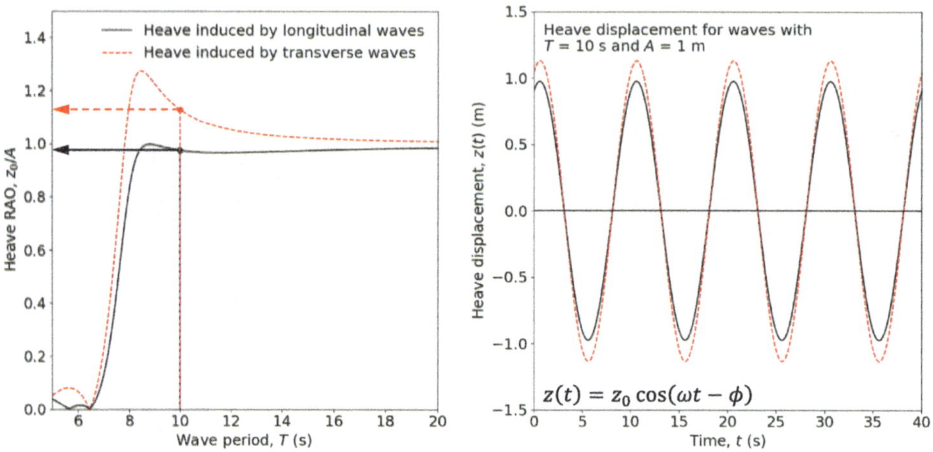

Fig. 19.9 Left: Heave RAO as a function of wave period. Right: Heave motion under waves with $T = 10$ s and $A = 1$ m approaching from the longitudinal ($\theta \approx 0°$) or transverse direction ($\theta \approx 90°$)

regular waves with period $T = 10$ s and an amplitude of $A = 1$ m. The amplitude of this motion corresponds to the RAO at $T = 10$ s (pertaining to its respective wave direction) multiplied by the wave amplitude A.

19.6 Dynamic Response in Roll

19.6.1 Froude-Krylov Force

Roll motions are induced by the y component of propagating waves with its dynamic behavior being stiffness-dominated. The Froude-Krylov moment amplitude in roll $F_{FK\theta_x}(t_0)$ is determined via the integration of p_0 corresponding to $t_0 = -T/4$ over the keel ($z = -D$) multiplied by the y-axis distance to the CoR. This is combined with the integration of p_0 along the longitudinal side walls (x_1 and x_2) multiplied with the z-axis distance to the COR. Mathematically, these operations are expressed as:

$$F_{FK\theta_x}(t_0) = \int_{-B_p/2}^{B_p/2} \int_{-C_p/2}^{C_p/2} p_0(x_\theta) y dx dy + \int_{-D}^{0} \int_{-C_p/2}^{C_p/2} \left[p_0(x_1) - p_0(x_2)\right]\left(r_{\theta_x} - z\right) dx dz \tag{19.38}$$

Here, the first term on the right-hand side depicts the moment caused by the dynamic pressure acting on the keel (where $z = -D$) while second term is the moment resulting from pressures on the left (x_1) and right (x_2) side walls of the pontoon (see Fig. 19.10).

Evaluation of the integrals yields:

$$F_{FK\theta_x}(t_0) = \rho g\left(\zeta_{\theta_x}\gamma_{FKy} + \zeta_{FKz}\gamma_{FK\theta_x}\right) \tag{19.39}$$

where:

$$\zeta_{\theta_x} = \frac{sech(kh)\{k\left(D + r_{\theta_x}\right)\sinh[k(D-h)] - \cosh[k(h-D)] + kr_{\theta_x}\sinh(kh)\} + 1}{k^2} \tag{19.40}$$

for all water depths, or:

$$\zeta_{\theta_x} = \frac{\exp(-kD)\left[\left(kr_{\theta_x} + 1\right)\exp(kD) - k\left(D + r_{\theta_x}\right) - 1\right]}{k^2} \tag{19.41}$$

for deep water conditions ($kh \gg 1$). Note that γ_{FKy} and ζ_{FKz} are given by Eqs. (19.25) and (19.32), respectively, and:

$$r_{\theta_x} = \overline{KR}_{\theta_x} - D \tag{19.42}$$

is the elevation of the CoR in roll above the waterplane. In addition:

$$\gamma_{FK\theta_x} = \frac{2s_C\left(B_p k c_B - 2s_B\csc\theta\right)}{k^3\sin\theta\cos\theta} \tag{19.43}$$

where s_B, s_C, and c_B are presented in Eq. (19.10).

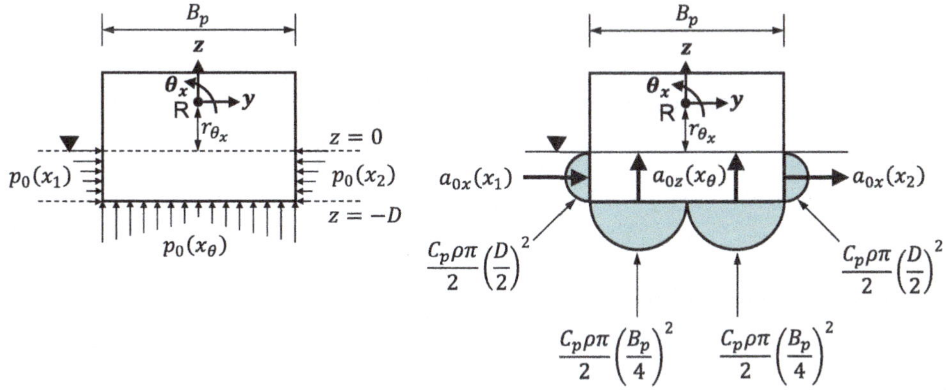

Fig. 19.10 Froude-Krylov (left) and added mass (right) components for roll

19.6.2 Added Mass Force

The added mass moment amplitude in roll stems from integration of the vertical acceleration $a_{0z}(x_\theta)$ longitudinally along the keel ($z = -D$) at $y = \pm B_p/4$ to coincide with the centroid of the semicircular prismatic volumes shown in Fig. 19.10. This acceleration is multiplied with the corresponding added mass (treated as a lumped mass) and its y-axis distance from the CoR ($B_p/4$) in order to yield the added mass moment in roll. To account for the added mass along the longitudinal side walls, the horizontal acceleration (a_{ox}) oriented towards the y direction is integrated along x_1 and x_2 at $z = -D/2$ and multiplied with the added mass and its z-axis distance to the CoR to produce the roll added mass moment induced by fluid acceleration along the side walls. Mathematically, the total added mass moment in roll is expressed as:

$$F_{AM\theta_x}(t_0) = \frac{\rho\pi B_p^2}{32}\left(\frac{B_p}{4}\right)\int_{-C_p/2}^{C_p/2}\left[a_{0z}\left(x\cos\theta + \frac{B_p}{4}\sin\theta\right) - a_{0z}\left(x\cos\theta - \frac{B_p}{4}\sin\theta\right)\right]dx$$
$$+ \frac{\rho\pi D^2}{8}\left(\frac{D}{2} + r_{\theta_x}\right)\int_{-C_p/2}^{C_p/2}[a_{0x}(x_1) + a_{0x}(x_2)]\sin\theta dx \tag{19.44}$$

Here, the first term on the right-hand side reflects the moment produced by the added mass on the keel and the second term accounts for the moment resulting from the added mass along the longitudinal side walls. Resolving the integrals yields:

$$F_{AM\theta_x}(t_0) = \frac{\rho\pi\omega^2}{8}\left[\frac{B_p^2}{4}\zeta_{AMz}\gamma_{AM\theta_x} + D^2\left(\frac{D}{2} + r_{\theta_x}\right)\zeta_{AMxy}\gamma_{AMy}\sin\theta\right] \tag{19.45}$$

where ζ_{AMxy}, ζ_{AMz}, and γ_{AMy} are given by Eqs. (19.20), (19.36), and (19.28), respectively. Finally:

$$\gamma_{AM\theta_x} = \frac{B_p s_C \sin\left(\frac{B_p k}{4}\sin\theta\right)}{k\cos\theta} \tag{19.46}$$

with s_C given in Eq. (19.10).

19.6.3 Response Amplitude Operator

Knowledge of $F_{FK\theta_x}(t_0)$ and $F_{AM\theta_x}(t_0)$ allows us to calculate the RAO in roll (with units of rad/m):

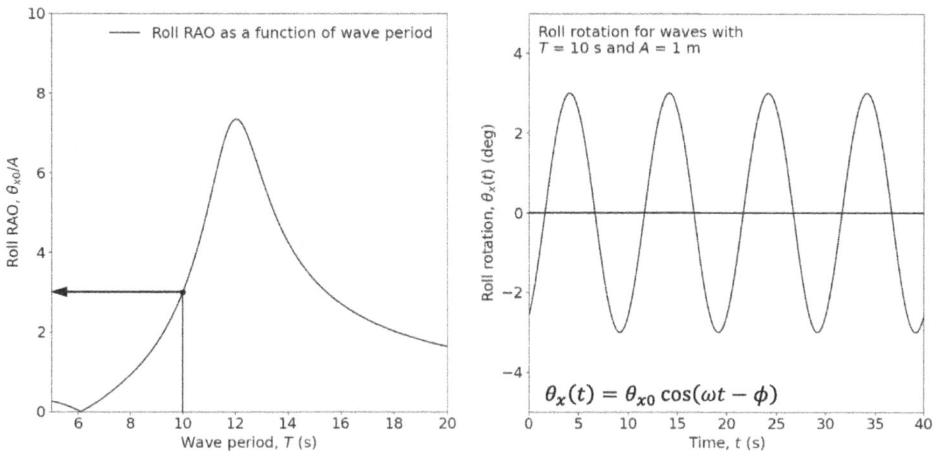

Fig. 19.11 Left: Roll RAO as a function of wave period. Right: Roll motion under waves with $T = 10\,\mathrm{s}$ and $A = 1$ m approaching from the transverse direction ($\theta \approx 90°$)

$$\mathrm{RAO}_{\theta_x} = \frac{\theta_{x0}}{A} = \frac{Q_{\theta_x}\left|F_{FK\theta_x}(t_0) + F_{AM\theta_x}(t_0)\right|}{K_{\theta_x}} \tag{19.47}$$

where θ_{x0} is the harmonic roll amplitude in radians. The response amplitude operator Q_{θ_x} and waterplane stiffness in roll K_{θ_x} are from Eqs. (19.3) and (18.27), respectively. Figure 19.11 illustrates the roll RAO in deg/m as a function of the wave period from 5 to 20 s for the floating structure described in Sect. 19.2 in response to waves traveling along the transverse direction ($\theta \approx 90°$). Note that a roll damping ratio of $\xi_{\theta_x} = 0.1$ is assumed per Wang [2]. Also shown is its rotation as a function of time about the x-axis for regular waves with period $T = 10\,\mathrm{s}$ and an amplitude of $A = 1$ m. The amplitude of this rotation corresponds to the RAO at $T = 10\,\mathrm{s}$ multiplied by the wave amplitude A.

19.7 Dynamic Response in Pitch

19.7.1 Froude-Krylov Force

Pitch motions are induced by the x component of propagating waves with its dynamic behavior being stiffness-dominated. The Froude-Krylov moment amplitude in pitch $F_{FK\theta_y}(t_0)$ is determined via the integration of p_0 corresponding to $t_0 = -T/4$ over the keel ($z = -D$) multiplied by the x-axis distance to the CoR. This is combined with the integration of p_0 along the transverse side walls (y_1 and y_2) multiplied with the z-axis distance to the COR (see Fig. 19.12). Mathematically, these operations are expressed as:

$$F_{FK\theta_y}(t_0) = \int_{-B_p/2}^{B_p/2} \int_{-C_p/2}^{C_p/2} -p_0(x_\theta)xdxdy + \int_{-D}^{0} \int_{-B_p/2}^{B_p/2} [-p_0(y_1) + p_0(y_2)](r_{\theta_y} - z)dydz \tag{19.48}$$

giving:

$$F_{FK\theta_y}(t_0) = -\rho g(\zeta_{\theta_y}\gamma_{FKx} + \zeta_{FKz}\gamma_{FK\theta_y}) \tag{19.49}$$

where:

$$\zeta_{\theta_y} = \frac{sech(kh)\{k(D + r_{\theta_y})\sinh[k(D-h)] - \cosh[k(h-D)] + kr_{\theta_y}\sinh(kh)\} + 1}{k^2} \tag{19.50}$$

for all water depths, or:

$$\zeta_{\theta_y} = \frac{\exp(-kD)[(kr_{\theta_y} + 1)\exp(kD) - k(D + r_{\theta_y}) - 1]}{k^2} \tag{19.51}$$

for deep water conditions ($kh \gg 1$). Note that γ_{FKx} and ζ_{FKz} are given by Eqs. (19.17) and (19.32), respectively, and:

$$r_{\theta_y} = \overline{KR}_{\theta_y} - D \tag{19.52}$$

is the elevation of the CoR in pitch above the waterplane. In addition:

$$\gamma_{FK\theta_y} = \frac{2s_B(C_pkc_C - 2s_C\sec\theta)}{k^3\sin\theta\cos\theta} \tag{19.53}$$

where s_B, s_C, and c_C are presented via Eq. (19.10).

19.7.2 Added Mass Force

For pontoons with $C_p/B_p \le 8/3$, such as that considered in this example, the added mass moment amplitude in pitch arise from the integration of the vertical acceleration $a_{0z}(x_\theta)$ transversely along the keel ($z = -D$) at $x = \pm C_p/4$ to coincide with the centroid of the semicircular prismatic volumes shown in Fig. 19.12. This acceleration is multiplied with the corresponding added mass (treated as a lumped mass) and its x-axis distance from the CoR ($C_p/4$) in order to yield the added mass moment in pitch. However, for more slender pontoons with $C_p/B_p > 8/3$, the vertical accelerations are integrated along the x-axis to coincide with the centroid of the longitudinal semicircular added mass at $y = 0$. To account for the added mass along the transverse side walls, the horizontal acceleration (a_{ox}) oriented towards the x direction is integrated along y_1 and y_2 at $z = -D/2$ and

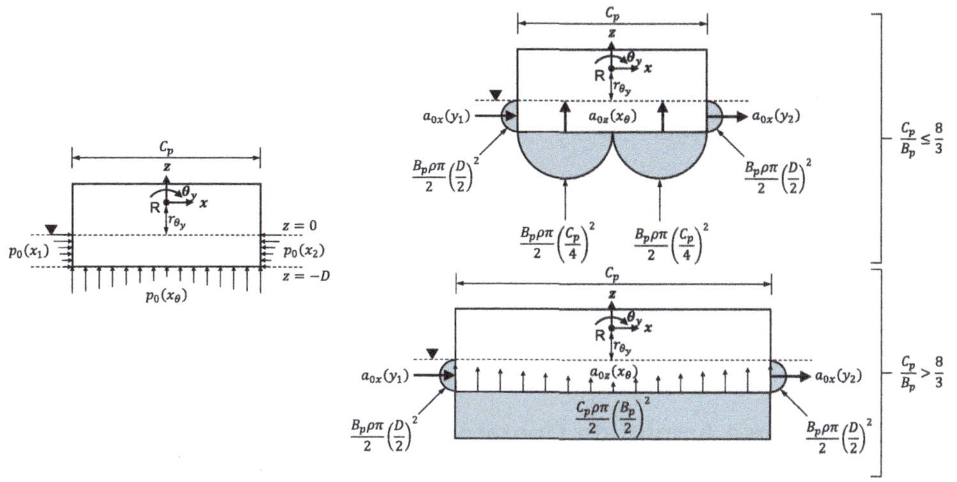

Fig. 19.12 Froude-Krylov (left) and added mass (right) components for pitch

multiplied with the added mass and its z-axis distance to the CoR to produce the pitch added mass moment induced by fluid acceleration along the side walls. Mathematically, these conditions are expressed as:

$$F_{AM\theta_y}(t_0) = \begin{cases} R_1 + R_3, & 1 \leq C_p/B_p \leq 8/3 \\ R_2 + R_3, & C_p/B_p > 8/3 \end{cases} \tag{19.54}$$

where:

$$\begin{aligned} R_1 &= \frac{\rho\pi C_p^2}{32}\left(\frac{C_p}{4}\right)\int_{-B_p/2}^{B_p/2}\left[a_{0z}\left(-\frac{C_p}{4}\cos\theta + y\sin\theta\right) - a_{0z}\left(\frac{C_p}{4}\cos\theta + y\sin\theta\right)\right]dy \\ &= -\frac{\rho\pi\omega^2}{32}C_p^2\zeta_{AMz}\gamma_{AM\theta_y} \end{aligned} \tag{19.55}$$

$$R_2 = -\frac{\rho\pi B_p^2}{8}\int_{-C_p/2}^{C_p/2} a_{0z}(x\cos\theta)x dx = -\frac{\rho\pi\omega^2}{8}B_p^2\zeta_{AMz}\gamma_{AM\theta_y} \tag{19.56}$$

denote the pitch moment induced by the added mass along the keel, and:

$$R_3 = -\frac{\rho\pi D^2}{8}\left(\frac{D}{2} + r_{\theta_y}\right)\int_{-B_p/2}^{B_p/2}\left[a_{0x}(y_1) + a_{0x}(y_2)\right]\cos\theta dy$$

$$= -\frac{\rho\pi\omega^2}{8}D^2\left(\frac{D}{2} + r_{\theta_y}\right)\zeta_{AMxy}\gamma_{AMx}\cos\theta \tag{19.57}$$

is the pitch moment produced by the added mass along the transverse side walls. Note that γ_{AMx}, ζ_{AMz}, and ζ_{AMxy} are given by Eqs. (19.21), (19.36), and (19.20), respectively, and:

$$\gamma_{AM\theta_y} = \begin{cases} \frac{C_p s_B \sin\left(\frac{C_p k}{4}\cos\theta\right)}{k \sin\theta}, & 1 \le C_p/B_p \le 8/3 \\ \frac{2s_C \sec\theta - C_p k c_C}{k^2 \cos\theta}, & C_p/B_p > 8/3 \end{cases} \tag{19.58}$$

with s_B, s_C, and c_C given in Eq. (19.10).

19.7.3 Response Amplitude Operator

Knowledge of $F_{FK\theta_y}(t_0)$ and $F_{AM\theta_y}(t_0)$ allows us to calculate the RAO in pitch (with units of rad/m):

$$\mathrm{RAO}_{\theta_y} = \frac{\theta_{y0}}{A} = \frac{Q_{\theta_y}\left|F_{FK\theta_y}(t_0) + F_{AM\theta_y}(t_0)\right|}{K_{\theta_y}} \tag{19.59}$$

where θ_{y0} is the harmonic pitch amplitude in radians. The response amplitude operator Q_{θ_y} and waterplane stiffness in pitch K_{θ_y} are from Eqs. (19.3) and (18.27), respectively. Figure 19.13 illustrates the pitch RAO in deg/m as a function of the wave period from 5 to 20 s for the floating structure described in Sect. 19.2 in response to waves traveling along the longitudinal direction ($\theta \approx 0°$). Note that a pitch damping ratio of $\xi_{\theta_y} = 0.1$ is assumed per Wang [2]. Also shown is its rotation as a function of time about the y-axis for regular waves with period $T = 10$ s and an amplitude of $A = 1$ m. The amplitude of this rotation corresponds to the RAO at $T = 10$ s multiplied by the wave amplitude A.

19.8 Dynamic Response in Yaw

19.8.1 Froude-Krylov Force

The yaw response (θ_z) can only be excited by waves approaching at an oblique (non-perpendicular) angle relative to the side walls. Similar to heave, the maximum yaw response occurs when a wave crest coincides with the z-axis such that $t_0 = 0$. The Froude-Krylov moment amplitude in yaw $F_{FK\theta_Z}$ arises from the integration of p_0 around the perimeter of the pontoon (x_1, x_2, y_1, and y_2) multiplied by the perpendicular lever arm (x or y) to the z-axis:

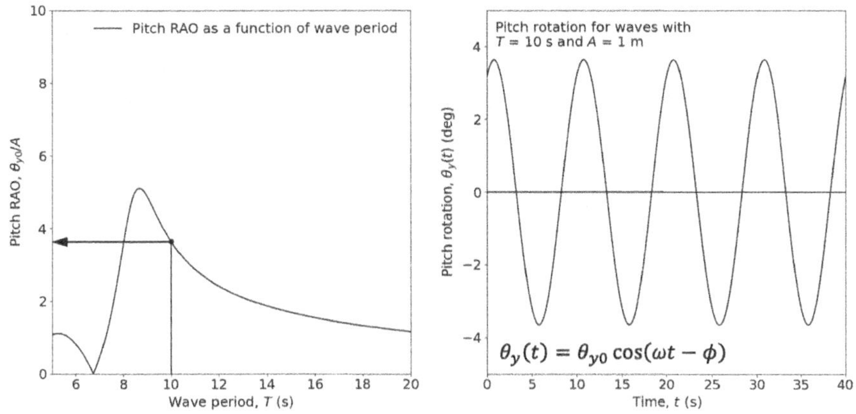

Fig. 19.13 Left: Pitch RAO as a function of wave period. Right: Pitch motion under waves with $T = 10$ s and $A = 1$ m approaching from the longitudinal direction ($\theta \approx 0°$)

$$F_{FK\theta_z}(t_0) = \int_{-D}^{0}\int_{-C_p/2}^{C_p/2}\big[p_0(x_1) - p_0(x_2)\big]x dx dz + \int_{-D}^{0}\int_{-B_p/2}^{B_p/2}\big[p_0(y_2) - p_0(y_1)\big]y dy dz \tag{19.60}$$

Here, the first and second terms on the right-hand side represent the dynamic wave pressure acting on the longitudinal and transverse side walls, respectively, and are visualized in Fig. 19.14. Resolving the integrals yields:

$$F_{FK\theta_z}(t_0) = \rho g \zeta_{FKxy}\left[\frac{2s_B\left(2s_C \sec\theta - C_p k c_C\right)}{k^2\cos\theta} + \frac{2s_C\left(B_p k c_B - 2s_B \csc\theta\right)}{k^2 \sin\theta}\right] \tag{19.61}$$

with ζ_{FKxy} given by Eq. (19.16) and s_B, s_C, c_B, and c_C are presented in Eq. (19.10).

19.8.2 Added Mass Force

Likewise, the yaw added mass moment amplitude results from the prismatic added mass distribution shown in Fig. 19.15 accelerated along the perimeter of the pontoon (normal to the side walls) multiplied by the perpendicular lever arm (x or y) to the z-axis:

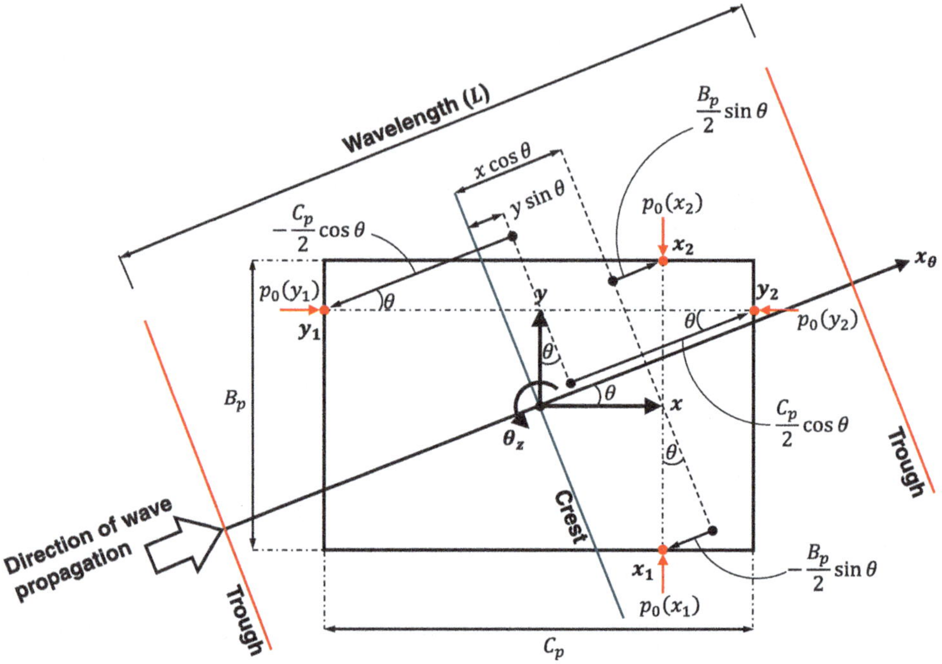

Fig. 19.14 Plan view of pontoon showing the direction of positive Froude-Krylov pressures on each side wall

$$F_{AM\theta_z}(t_0) = \frac{\rho\pi D^2}{4} \int_{-C_p/2}^{C_p/2} [a_{0x}(x_1) + a_{0x}(x_2)]x \sin\theta dx - \frac{\rho\pi D^2}{4} \int_{-B_p/2}^{B_p/2} \left[a_{0x}(y_1) + a_{0x}(y_2)\right] y \cos\theta dy \tag{19.62}$$

Here, the first and second terms on the right-hand side represent the yaw moment induced by the added mass on the longitudinal and transverse side walls, respectively. Evaluating the integrals at $z = -D/2$ yields:

$$F_{AM\theta_z}(t_0) = \frac{\rho\pi D^2}{4}\omega^2 \zeta_{AMxy}\left[\frac{2c_B\left(2s_C \sec\theta - C_p k c_C\right)}{k^2 \cot\theta} - \frac{2c_C\left(2s_B \csc\theta - B_p k c_B\right)}{k^2 \tan\theta}\right] \tag{19.63}$$

with ζ_{AMxy} given in Eq. (19.20) and s_B, s_C, c_B, and c_C are presented in Eq. (19.10).

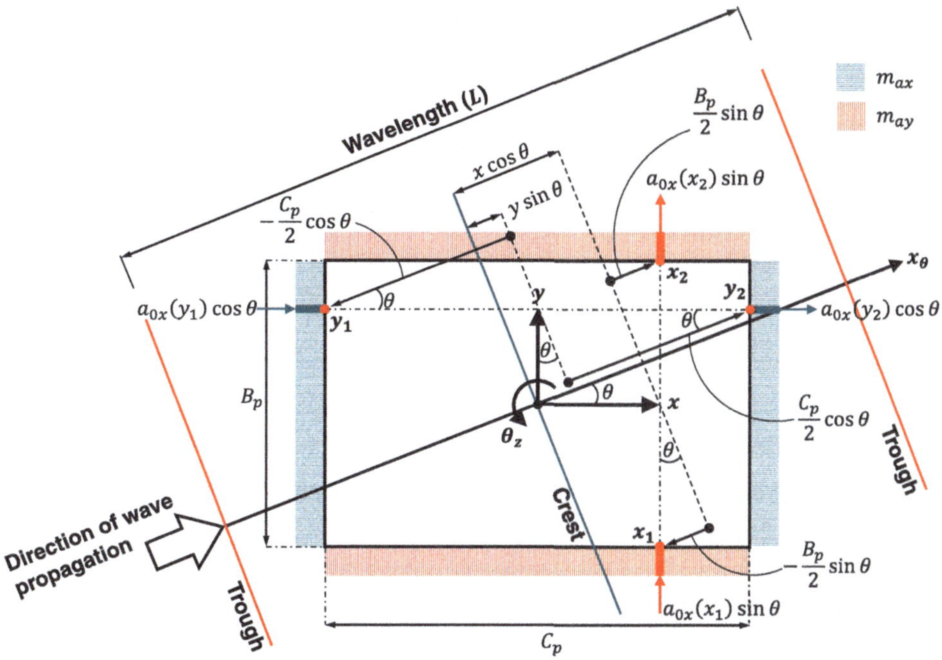

Fig. 19.15 Plan view of pontoon showing the direction of positive added mass accelerations on each side wall. Shaded areas represent added masses in surge (m_{ax}) and sway (m_{ay})

19.8.3 Response Amplitude Operator

Knowledge of $F_{FK\theta_z}(t_0)$ and $F_{AM\theta_z}(t_0)$ allows us to calculate the RAO in yaw (with units of rad/m):

$$\mathrm{RAO}_{\theta_z} = \frac{\theta_{z0}}{A} = \frac{\left|F_{FK\theta_z}(t_0) + F_{AM\theta_z}(t_0)\right|}{M_{\theta_z}\omega^2} \qquad (19.64)$$

where θ_{z0} denotes the harmonic yaw amplitude in radians and M_{θ_z} can be obtained using Eq. (18.48).

Figure 19.16 illustrates the yaw RAO in deg/m for wave directions $\theta = 15°$ and $30°$ as a function of the wave period from 5 to 20 s for the floating structure described in Sect. 19.2. The harmonic yaw rotations about the z-axis for $T = 10$ s and $A = 1$ m are also shown. Note that if $B_p = C_p$ and $\theta = 45°$, then $F_{0z} = 0$ since the accelerations and pressures on opposite side walls will cancel out and the RAO will be zero for yaw. Likewise, yaw will be zero when $\theta = 0°$ and $90°$ for all rectangular pontoons.

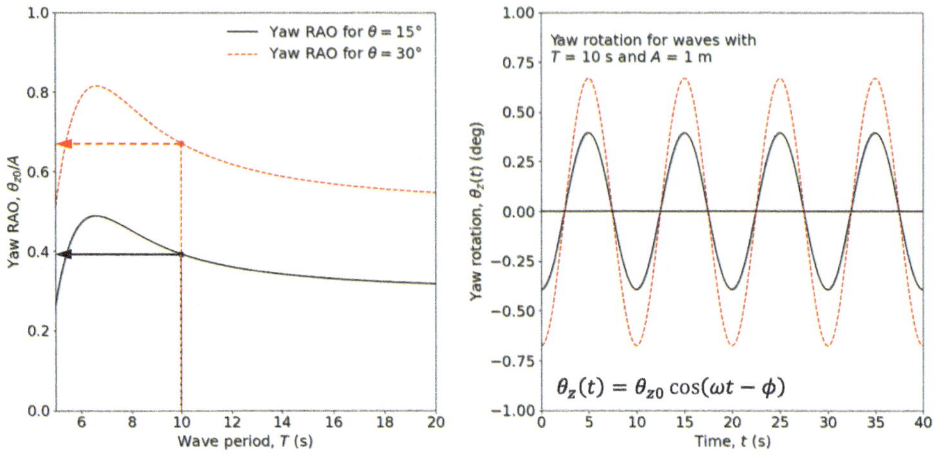

Fig. 19.16 Left: Yaw RAO as a function of wave period. Right: Yaw motion under waves with $T = 10\,\text{s}$ and $A = 1\,\text{m}$ with $\theta = 15°$ and $\theta = 30°$

19.9 Summary: Rigid-Body Dynamics Under Regular Waves

The determination of RAOs pertaining to each DoF R for a symmetrically loaded rectangular pontoon via the analytical approach presented in this chapter is summarized in Fig. 19.17. Relevant equations giving the dynamic properties used in the calculations are listed in Table 19.4. After identifying the structural and wave characteristics, dynamic properties independent of wave directionality encompassing the waterplane stiffness (K_R), total hydrodynamic mass (M_R), and the dynamic amplification factor (Q_R) may be obtained (step 1). Note that K_R and Q_R are only relevant for stiffness-dominated DoFs where $R = \{z, \theta_x, \theta_y\}$. Next, the total forcing amplitude, comprising Froude-Krylov (F_{FKR}) and added mass (F_{AMR}) components, are computed which are dependent upon the direction of wave propagation (θ) for each DoF R (step 2). After solving for the quantities listed in steps 1 and 2, the RAO (as a function of the wave frequency ω or period T) for each DoF R may be determined using the equations appropriate for mass (surge, sway, and yaw) or stiffness (heave, roll, and pitch) dominated DoFs as specified in step 3. The response as a function of time t is then:

$$R(t) = A \times \text{RAO}_R \cos(\omega t - \phi) \tag{19.65}$$

where A is the wave amplitude and ϕ is the phase lag corresponding to DoF R from Eq. (19.6).

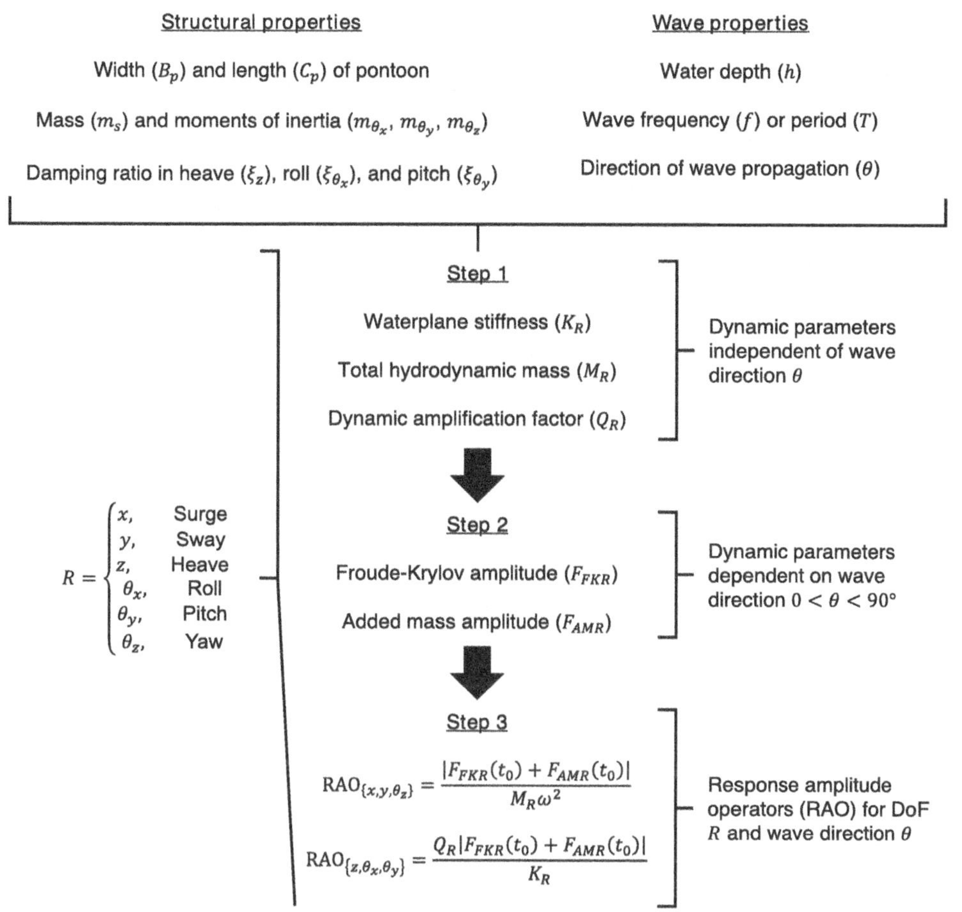

Fig. 19.17 RAO calculation flowchart for DoFs $R = \{x, y, z, \theta_x, \theta_y, \theta_z\}$

19.10 Internal Pontoon Forces Induced by Regular Waves

Just like the hydrostatic scenario, it is necessary to obtain the global bending and shear response of floating pontoons subject to wave excitation for purposes of structural design under dynamic conditions. For a platform with length $C_p > B_p$, the maximum bending and shear demands are typically associated with the heave response excited by waves propagating in the longitudinal direction (x-axis) parallel to C_p due to its longer dimension. We need therefore consider the response when $R = z$ (motion along the z-axis) in order to determine the maximum bending and shear forces induced by waves on a floating pontoon [1].

Table 19.4 Equations used to determine the RAO across all DoFs

Dynamic property	Degree of freedom $R =$					
	x (Surge)	y (Sway)	z (Heave)	θ_x (Roll)	θ_y (Pitch)	θ_z (Yaw)
K_R	–	–	Eq. (18.27)	Eq. (18.27)	Eq. (18.27)	–
M_R	Eq. (18.48)	Eq. (18.48)	Eq. (18.48)	Eq. (18.48)	Eq. (18.48)	Eq. (18.48)
Q_R	–	–	Eq. (19.3)	Eq. (19.3)	Eq. (19.3)	–
F_{FKR}	Eq. (19.13)	Eq. (19.24)	Eq. (19.31)	Eq. (19.39)	Eq. (19.49)	Eq. (19.61)
F_{AMR}	Eq. (19.19)	Eq. (19.27)	Eq. (19.35)	Eq. (19.45)	Eq. (19.54)	Eq. (19.63)
RAO_R	Eq. (19.22)	Eq. (19.29)	Eq. (19.37)	Eq. (19.47)	Eq. (19.59)	Eq. (19.64)

Bending and shear result from differences in the longitudinal distribution of net forces acting over the pontoon at each instant in time. The pontoon can be treated as a simply supported beam where the maximum positive (upwards) and negative (downwards) bending is governed by the time t_0:

$$t_0 = \begin{cases} 0, & \text{Maximum positive bending in heave} \\ T/2, & \text{Maximum negative bending in heave} \end{cases} \tag{19.66}$$

where $t_0 = 0$ and $T/2$ respectively signifies a wave crest and trough below the pontoon (T = wave period). A visual description of positive and negative bending of a floating pontoon under heave is presented by Fig. 19.18.

To quantify the distribution of shear and bending moments at t_0, we first discretize the submerged portion of the pontoon (displacing the same mass of water as the structural mass m) into an arbitrary number of N slices along the x-axis each with width δx as shown in Fig. 19.19. The width of each slice is therefore:

$$\delta x = \frac{C_p}{N} \tag{19.67}$$

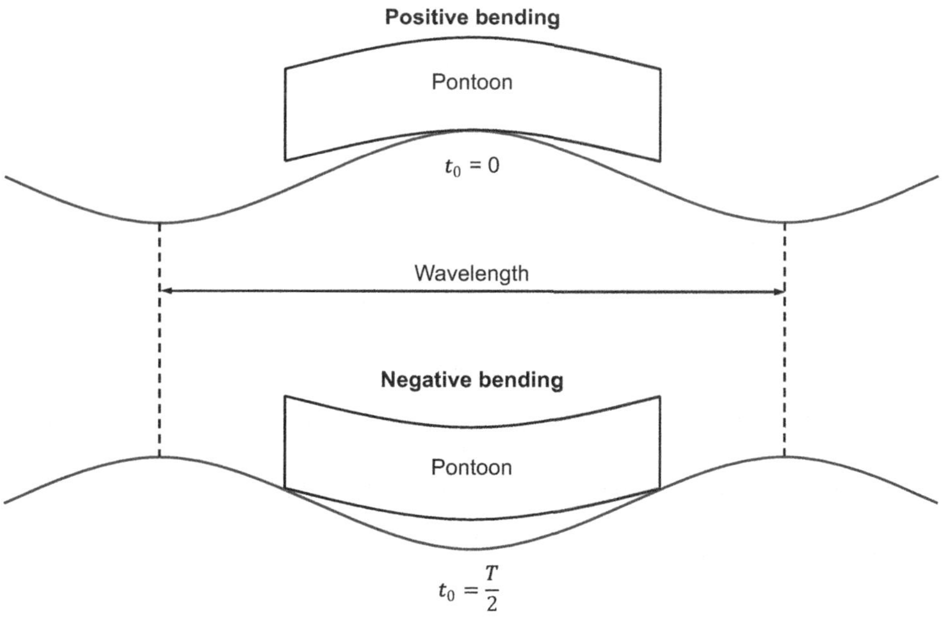

Fig. 19.18 Maximum positive and negative bending of a pontoon under heave

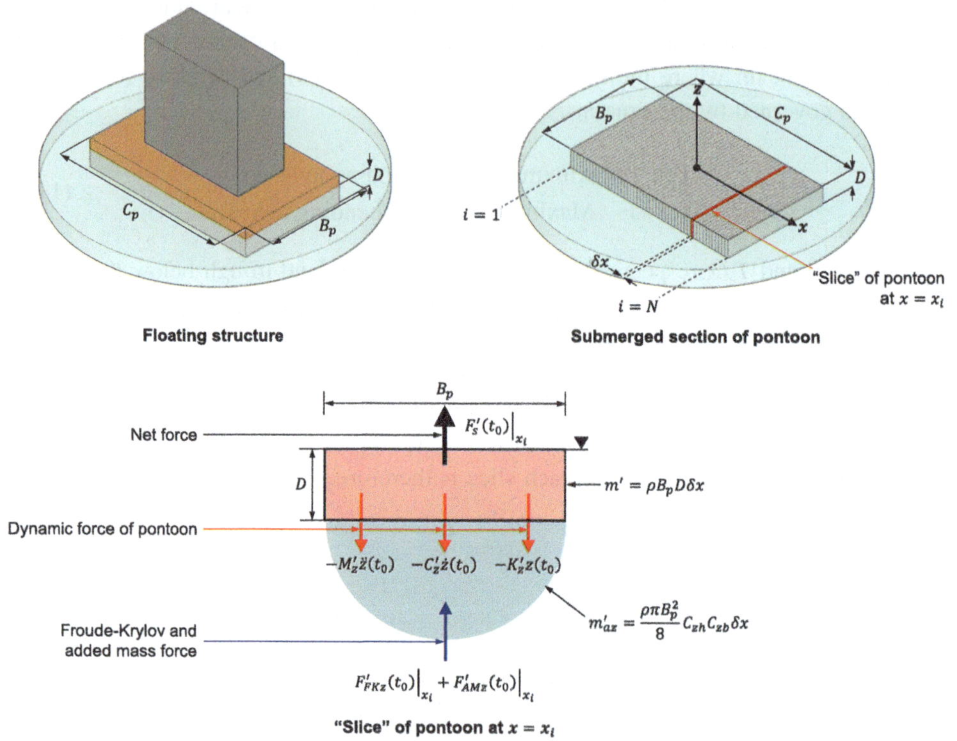

Fig. 19.19 Discretization of pontoon into slices of width δx with force components shown

The centroid of each slice i along the x-axis is denoted x_i and may be determined for a rectangular pontoon via:

$$x_i = \frac{-C_p + \delta x}{2} + (i - 1)\delta x, \quad i \in [1, 2, 3, \ldots, N] \tag{19.68}$$

Returning to our mass-spring-damper model:

$$F(t_0) = M\ddot{R}(t_0) + C\dot{R}(t_0) + KR(t_0) \tag{19.69}$$

we can now construct a free-body diagram for each slice (at time t_0) based on Eq. (19.69) while including an additional force $F'_S(t_0)\big|_{x_i}$ which reflects the net force acting on slice i that must be supported by the stiffness of the structure. Equation (19.69) can therefore be expressed as:

$$F'_{FKz}(t_0)\big|_{x_i} + F'_{AMz}(t_0)\big|_{x_i} = M'_z\ddot{z}(t_0) + C'_z\dot{z}(t_0) + K'_z z(t_0) + F'_S(t_0)\big|_{x_i} \tag{19.70}$$

where x_i is a vector along the x-axis from $x = 0$ (located at the center of floatation) to the center of slice i. $F'_{FKz}(t_0)\big|_{x_i}$ and $F'_{AMz}(t_0)\big|_{x_i}$ are the Froude-Krylov force and added mass force in heave at location x_i and time t_0. Equation (19.70) can subsequently be rearranged for the net force (which we are trying to solve for) on each slice:

$$F'_S(t_0)\big|_{x_i} = F'_{FKz}(t_0)\big|_{x_i} + F'_{AMz}(t_0)\big|_{x_i} - M'_z\ddot{z}(t_0) - C'_z\dot{z}(t_0) - K'_z z(t_0) \tag{19.71}$$

Note that the prime implies that we are only considering forces acting on a single slice i located at x_i. The mass (M'_z), damping (C'_z), and stiffness (K'_z) is:

$$M'_z = m' + m'_{az} = \rho B_p D\delta x + \frac{\rho\pi B_p^2}{8} C_{zh} C_{zb}\delta x \tag{19.72}$$

$$C'_z = 2\xi_z\sqrt{M'_z K'_z} \tag{19.73}$$

$$K'_z = \rho g B_p \delta x \tag{19.74}$$

where m' and m'_{az} denote the mass and added mass of each slice, respectively. The structure's vertical displacement, velocity, and acceleration per unit wave amplitude ($A = 1$) at t_0 is:

$$z(t_0) = \mathrm{RAO}_z \cos(\omega t_0 - \phi) \tag{19.75}$$

$$\dot{z}(t_0) = -\omega \mathrm{RAO}_z \sin(\omega t_0 - \phi) \tag{19.76}$$

$$\ddot{z}(t_0) = -\omega^2 \mathrm{RAO}_z \cos(\omega t_0 - \phi) \tag{19.77}$$

with RAO_z determined via Eq. (19.37) and ϕ is from (19.6). The Froude-Krylov and added mass force per unit wave amplitude for each slice is:

$$F'_{FKz}(t_0)\big|_{x_i} = p(t_0)|_{x_i} B_p \delta x = B_p \rho g \left\{ \frac{\cosh[k(h-D)]}{\cosh(kh)} \right\} \cos(kx_i - \omega t_0)\delta x \tag{19.78}$$

$$F'_{AMz}(t_0)\big|_{x_i} = m'_{az} a_z(t_0)|_{x_i} = -\frac{\rho\pi B_p^2}{8} C_{zh} C_{zb}\omega^2 \left\{ \frac{\sinh[k(h-D)]}{\sinh(kh)} \right\} \cos(kx_i - \omega t_0)\delta x \tag{19.79}$$

The distribution of net forces as a function of x_i can now be determined by substituting (19.72) to (19.79) into (19.71). The shear force at x_i arises from the summation of $F'_S(t_0)\big|_{x_j}$ from $j = 1$ to $j = i$:

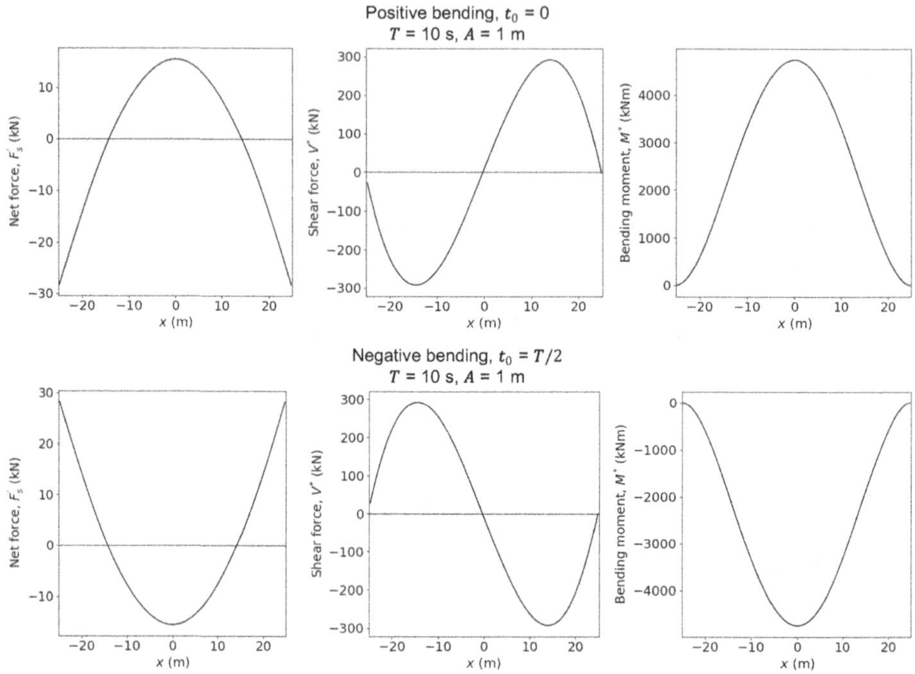

Fig. 19.20 Distribution of longitudinal net force, shear force, and bending moment for maximum positive (top) and negative (bottom) bending induced by regular waves with $T = 10\,\mathrm{s}$ and $A = 1\,\mathrm{m}$

$$V^*(t_0)\big|_{x_i} = A\sum_{j=1}^{i} F_S'(t_0)\big|_{x_j} \tag{19.80}$$

Similarly, the moment at x_i is:

$$M^*(t_0)\big|_{x_i} = A\sum_{j=1}^{i}\left[F_S'(t_0)\big|_{x_j} \times \left(x_j - x_i\right)\right] \tag{19.81}$$

Here, we have included the wave amplitude A in (19.80) and (19.81) since $F_S'(t_0)$ describes the force per unit wave amplitude. The distribution of net force, shear, and bending moment along the length (x) of the pontoon for our example structure (see Sect. 19.2) is plotted in Fig. 19.20 for both positive ($t_0 = 0$) and negative ($t_0 = T/2$) bending. Regular waves with period $T = 10\,\mathrm{s}$ propagating with amplitude $A = 1\,\mathrm{m}$ are assumed. The total shear and moment demands are produced from the summation of the global dynamic and static (e.g., Fig. 17.22) components.

References

1. Barltrop, N.D.P., 1998. Floating structures: a guide for design and analysis. The Centre for Marine and Petroleum Technology.
2. Wang, S., 2022. Analytical solutions for the dynamic analysis of a modular floating structure for urban expansion. Ocean Engineering 266.

Dynamics of Floating Structures Under Irregular Waves

20

In Chap. 19, we have examined the response $R(t)$ of a floating structure about each of its six degrees of freedom when excited by regular waves with a single frequency f (or period $T = 1/f$) and amplitude A approaching from any arbitrary direction θ. Clearly, this behavior is not representative of real-world conditions where incoming waves exhibit a wide range of frequencies and amplitudes. To facilitate the design of floating structures that guarantee the comfort of its occupants, the maximum response and accelerations excited by irregular waves must be determined [21].

20.1 Dynamic Response Time History

From our discussions into irregular waves presented in Chap. 6, we know that realistic sea states can be described using a wave spectrum where the spectral density $S(f)$ quantifying the wave energy is plotted against the wave frequency f. Also recall that the wave spectrum can be split into an arbitrarily large number of N individual wave components (ideally $N > 200$), each with a different frequency f_i. The sea surface elevation η for irregular waves arises from the linear superposition (summation) of each frequency component f_i within the wave spectrum:

$$\eta(x, t) = \sum_{i=1}^{N} A_i \cos\left(k_i x - \omega_i t + \widetilde{\phi}_i\right) \tag{20.1}$$

where k_i is the wavenumber associated with frequency f_i determined using the dispersion relation:

S. Wang, *Applied Ocean Mechanics for Structural Engineering and Climate Adaptation*, Synthesis Lectures on Ocean Systems Engineering,
https://doi.org/10.1007/978-3-032-18199-2_20

$$\omega_i = \sqrt{gk_i \tanh(k_i h)} \tag{20.2}$$

and $\tilde{\phi}_i$ is the random phase which follows a uniform distribution between $-\pi$ and π. A_i is the wave amplitude as determined from the spectral density $S(f_i)$:

$$A_i = \sqrt{2S(f_i)\Delta f} \tag{20.3}$$

and Δf is the width of each frequency component f_i. Figure 20.1 demonstrates how Δf is determined for a wave spectrum discretized into N frequency bands distributed between a minimum (f_{min}) and maximum (f_{max}) frequency threshold where $S(f) \approx 0$. Due to the property of linearity as seen in Eq. (20.1), we can also treat the structural response R_i induced by each frequency f_i separately via the RAO:

$$R_i(t) = A_i \mathrm{RAO}_R(f_i) \cos\left(\omega_i t - \phi(f_i) + \tilde{\phi}_i\right) \tag{20.4}$$

Note that Eq. (20.4) gives the response (translation or rotation depending on the DoF under consideration) for only one wave frequency f_i. The frequency-dependent phase lag $\phi(f_i)$ for stiffness and mass dominated DoFs pertaining to this frequency is:

$$\phi(f_i) = \begin{cases} \arctan\left(\dfrac{2\xi_R \frac{f_i}{f_{nR}}}{1 - \frac{f_i^2}{f_{nR}^2}}\right), & R = \left\{z, \theta_x, \theta_y\right\} \\ \pi, & R = \{x, y, \theta_z\} \end{cases} \tag{20.5}$$

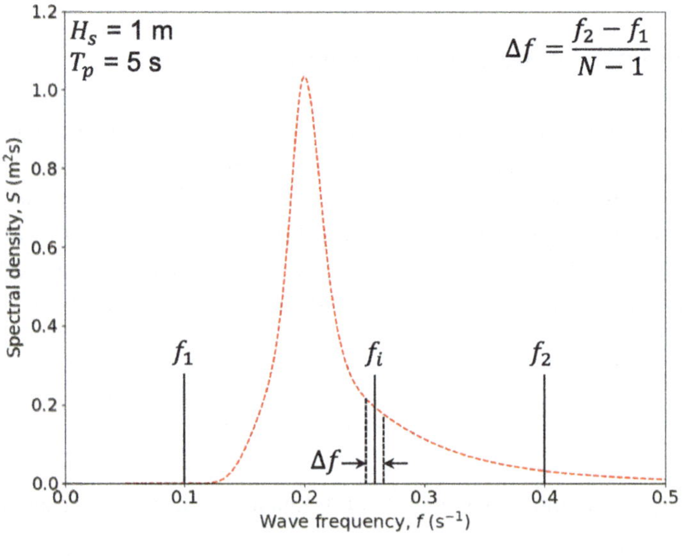

Fig. 20.1 Determination of Δf from a wave spectrum with N frequency bands

where ξ_R is the damping ratio for DoF R and its natural frequency f_{nR} is given by Eq. (19.4). When accounting for the entire distribution of frequencies within a wave spectrum, the response $R(t)$ is simply the summation of each individual component R_i:

$$R(t) = \sum_{i=1}^{N} A_i \mathrm{RAO}_R(f_i) \cos\left(\omega_i t - \phi(f_i) + \tilde{\phi}_i\right) \tag{20.6}$$

As an example, Fig. 20.2 shows the free surface elevation (η) from Eq. (20.1), and surge (x), heave (z), and pitch (θ_y) response from Eq. (20.6) of the floating structure depicted in Fig. 19.3 over a period of 3 min (180 s). Waves are taken to propagate along the longitudinal direction (x-axis), hence only surge, heave, and pitch DoFs are activated. Likewise, Fig. 20.3 shows the free surface elevation (η), sway (x), heave (z), and roll (θ_x) response excited by waves propagating along the transverse direction (y-axis). The structure is being excited by irregular waves with $H_s = 1$ m and $T_p = 5$ s in accordance with the JONSWAP wave spectrum (see Fig. 20.1) with its spectral density given by Goda [6]:

$$S(f) = \beta_J H_s^2 \left(\frac{f_p^4}{f^5}\right) \exp\left[-1.25\left(\frac{f_p}{f}\right)^4\right] \gamma^{\exp\left[\frac{-\left(\frac{f}{f_p}-1\right)^2}{2\sigma^2}\right]} \tag{20.7}$$

where the JONSWAP parameters are:

$$\beta_J = \frac{0.0624\left[1.094 - 0.01915\ln(\gamma)\right]}{0.230 + 0.0336\gamma - 0.185(1.9+\gamma)^{-1}}, \quad \gamma = 3.3 \tag{20.8}$$

$$\sigma = \begin{cases} 0.07, & f \le f_p \\ 0.09, & f > f_p \end{cases} \tag{20.9}$$

and $f_p = 1/T_p$ is the frequency of the spectral peak.

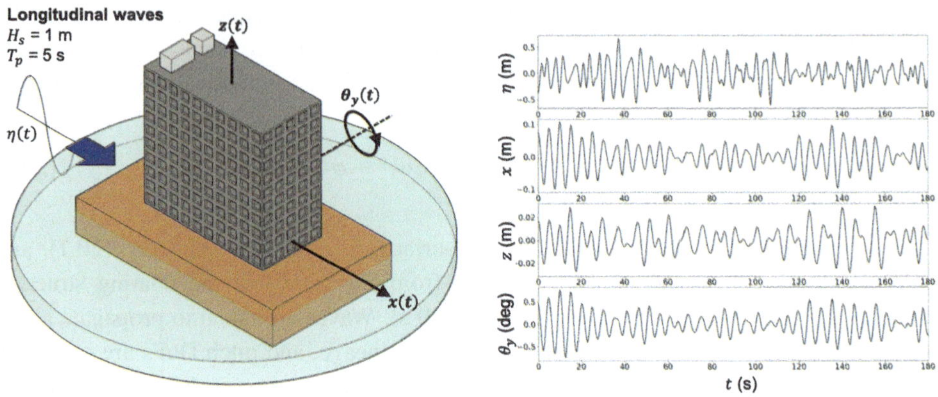

Fig. 20.2 Dynamic motions induced on a floating structure by irregular waves with $H_s = 1$ m and $T_p = 5$ s propagating along the longitudinal ($\theta \approx 0°$) direction. Structural properties are summarized in Tables 19.1, 19.2 and 19.3

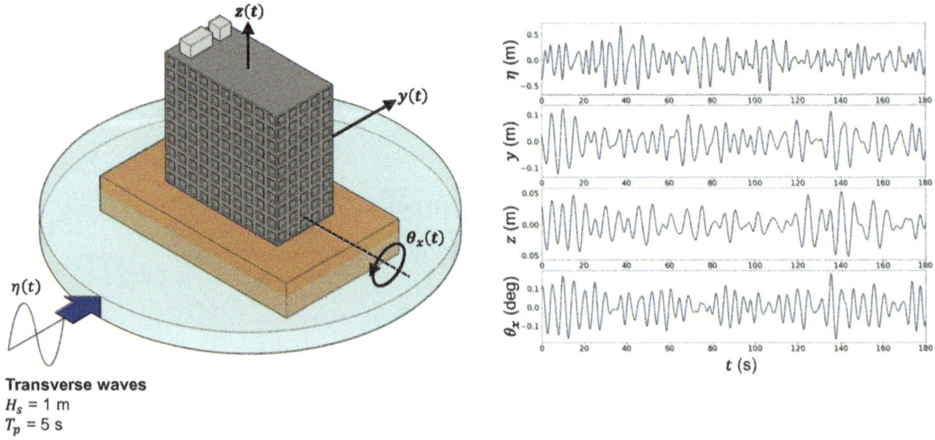

Fig. 20.3 Dynamic motions induced on a floating structure by irregular waves with $H_s = 1$ m and $T_p = 5$ s propagating along the transverse ($\theta \approx 90°$) direction. Structural properties are summarized in Tables 19.1, 19.2 and 19.3

20.2 Most Probable Maximum Response

During storm conditions, it is of interest to predict the maximum response of a floating structure excited by irregular waves described by a wave spectrum. However, due to the inherently random nature of realistic sea states, we need to first calculate the variance of the response [5]. For irregular waves, the variance is simply given by the area underneath the wave spectrum $S(f)$, which is also known as its zeroth-moment m_0. The variance of the response of a floating structure pertaining to DoF R placed within this sea state must also take into account its RAO, yielding:

$$\sigma_R^2 = \int_0^\infty S(f)\left[\mathrm{RAO}_R(f)\right]^2 df = \sum_{i=1}^{N} S(f_i)\left[\mathrm{RAO}_R(f_i)\right]^2 \Delta f \tag{20.10}$$

where, from Eq. (20.3):

$$S(f_i)\Delta f = \frac{A_i^2}{2} \tag{20.11}$$

and:

$$\mathrm{RAO}_R(f_i) = \frac{R_i}{A_i} \tag{20.12}$$

is ratio between the response amplitude of DoF R and the wave amplitude, with both pertaining to frequency component f_i within the spectrum. We can now use the variance per Eq. (20.10) to determine the most probable largest value of the response R_{max} over a set period of time t (e.g., the duration of a storm) when assuming a Rayleigh distribution. This quantity can be interpreted as the average of the largest response values taken from a large number of sea states that all have mean a period of T_m and span a duration of t. The value of R_{max} is determined via:

$$R_{max} = \sqrt{2\sigma_R^2 \ln\left(\frac{t}{T_m}\right)} \tag{20.13}$$

where T_m is the mean period of the wave spectrum:

$$T_m = \sqrt{\frac{m_0}{m_2}} \tag{20.14}$$

with m_0 and m_2 denoting the zeroth and second-order moments, respectively:

$$m_0 = \int_0^\infty S(f)df = \sum_{i=1}^{N} S(f_i)\Delta f \tag{20.15}$$

$$m_2 = \int_0^\infty S(f)f^2 df = \sum_{i=1}^{N} S(f_i)f_i^2 \Delta f \tag{20.16}$$

Note that m_0 (i.e., variance of the wave spectrum) can also be estimated using the significant wave height via $m_0 = (H_s/4)^2$. Figure 20.4 shows the RAO plotted alongside the irregular wave spectrum (from Fig. 20.1) for longitudinal and transverse waves. Determination of the variance via Eq. (20.10) for each DoF R involves multiplying the

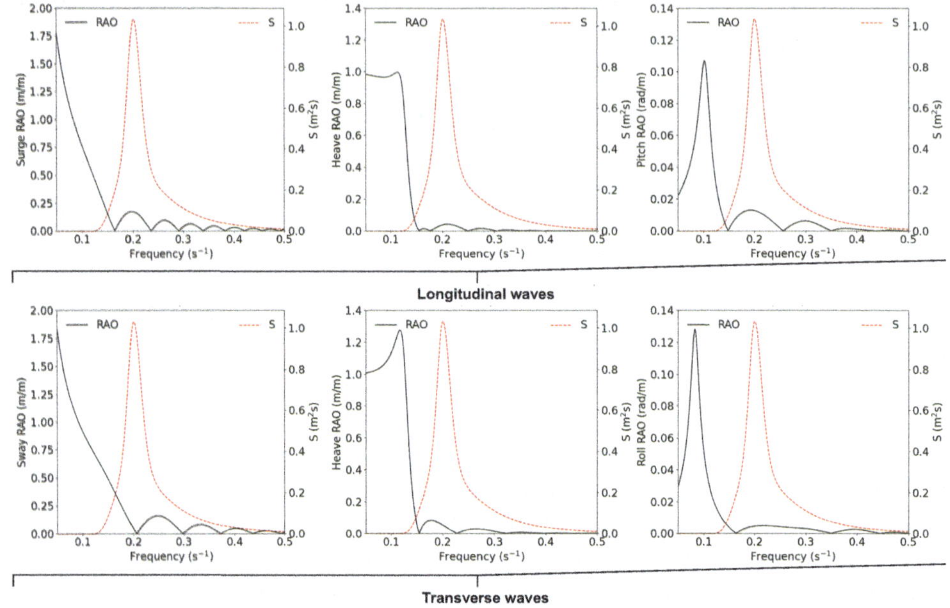

Fig. 20.4 RAO overlaid with the JONSWAP wave spectrum for DoFs excited by longitudinal (top) and transverse (bottom) waves

area of each frequency band on the spectrum, given by $S(f_i)\Delta f$, by the square of the RAO associated with the same frequency f_i, given by $\left[\text{RAO}_R(f_i)\right]^2$ then summing them all together over the entire frequency range. More overlap between the RAO and spectral density curves thus increases the overall variance σ_R^2 which translates to a larger maximum response R_{max}.

Table 20.1 shows how the most probable maximum response for surge (x_{max}), heave (z_{max}), and pitch ($\theta_{y,max}$) of the floating structure in Fig. 20.3 changes as the time period is increased from $t = 410$ s (100 times the mean period T_m) to 41,000 s (10,000 times the mean period T_m). The mean period associated with the spectrum shown in Fig. 20.1 was determined to be 4.1 s using Eq. (20.14). Likewise, Table 20.2 shows the most probable maximum response for sway (y_{max}), heave (z_{max}), and roll ($\theta_{x,max}$) over the same three time periods. As expected, we notice that longer durations correspond to an increase in the most probable maximum response.

Table 20.1 Most probable maximum response for longitudinal waves

T_m	t/T_m	t (s)	x_{max} (m)	z_{max} (m)	$\theta_{y,max}$ (°)
4.1 s	100	410	0.094	0.031	0.43
	1000	4100	0.11	0.038	0.53
	10,000	41,000	0.13	0.043	0.61

Table 20.2 Most probable maximum response for transverse waves

T_m	t/T_m	t (s)	y_{max} (m)	z_{max} (m)	$\theta_{x,max}$ (°)
4.1 s	100	410	0.12	0.05	0.18
	1000	4100	0.15	0.06	0.22
	10,000	41,000	0.17	0.07	0.25

20.3 Structural Accelerations Under Irregular Waves

The evaluation of human comfort under wave excitation involves the quantification of root mean square (RMS) horizontal (x and y) and vertical (z) building accelerations. For a pontoon-supported building such as that shown in Fig. 19.3, the RMS horizontal acceleration will be largest at roof level ($G_{x,max}$ and $G_{y,max}$—see Fig. 20.5), whereas the critical RMS vertical acceleration will be largest at the outer edge of the pontoon ($G_{zx,max}$ and $G_{zy,max}$—see Fig. 20.5) induced by waves propagating in the longitudinal (x) and transverse (y) direction. Such accelerations can be approximated as the square root of sum squared RMS accelerations pertaining to the surge, sway or heave taken about the CoR and the linear component induced by the roll for pitch acceleration at the desired location. RMS horizontal and vertical accelerations induced by longitudinal waves (left schematic of Fig. 20.5) are determined through:

$$G_{x,max} = \sqrt{G_x^2 + \left[G_{\theta_y}\left(D_p + D_b - \overline{KR}_{\theta_y}\right)\right]^2}, \quad \theta = 0° \tag{20.17}$$

$$G_{zx,max} = \sqrt{G_z^2 + \left(\frac{G_{\theta_y} C_p}{2}\right)^2}, \quad \theta = 0° \tag{20.18}$$

whereas for transverse waves (right schematic of Fig. 20.5):

$$G_{y,max} = \sqrt{G_y^2 + \left[G_{\theta_x}\left(D_p + D_b - \overline{KR}_{\theta_x}\right)\right]^2}, \quad \theta = 90° \tag{20.19}$$

$$G_{zy,max} = \sqrt{G_z^2 + \left(\frac{G_{\theta_x} B_p}{2}\right)^2}, \quad \theta = 90° \tag{20.20}$$

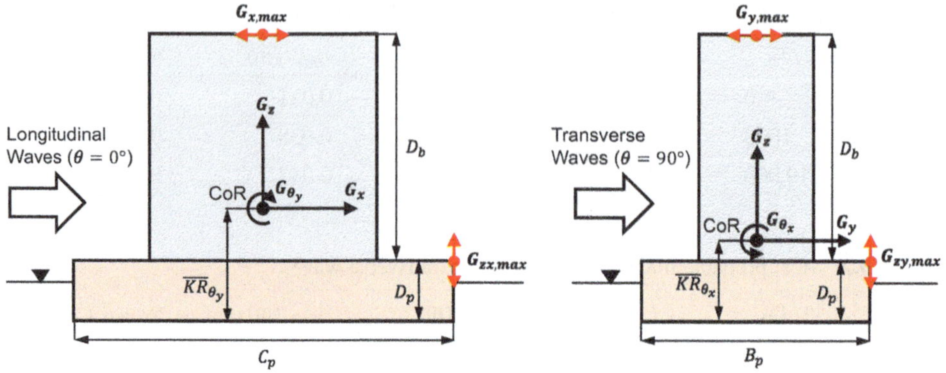

Fig. 20.5 Maximum RMS horizontal and vertical accelerations for longitudinal (left) and transverse (right) waves

where G_x, G_y, G_z, G_{θ_x}, and G_{θ_y} denote the RMS acceleration about the CoR in surge, sway, heave, roll, and pitch, respectively. For oblique waves ($0° < \theta < 90°$), however, yaw motions will produce maximum horizontal accelerations at the corners of the superstructure. When combined with accelerations along the x and y axes at roof level, the overall horizontal RMS acceleration becomes:

$$G_{xy,max} = \sqrt{G_{x,max}^2 + G_{y,max}^2 + \left(G_{\theta_z} r_b\right)^2}, \quad 0° < \theta < 90° \tag{20.21}$$

where $G_{x,max}$ and $G_{y,max}$ are the RMS longitudinal and transverse accelerations at roof level from Eqs. (20.17) and (20.19), respectively, when computed using a wave incidence angle of θ. In addition, G_{θ_z} denotes the RMS angular acceleration in yaw about the CoR, and:

$$r_b = \sqrt{\left(\frac{C_b}{2}\right)^2 + \left(\frac{B_b}{2}\right)^2} \tag{20.22}$$

is the distance between the CoR and each corner of the superstructure as shown in Fig. 20.6. The critical vertical acceleration for oblique waves will be present at each corner of the pontoon given by:

$$G_{z,max} = \sqrt{G_z^2 + \left(\frac{G_{\theta_y} C_p}{2}\right)^2 + \left(\frac{G_{\theta_x} B_p}{2}\right)^2}, \quad 0° < \theta < 90° \tag{20.23}$$

Note that the RMS acceleration G_R for any DoF R featured in Eqs. (20.17)–(20.23) can be obtained using the wave spectrum and the response amplitude operator (RAO) corresponding to that particular DoF via:

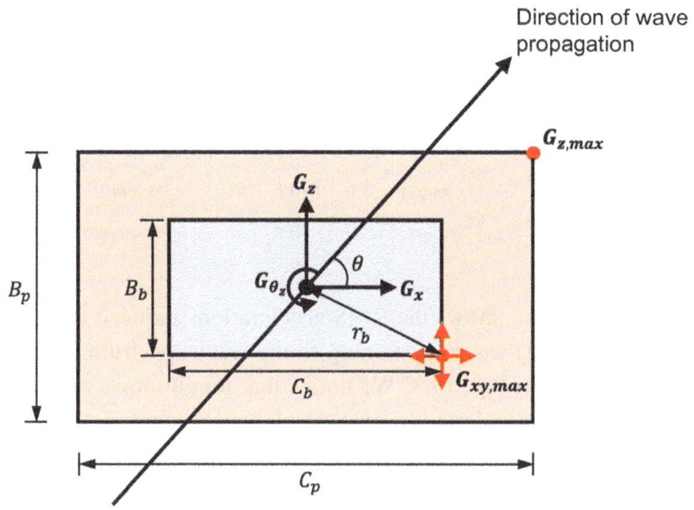

Fig. 20.6 Maximum RMS horizontal and vertical accelerations due to oblique waves

$$G_R = \sqrt{\sum_{i=1}^{N} S(f_i)\left[\omega_i^2 \mathrm{RAO}_R(f_i)\right]^2 \Delta f}, \quad R = \begin{cases} x, & \text{Surge} \\ y, & \text{Sway} \\ z, & \text{Heave} \\ \theta_x, & \text{Roll} \\ \theta_y, & \text{Pitch} \\ \theta_z, & \text{Yaw} \end{cases} \tag{20.24}$$

with the equations used to determine the RAOs being presented in Table 19.4. Equation (20.24) is analogous to the square root of the variance of the acceleration, thus yielding its standard deviation. Since the mean acceleration is taken to be zero, the standard deviation will be identical to the root mean square.

Equation (20.21) essentially presents the general form of the critical horizontal RMS acceleration which reduces to Eqs. (20.17) or (20.19) when θ is 0° or 90°, respectively. Likewise, the general form of the critical vertical acceleration is given by (20.23), which becomes Eqs. (20.18) or (20.20) for $\theta = 0°$ or 90°, respectively. If we assume that waves can approach the pontoon from any arbitrary direction, the maximum horizontal (G_h) and vertical RMS (G_v) accelerations for all θ are:

$$G_h = \max_{\theta}\left(G_{xy,max}\right) \tag{20.25}$$

$$G_v = \max_{\theta}\left(G_{z,max}\right) \tag{20.26}$$

Table 20.3 RMS accelerations induced by longitudinal and transverse waves

Longitudinal waves		Transverse waves	
G_x	0.052 m s^{-2}	G_y	0.060 m s^{-2}
G_z	0.013 m s^{-2}	G_z	0.019 m s^{-2}
G_{θ_y}	0.004 rad s^{-2}	G_{θ_x}	0.002 rad s^{-2}
$G_{x,max}$	0.13 m s^{-2}	$G_{y,max}$	0.085 m s^{-2}
$G_{zx,max}$	0.11 m s^{-2}	$G_{zy,max}$	0.036 m s^{-2}

As an example, Table 20.3 shows the RMS accelerations induced by longitudinal ($\theta = 0°$) and transverse ($\theta = 90°$) waves pertaining to the wave spectrum in Fig. 20.1 applied to the floating structure from Fig. 19.3. We notice that longitudinal waves result in lower surge (G_x) and heave (G_z) accelerations at the CoR when compared with the sway (G_y) and heave (G_z) accelerations induced by transverse waves. However, larger accelerations are observed at roof level and at the pontoon's edge as a result of waves propagating in the longitudinal direction. This result is due to the higher pitch accelerations at the CoR when compared to roll (i.e., $G_{\theta_y} > G_{\theta_x}$).

20.4 Acceleration Thresholds for Occupant Comfort

To evaluate the feasibility of a particular design, it is necessary to compare the RMS accelerations calculated in Sect. 20.3 against the maximum acceleration limits recommended by international serviceability standards. While dedicated serviceability guidelines pertaining to floating structures for permanent human habitation do not exist, the lateral acceleration limits specified for tall buildings undergoing wind-induced vibrations can be utilized as a general guide [16].

20.4.1 ISO 6897

The most widely used benchmarks for the evaluation of acceptable horizontal acceleration thresholds during windstorms were published by the International Organization for Standardization (ISO) in the form of ISO 6897 [13] and ISO 10137 [12]. First released in 1984, ISO 6897 proposes maximum horizontal acceleration limits as a function of the frequency (between 0.063 and 1 Hz) of vibration for two types of structures:

- General purpose buildings built on land
- Fixed offshore structures.

The development of these acceleration thresholds were formulated from the results of motion chamber experiments conducted by Irwin [10, 11], Chen and Robertson [4], and Hansen et al. [9]. These experiments involve placing human subjects into an enclosed room undergoing sinusoidal oscillations with gradually increasing acceleration at a constant frequency, then asking the subjects to indicate when motion is perceived [19]. Using this information, ISO 6897 specifies the maximum threshold in terms of the RMS acceleration at a given frequency over the worst 10 min of a windstorm that has a return period of at least 5 years.

20.4.2 ISO 10137

In 2007, ISO 6897 was superseded with the release of ISO 10137, which proposes different levels of tolerance based on the following two types of structures:

- Office buildings
- Residential buildings.

The formulation of ISO 10137 was based upon AIJ-GBV-2004 [1] published by the Architectural Institute of Japan (AIJ) after consulting motion simulator experiments from Goto [7, 8], Kanda et al. [14], and Shioya et al. [18]. Note that AIJ-GBV-2004 takes a probabilistic approach to motion perceptibility by delineating thresholds at which a certain percentage of people would be expected to notice accelerations of a given magnitude. In ISO 10137, the maximum acceleration threshold for residential buildings was taken as the acceleration where 90% of people are expected to sense movement if exposed to horizontal accelerations of that particular magnitude. This acceleration threshold was then increased by a factor of 1.5 for office buildings to reflect a higher tolerance to motion for people working in high-rise offices. Notably, ISO 10137 expresses all acceleration thresholds in terms of the peak acceleration (a_{peak}) over a duration of 10 min, which contrasts with the RMS accelerations (a_{rms}) given in ISO 6897. The relationship between peak and RMS accelerations can be approximated using:

$$a_{peak} \approx g_f a_{rms} \tag{20.27}$$

where g_f is known as the "peak factor" and is commonly taken as 3.5 for random vibrational waveforms such as that produced by wind or irregular waves [3].

20.4.3 Netherlands Organization for Applied Scientific Research

In addition to ISO 6897 and ISO 10137, the Netherlands Organization for Applied Scientific Research (TNO) has also released guidelines pertaining to the perceptibility of environmental vibrations [20]. Here, RMS accelerations are classified into the following different zones based on their noticeability and acceptability to building occupants [15]:

- Zone D: acceleration is clearly noticeable and motion sickness may occur
- Zone E: acceleration is noticeable but may be acceptable for short durations in houses
- Zone F: acceleration is hardly noticeable and will be acceptable for long durations in houses
- Zone G: acceleration is not noticeable.

20.4.4 NordForsk

As supplementary to guidelines intended for buildings, the NordForsk organization under the Nordic Council of Ministers has also proposed acceptable acceleration thresholds for docked cruise ships [17]. A constant RMS acceleration threshold is proposed for both horizontal and vertical components that is independent of the vibration frequency, with acceptable magnitudes being:

- Horizontal acceleration below 0.2 m s^{-2}
- Vertical acceleration below 0.3 m s^{-2}.

20.4.5 Summary of Acceleration Limits

The recommended horizontal acceleration thresholds aggregating these aforementioned guidelines are presented in Fig. 20.7. For RMS accelerations produced by irregular waves, the frequency may be taken as the mean frequency $f_m = 1/T_m$, where the mean period T_m can be determined from Eq. (20.14). Since ISO 10137 specifies the acceleration in terms of the peak acceleration, the thresholds shown in Fig. 20.7 represent their conversion into the RMS acceleration via Eq. (20.27). Figure 20.8 additionally presents the recommended thresholds for vertical acceleration. Here, only the TNO and NordForsk limits are presented since the ISO standards were developed only for horizontal accelerations. In addition, recommended thresholds pertaining to rotational accelerations (roll, pitch, and yaw) do not currently exist. Research by Allred and Clark [2] suggests that human sensitivity to rotational motion increases with frequency between 0.1 and 0.5 Hz. However, such knowledge has yet to be synthesized into practical serviceability guidelines applicable towards floating structures.

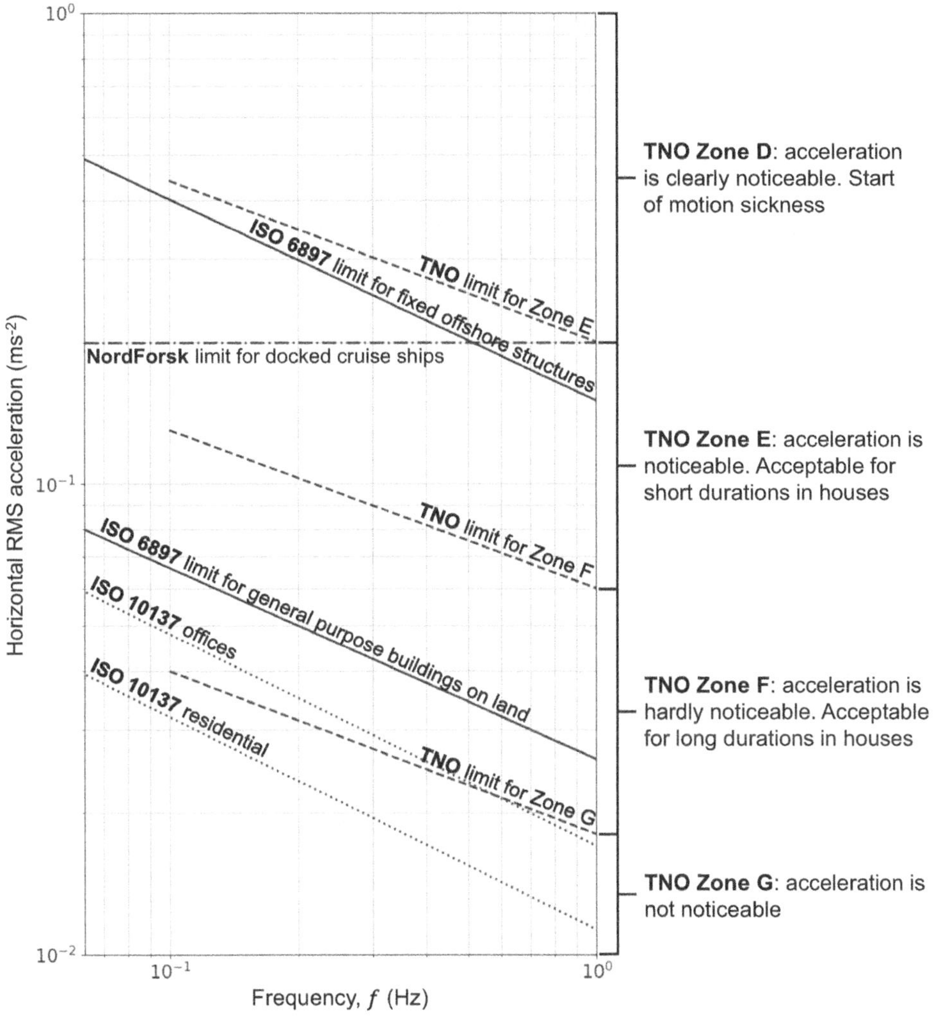

Fig. 20.7 Recommended horizontal acceleration thresholds as a function of frequency

20.5 Summary: Comfort Assessment of Floating Structures for Human Habitation

The procedure used to determine the serviceability of floating structures from the perspective of human comfort under irregular waves is summarized in Table 20.4. This methodology assumes that the structure consists of a rectangular pontoon supporting a symmetrical superstructure such as that shown in Fig. 19.3. The following parameters are required as input:

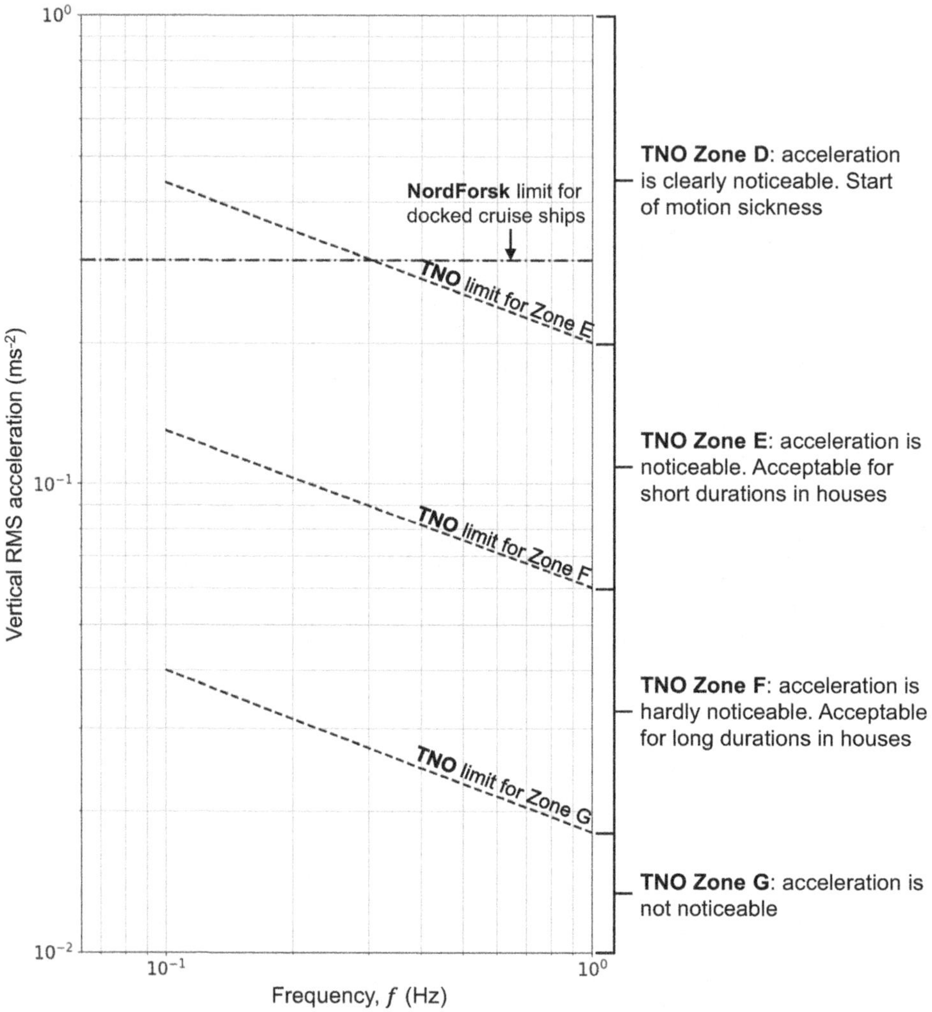

Fig. 20.8 Recommended vertical acceleration thresholds as a function of frequency

- Dimensions and mass of the pontoon (e.g., C_p, B_p, D_p, and m_p) and superstructure (e.g., C_b, B_b, D_b, and m_b)
- The significant wave height (H_s) and peak or mean wave period (T_p or T_m) describing the design sea state at the site.

Additional details of the structural parameters required for dynamic analysis can also be found in Sect. 19.2.

Table 20.4 Comfort assessment of floating structures under irregular waves

Step 1	Plot the wave spectrum $S(f)$ then discretize into $N > 200$ frequency bands of Δf between f_{max} and f_{min}	$\Delta f = \frac{f_{max}-f_{min}}{N-1}$
Step 2	For a given angle of wave approach θ, determine the RAO for each DoF R via the equations in Table 19.4	$\text{RAO}_R = \begin{cases} \frac{Q_R F_{0R}}{K_R}, & R = \{z, \theta_x, \theta_y\} \\ \frac{F_{0R}}{M_R \omega^2}, & R = \{x, y, \theta_z\} \end{cases}$
Step 3	Determine the RMS acceleration for each DoF R via (20.24)	$G_R = \sqrt{\sum_{i=1}^{N} S(f_i)\left[\omega_i^2 \text{RAO}_R(f_i)\right]^2 \Delta f}$
Step 4	Determine the maximum horizontal ($G_{xy,max}$) and vertical ($G_{z,max}$) RMS acceleration via (20.21) and (20.23), respectively	$G_{xy,max} = \sqrt{G_{x,max}^2 + G_{y,max}^2 + \left(G_{\theta_z} r_b\right)^2}$ $G_{z,max} = \sqrt{G_z^2 + \left(\frac{G_{\theta_y} C_p}{2}\right)^2 + \left(\frac{G_{\theta_x} B_p}{2}\right)^2}$
Step 5	Repeat Steps 2–4 across $0° \leq \theta \leq 90°$ to identify the maximum horizontal (G_h) and vertical (G_v) RMS accelerations via (20.25) and (20.26) for all θ	$G_h = \max_{\theta}\left(G_{xy,max}\right), \quad 0° \leq \theta \leq 90°$ $G_v = \max_{\theta}\left(G_{z,max}\right), \quad 0° \leq \theta \leq 90°$
Step 6	Compare G_h and G_v against recommended acceleration limits at the mean frequency $f_m = 1/T_m$	Refer to Fig. 20.7 for horizontal accelerations and Fig. 20.8 for vertical accelerations

References

1. AIJ, 2004. AIJ-GBV-2004. Guidelines for the evaluation of habitability to building vibration. Architectural Institute of Japan, Tokyo.
2. Allred, A.R., Clark, T.K., 2023. Vestibular perceptual thresholds for rotation about the yaw, roll, and pitch axes. Experimental Brain Research 241, 1101–1115.
3. Boggs, D., 1997. Acceleration Indexes for Human Comfort in Tall Buildings - Peak or RMS?,

in: CTBUH Monograph: Motion Perception Tolerance and Mitigation. Council on Tall Buildings and Urban Habitat.
4. Chen, P.W., Robertson, L.E., 1972. Human perception thresholds of horizontal motion. Journal of the Structural Division (ASCE) 98, 1681–1695.
5. Faltinsen, O.M., 1990. Sea loads on ships and offshore structures. Cambridge University Press.
6. Goda, Y., 2000. Random seas and design of maritime structures. World Scientific Publishing Co., Singapore.
7. Goto, T., 1983. Studies on wind-induced motion of tall buildings based on occupants' reactions. Journal of Wind Engineering and Industrial Aerodynamics 13, 241–252.
8. Goto, T., 1975. The criteria to motions in tall buildings, (Part 1) Factors affecting human perception and tolerance of motion. Journal of Architecture, Transactions of the Architectural Institute of Japan 237, 109–119.
9. Hansen, R.J., Reed, J.W., Vanmarcke, E.H., 1973. Human response to wind-induced motion of buildings. Journal of the Structural Division (ASCE) 99, 66–74.
10. Irwin, A.W., 1981. Perception, comfort and performance criteria for human beings exposed to whole body pure yaw vibration and vibration containing yaw and translational components. Journal of Sound and Vibration 76, 481–497.
11. Irwin, A.W., 1978. Human response to dynamic motion of structures. The Structural Engineer 56A, 237–244.
12. ISO, 2007. ISO 10137:2007 Bases for design of structures - Serviceability of buildings and walkways against vibrations. International Organization for Standardization, Geneva, Switzerland.
13. ISO, 1984. ISO 6897:1984 Guidelines for the evaluation of the response of occupants of fixed structures, especially buildings and off-shore structures, to low-frequency horizontal motion (0.063 to 1 Hz). International Organization for Standardization, Geneva, Switzerland.
14. Kanda, J., Tamura, Y., Fujii, K., 1988. Probabilistic criteria for human perception of low-frequency horizontal motions, in: Proceedings of Symposium/Workshop on Serviceability of Buildings. Ottawa, Canada, pp. 260–269.
15. Ko, K., 2015. Realising a floating city: a feasibility study of the construction of a floating city (MSc Thesis). Delft University of Technology, Delft, Netherlands.
16. Kwok, K.C.S., Hitchcock, P.A., Burton, M.D., 2009. Perception of vibration and occupant comfort in wind-excited tall buildings. Journal of Wind Engineering and Industrial Aerodynamics 97, 368–380.
17. NordForsk, 1987. Assessment of Ship Performance in a Seaway: The Nordic Co-operative Project: "Seakeeping Performance of Ships." NordForsk, Copenhagen, Denmark.
18. Shioya, K., Kanda, J., Tamura, Y., Fujii, K., 1992. Human perception thresholds of two dimensional horizontal motion, in: Proceedings of the Structures Congress. American Society of Civil Engineers, San Antonio, Texas, pp. 480–483.
19. Tamura, Y., Kawana, S., Nakamura, O., Kanda, J., Nakata, S., 2006. Evaluation perception of wind-induced vibration in buildings. Structures and Buildings 159, 283–293.
20. van Koten, H., 1967. Grenzen voor dynamische beweging (boundaries for dynamic movement). TNO-IBBC rapport BI-67-107.
21. Wang, G., Rosenfeld, Y., Drimer, N., Goldfeld, Y., 2021. Occupant comfort analysis for rigid floating structures – methodology and design assessment for offshore dwelling module. Ships and Offshore Structures 16, 184–199.

21 Floating Breakwaters for Wave Attenuation

The previous chapters approach floating structures largely from the perspective of land creation for ocean colonization. Yet, they can also serve as breakwaters to shelter coastal or floating communities from large waves. This chapter introduces simplified methods for the analysis of floating breakwaters and their mooring systems.

21.1 Overview of Floating Breakwaters

The simplicist description of a floating breakwater consists of a rectangular pontoon anchored to the seabed via mooring lines such as that illustrated in Fig. 21.1. The main advantage of floating breakwaters over traditional bottom-founded structures (e.g., Fig. 11.4) lies in their ease of installation and lower cost when deployed in deeper waters. Since most wave energy is concentrated near the free surface, the use of floating breakwaters naturally presents an attractive alternative to concrete caissons or rubble mounds which often occupy significant space on the seabed. As a consequence, bottom-founded breakwaters can block water circulation and exacerbate pollution or sediment issues [3]. Similar to that of vegetation (see Chaps. 15 and 16), floating breakwaters can therefore provide an environmentally friendly solution to the protection of coastlines from large waves. The ease at which floating breakwaters can be transported and relocated has even resulted in their use during military operations such as the WWII Normandy landings in June 1944 [11]. During this operation (codenamed "Overlord"), floating breakwaters referred to as the "Bombardon" were used to create two artificial harbors to facilitate the deployment of Allied troops on Normandy's beaches (Fig. 21.2).

S. Wang, *Applied Ocean Mechanics for Structural Engineering and Climate Adaptation*, Synthesis Lectures on Ocean Systems Engineering,
https://doi.org/10.1007/978-3-032-18199-2_21

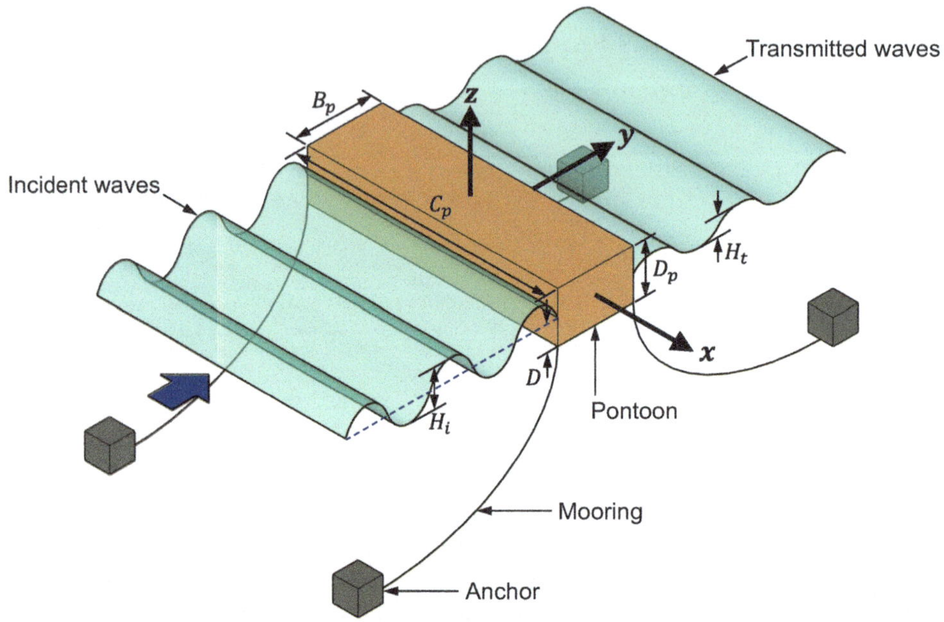

Fig. 21.1 Simplified representation of a rectangular pontoon acting as floating breakwater

Floating breakwaters can also exhibit numerous co-benefits that serve coastal communities beyond their ability to attenuate waves. One such example is the floating breakwater in Condamine Harbor, Monaco shown in Fig. 21.3 [1]. Measuring over 350 m in length, the breakwater consists of a hollow concrete caisson connected to land via a ball-and-socket joint while being anchored to the seabed via 8 mooring lines on its opposite end. As seen in Fig. 21.3, two 8 m long fins are attached at the base of the rectangular caisson to provide additional stability. This added stability is necessary since the entire structure is designed to serve as a parking terminal for vehicles while hosting numerous shops and promenades on its upper deck. In addition, the breakwater acts as a floating terminal for boats to dock on its leeward side. To provide an additional layer of safety for its occupants, the caisson features a double hull structural system to shield the internal parking compartments from a potential breach of the outermost walls.

Like most floating structures, floating breakwaters must be anchored in place to prevent excessive lateral displacements during wave action. As shown in Fig. 21.1, these anchoring systems commonly take the form of slender mooring lines attached to an anchor at the seabed. To further limit the motion of floating breakwaters, rigid mooring dolphins can be used in lieu of flexible lines. An example of mooring dolphins can be found in the Kan-on floating pontoon breakwaters installed in Hiroshima Port at the mouth of the Tenma river [10]. Figure 21.4 presents an illustration of a pontoon comprising the Kan-on breakwater which measures 97.8 m long, 20 m wide, and 3.5 m deep, and features

Fig. 21.2 Floating breakwaters at Gold Beach (near Arromanches-les-Bains, France) during the Normandy landings in 1944 (courtesy of the UK Army Film and Photographic Unit)

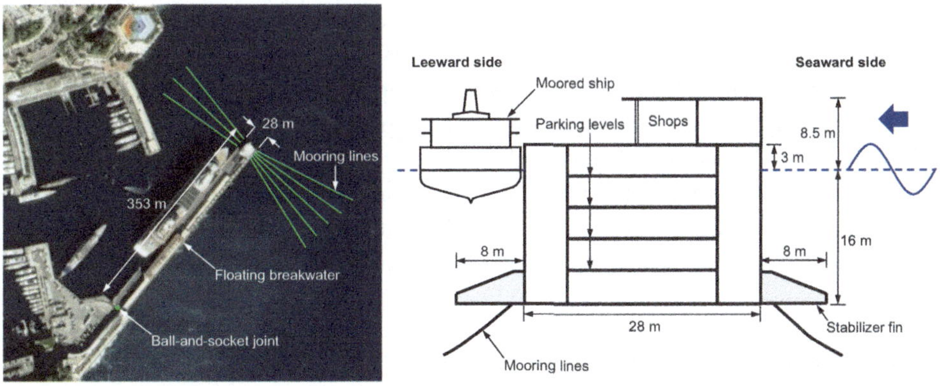

Fig. 21.3 Floating breakwater in Condamine Harbor, Monaco. Satellite image courtesy of USGS

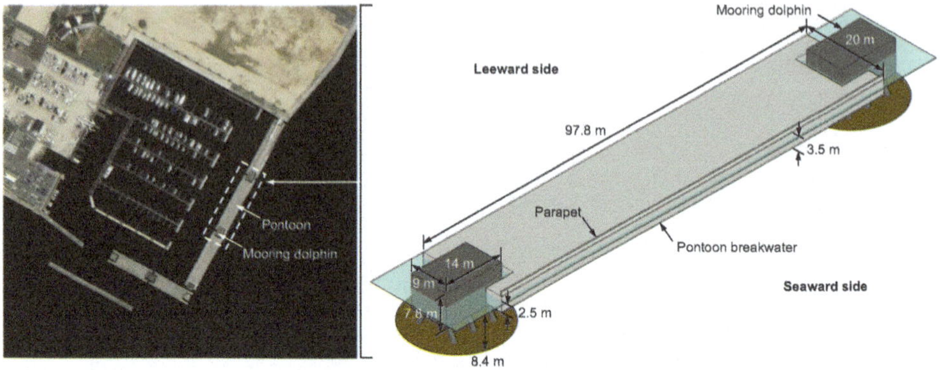

Fig. 21.4 Kan-on floating breakwater in Hiroshima Port, Japan. Satellite image courtesy of USGS

a 1 m high parapet on the seaward side to mitigate wave overtopping. Constructed from reinforced concrete surrounding a hollow steel core with vertical bulkheads, the pontoons are connected at each end to a concrete block anchored into the seabed via piles. These blocks effectively limit the lateral motion of the pontoon while still permitting vertical movement to accommodate rising or falling water levels within the harbor.

Despite the advantages of floating breakwaters, it is important to realize that such structures are generally only effective at attenuating moderate waves (< 2 m wave height) that have relatively short periods of less than 6 s [7]. In addition, floating breakwaters and their mooring systems are susceptible to failure during severe storms and require increased maintenance when compared to traditional bottom-founded structures. Beyond the prismatic rectangular pontoons shown in Figs. 21.1 and 21.4, many alternative designs have been proposed which include the use of double pontoons [8], energy dissipating cages [9], and their integration with wave energy converters (WECs) for wave energy harvesting [14], which are all illustrated in Fig. 21.5. Even agglomerations of bundled scrap tires have been trialed as a more economical option to traditional floaters [2]. While many more innovative designs exist that aim to improve the wave attenuating properties of floating breakwaters, this chapter will focus on conventional box-type pontoons since simplified analytical solutions can be employed to analyze their performance. The reader is directed to [3] and [15] for a more detailed discussion into different types of floating breakwaters and hybrid floating breakwater-WEC solutions.

21.2 Transmission Coefficient Under Regular Waves

Since floating breakwaters are designed to shield harbors from large waves, their performance is primarily quantified through the wave transmission coefficient K_t given as:

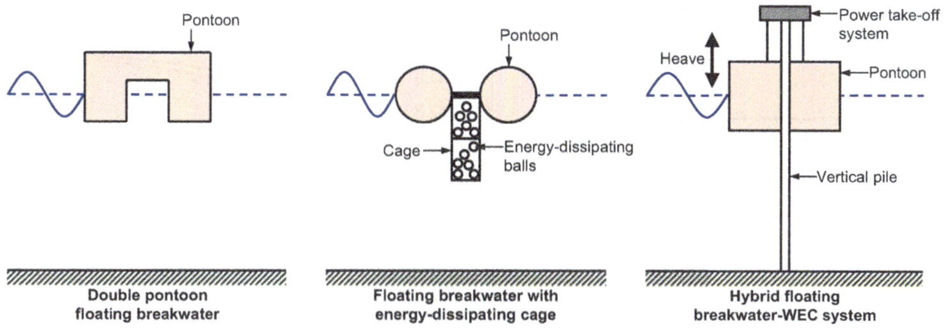

Fig. 21.5 Examples of different floating breakwater designs

$$K_t = \frac{H_t}{H_i} \tag{21.1}$$

where H_i and H_t denote the height of the incident and transmitted waves, respectively, as illustrated in Fig. 21.1. Breakwaters exhibiting lower transmission coefficients signify an increased capacity to suppress the height of incoming waves and would therefore be more effective at fulfilling their intended function. Floating breakwaters work by reflecting incoming waves in addition to dissipating wave energy via turbulence induced by its motion. However, this motion will also lead to wave radiation which could adversely contribute to generation of unwanted waves on the leeward (i.e., sheltered) side. The value of K_t strongly depends on multiple parameters including the wave period, breakwater geometry, and mooring solution. Hence, obtaining an accurate value for the transmission coefficient, especially for unconventional floater geometries, often necessitates complex methods of analyses and is often achieved through experimental testing [12] or advanced numerical simulations [4]. Yet, a simplified analytical solution for K_t was proposed by Macagno [13] for rectangular floating breakwaters when assuming that the pontoon is rigidly fixed in place and no wave overtopping occurs:

$$K_t = \frac{1}{\sqrt{1 + \frac{kB_p \sinh(kh)}{2\cosh[k(h-D)]}}} \tag{21.2}$$

Here, B_p denotes the pontoon width, h is the water depth, $k = 2\pi/L$ is the wavenumber, and D is the draft given by:

$$D = \frac{m'_p}{\rho B_p} \tag{21.3}$$

where m'_p is the mass per unit length of the pontoon and ρ is the fluid density. Note that while Eq. (21.2) relies upon significant simplifications that may not reflect the actual behavior of floating breakwaters, the formula is often useful when seeking an initial

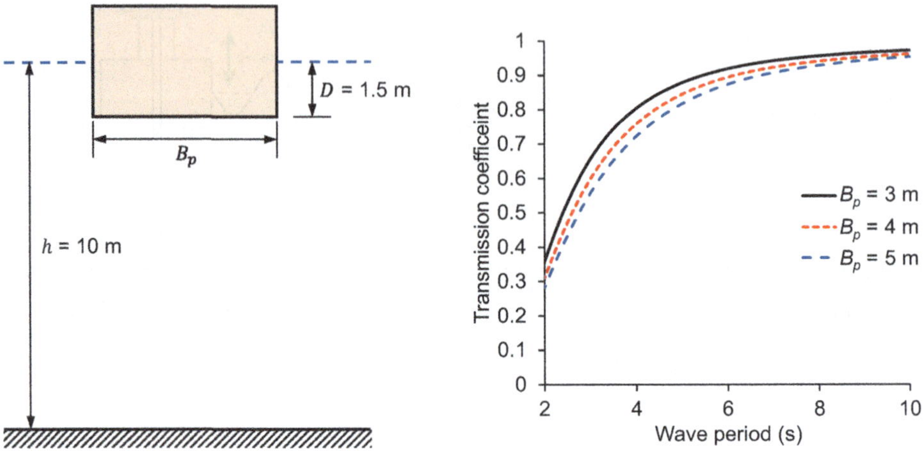

Fig. 21.6 Wave transmission coefficient predicted using Macagno's formula

prediction for preliminary design purposes. As an example, Fig. 21.6 presents the transmission coefficient predicted using Macagno's formula for a rectangular breakwater with $D = 1.5\,\text{m}$ in water $h = 10\,\text{m}$ deep. Here, K_t is plotted against the wave period T for three different pontoon widths of $B_p = 3, 4,$ and $5\,\text{m}$. We clearly see a decrease in the transmission coefficient for smaller wave periods and wider pontoons, which is consistent with the expected behavior of floating breakwaters. If a width of $B_p = 5\,\text{m}$ were to be adopted, incident waves with a period of $T = 4\,\text{s}$ would result in a K_t of approximately 0.7, thereby signifying a 30% reduction in the transmitted wave height on the leeward side. However, if the wave period was increased to 7 s, the reduction in wave height would only be approximately 10% to reflect a K_t of 0.9.

21.3 Transmission Coefficient Under Irregular Waves

Note that Eq. (21.2) is only intended for regular waves propagating with a constant wave height and frequency. The transmission coefficient for irregular waves described using a wave spectrum (e.g., JONSWAP) is given by:

$$K_{t,s} = \sqrt{\frac{m_{0,t}}{m_{0,i}}} \tag{21.4}$$

where:

$$m_{0,i} = \int_0^{\infty} S_i(f)df \tag{21.5}$$

$$m_{0,t} = \int_0^\infty S_t(f)df \tag{21.6}$$

denote the area underneath the wave spectrum pertaining to the incident and transmitted waves, respectively. The incident wave spectrum $S_i(f)$ can be obtained from the significant wave height as demonstrated in Eq. (20.7). The transmitted wave spectrum is:

$$S_t(f) = \left[K_t(f)\right]^2 S_i(f) \tag{21.7}$$

where $K_t(f)$ is the transmission coefficient for a regular wave propagating with frequency f which can be estimated using Eq. (21.2). The significant wave height of the transmitted waves is then:

$$H_{t,s} = 4\sqrt{m_{0,t}} \tag{21.8}$$

or:

$$H_{t,s} = K_{t,s} H_{i,s} \tag{21.9}$$

where $H_{i,s}$ denotes the significant wave height of the incident waves.

To demonstrate the application of Eqs. (21.4) to (21.9), let us examination the attenuation of irregular waves with significant wave height $H_{i,s} = 1.2\,\text{m}$ propagating towards the breakwater shown in Fig. 21.6 where its width is $B_p = 5\,\text{m}$. We can obtain the wave spectrum of the incident waves $S_i(f)$ using the JONSWAP spectrum given by Goda [5] as:

$$S_i(f) = \beta_J H_{i,s}^2 \left(\frac{f_p^4}{f^5}\right) \exp\left[-1.25\left(\frac{f_p}{f}\right)^4\right] \gamma^{\exp\left[\frac{-\left(\frac{f}{f_p}-1\right)^2}{2\sigma^2}\right]} \tag{21.10}$$

where:

$$\beta_J = \frac{0.0624\left[1.094 - 0.01915\ln(\gamma)\right]}{0.230 + 0.0336\gamma - 0.185(1.9+\gamma)^{-1}} \tag{21.11}$$

$$\sigma = \begin{cases} 0.07, & f \le f_p \\ 0.09, & f > f_p \end{cases} \tag{21.12}$$

The peak frequency f_p of the incident waves can be approximated from the significant wave height via:

$$f_p \approx \frac{1 - 0.132(\gamma + 0.2)^{-0.559}}{3.3H_{i,s}^{0.63}} \tag{21.13}$$

which when adopting a peak enhancement factor of $\gamma = 3.3$, yields $f_p = 0.252\,\text{Hz}$. Figure 21.7 plots the spectral density of the incident waves $S_i(f)$ from Eq. (21.10) in addition to that of the transmitted waves $S_t(f)$ obtained from (21.7). The area under both spectra is next calculated to be:

$$m_{0,i} = \int_0^\infty S_i(f)df = 0.0911\,\text{m}^2$$

$$m_{0,t} = \int_0^\infty S_t(f)df = 0.0409\,\text{m}^2$$

which gives a transmission coefficient of:

$$K_{t,s} = \sqrt{\frac{m_{0,t}}{m_{0,i}}} = \sqrt{\frac{0.0409}{0.0911}} = 0.67$$

The significant wave height of the transmitted waves is then:

$$H_{t,s} = K_{t,s}H_{i,s} = 0.67 \times 1.2 = 0.804\,\text{m}$$

Note that a very similar result can also be directly calculated from zeroth-moment (m_0) of the transmitted wave spectrum:

$$H_{t,s} = 4\sqrt{m_{0,t}} = 4\sqrt{0.0409} = 0.809\,\text{m}$$

which differs from the solution obtained using Eq. (21.9) by less than 1%.

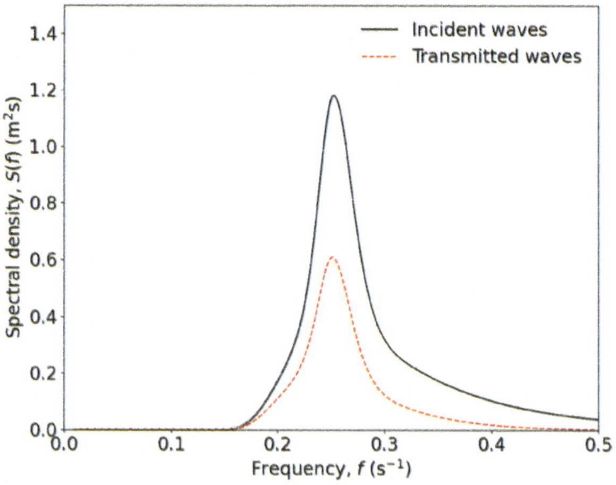

Fig. 21.7 Spectral density of incident and transmitted waves

21.4 Wave Forces on Floating Pontoon

In addition to the transmission coefficient, it is of great importance to determine the maximum wave force acting on a floating breakwater when designing the mooring and anchoring system. In this regard, we are most interested in the force acting along the sway (y) direction as shown in Figs. 21.1 and 21.8. From Chaps. 18 and 19, we know that such forces arise from the summation of the Froude-Krylov and added mass components. When following the procedure described in Sect. 19.4, the maximum force in sway from waves approaching with incident angle θ (relative to the x-axis) is given by:

$$F_{0y} = A\left[F_{FKy}(t_0) + F_{AMy}(t_0)\right] \tag{21.14}$$

where A denotes the wave amplitude. The Froude-Krylov force per unit wave amplitude is:

$$F_{FKy}(t_0) = \rho g \zeta_{FKxy} \gamma_{FKy} \tag{21.15}$$

with:

$$\zeta_{FKxy} = \begin{cases} \frac{\operatorname{sech}(kh)\{\sinh[k(D-h)]+\sinh(kh)\}}{k}, & \text{All depths} \\ \frac{1-\exp(-kD)}{k}, & \text{Deep water } (kh \gg 1) \end{cases} \tag{21.16}$$

$$\gamma_{FKy} = \frac{4\sec\theta}{k}\sin\left(\frac{B_p k}{2}\sin\theta\right)\sin\left(\frac{C_p k}{2}\cos\theta\right) \tag{21.17}$$

where B_p and C_p refers to the width and length of the breakwater, respectively. Likewise, the added mass force per unit wave amplitude is:

$$F_{AMy}(t_0) = \frac{\rho\pi D^2}{4}\omega^2 \zeta_{AMxy}\gamma_{AMy}\sin\theta \tag{21.18}$$

where:

$$\zeta_{AMxy} = \begin{cases} \frac{\cosh\left[k\left(h-\frac{D}{2}\right)\right]}{\sinh(kh)}, & \text{All depths} \\ \exp\left(-\frac{kD}{2}\right), & \text{Deep water } (kh \gg 1) \end{cases} \tag{21.19}$$

$$\gamma_{AMy} = \frac{4\sec\theta}{k}\cos\left(\frac{B_p k}{2}\sin\theta\right)\sin\left(\frac{C_p k}{2}\cos\theta\right) \tag{21.20}$$

In addition, the wave incidence angle should be taken as:

$$\theta = \begin{cases} \varepsilon, & \theta = 0^\circ \\ 90^\circ - \varepsilon, & \theta = 90^\circ \\ \theta, & 0^\circ < \theta < 90^\circ \end{cases} \tag{21.21}$$

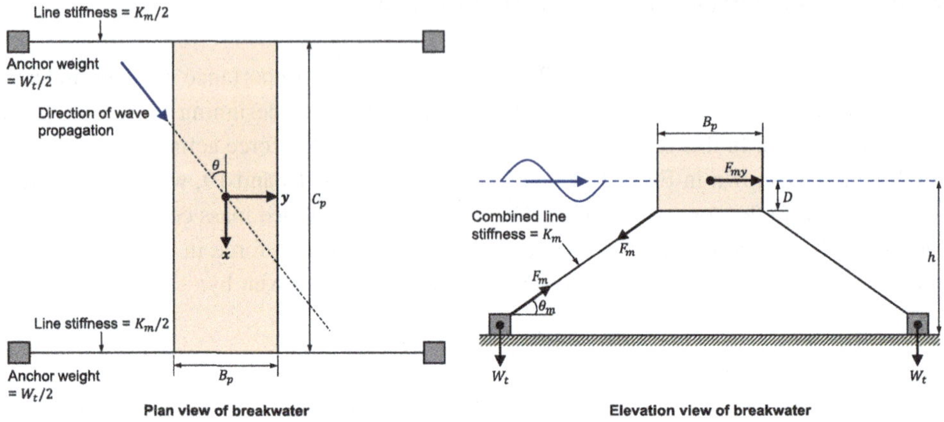

Fig. 21.8 Floating box-type breakwater with taut moorings

where ε represents an arbitrarily small positive quantity (e.g., $\varepsilon = 0.1°$) to avoid divisions by zero.

21.5 Mooring Forces

It is important to understand that the force given in Eq. (21.14) assumes that the breakwater is rigidly moored in place. To determine the tension force in the mooring line, we must consider for the natural frequency of the breakwater in sway (f_{ny}) resulting from the line stiffness (K_m). This involves calculating the dynamic amplification factor Q_y:

$$Q_y = \frac{1}{\sqrt{\left(1 - \frac{f^2}{f_{ny}^2}\right)^2 + \left(2\xi_y \frac{f}{f_{ny}}\right)^2}} \tag{21.22}$$

where ξ_y is the damping ratio in sway. The sway natural frequency is given by:

$$f_{ny} = \frac{1}{2\pi}\sqrt{\frac{K_{my}}{M_y}} \tag{21.23}$$

Here, K_{my} denotes the combined stiffness of all mooring lines on one side of the breakwater in the sway direction (i.e., along the y-axis). For taut mooring lines with a combined axial stiffness of K_m that are inclined at an angle θ_m at the anchor location (Fig. 21.8), their stiffness in the direction of sway would be:

$$K_{my} = K_m \cos^2 \theta_m \tag{21.24}$$

In addition, M_y refers to the combined structural and added mass of the breakwater in sway:

$$M_y = m_p + \frac{1}{2}\rho\pi D^2 C_p \tag{21.25}$$

where m_p is the mass of the pontoon. After obtaining Q_y via Eq. (21.22), the total force in the mooring lines acting along the sway direction (y-axis) is:

$$F_{my} = Q_y F_{0y} \tag{21.26}$$

This force can then be used to determine the tension force in the mooring lines (F_m):

$$F_m = \frac{F_{my}}{\cos\theta_m} \tag{21.27}$$

The required total weight of the anchors on each side of the breakwater can then be determined per the simplified method given by Hales [6]:

$$W_t = F_S\left[\frac{F_{my}}{\mu\left(1-\frac{\rho}{\rho_s}\right)} + F_m \sin\theta_m\right] \tag{21.28}$$

where μ is the coefficient of static friction between the anchor block and the seabed and F_S is the desired factor of safety which must be greater than 1. The density of water and the material used for the anchor block is given by ρ and ρ_s, respectively. Note that if more than one mooring line is present, W_t will refer to the combined mass of all anchor blocks on each side as shown in Fig. 21.8.

21.6 Example: Simplified Analysis of a Box-Type Floating Breakwater

We will now demonstrate how the simplified equations presented in this chapter can be used to analyze the performance of a box-type floating breakwater. For this example, we will consider a pontoon with length $C_p = 20\,\text{m}$, width $B_p = 5\,\text{m}$, and mass $m_p = 154 \times 10^3\,\text{kg}$ located in water with depth $h = 10\,\text{m}$. Regular waves with height $H = 1.2\,\text{m}$ and period $T = 4\,\text{s}$ are propagating towards the breakwater along a direction perpendicular to its longitudinal axis (i.e., $\theta = 90°$). In addition, let us assume that the pontoon is anchored to the seabed via taut mooring lines attached to each of its four corners in a similar manner to that shown in Fig. 21.8. Each line has an axial stiffness of 15,000 N/m and makes an angle of $\theta_m = 30°$ at the anchors. In this example, we are primarily interested in the following:

- Height of the transmitted waves.
- Maximum wave force on the breakwater.
- Maximum tension force in each of the mooring lines.
- Required weight of the anchor blocks.

21.6.1 Transmitted Wave Height

We begin our analysis by determining the draft D from the pontoon mass and dimensions. From Archimedes' principle, the draft is:

$$D = \frac{m_p}{\rho B_p C_p} = \frac{154{,}000}{1025 \times 5 \times 20} = 1.5\,\text{m}$$

The wavelength L and wavenumber k can be calculated next from the dispersion relation. When adopting the approximate solution, we obtain:

$$L \approx \frac{gT^2}{2\pi}\sqrt{\tanh\left(\frac{4\pi^2 h}{T^2 g}\right)} = \frac{9.81 \times 4^2}{2\pi}\sqrt{\tanh\left(\frac{4\pi^2 \times 10}{4^2 \times 9.81}\right)} = 24.8\,\text{m}$$

which leads to:

$$k = \frac{2\pi}{L} = \frac{2\pi}{24.8} = 0.253\,\text{m}^{-1}$$

The wave attenuation coefficient can now be estimated using Eq. (21.2):

$$K_t = \frac{1}{\sqrt{1 + \frac{kB_p \sinh(kh)}{2\cosh[k(h-D)]}}} = \frac{1}{\sqrt{1 + \frac{0.253 \times 5 \sinh(0.253 \times 10)}{2\cosh[0.253(10-1.5)]}}} = 0.72$$

which implies a transmitted wave height of:

$$H_t = K_t H_i = 0.72 \times 1.2 = 0.86\,\text{m}$$

on the leeward side of the breakwater.

21.6.2 Maximum Wave Force on Breakwater

To determine the maximum force on the breakwater due to the incident waves, we must first obtain its Froude-Krylov and added mass components. Considering first the Froude-Krylov force terms:

$$\zeta_{FKxy} = \frac{\text{sech}(kh)\{\sinh[k(D-h)] + \sinh(kh)\}}{k}$$

$$= \frac{\mathrm{sech}(0.253 \times 10)\{\sinh[0.253(1.5 - 10)] + \sinh(0.253 \times 10)\}}{0.253}$$
$$= 1.252\,\mathrm{m}$$

$$\begin{aligned}\gamma_{FKy} &= \frac{4\sec\theta}{k}\sin\left(\frac{B_p k}{2}\sin\theta\right)\sin\left(\frac{C_p k}{2}\cos\theta\right)\\ &= \frac{4\sec 89.9^\circ}{0.253}\sin\left(\frac{5\times 0.253}{2}\sin 89.9^\circ\right)\sin\left(\frac{20\times 0.253}{2}\cos 89.9^\circ\right)\\ &= 23.66\,\mathrm{m}\end{aligned}$$

Note that we have used $\theta = 89.9^\circ$ instead of 90° to avoid divisions by zero. The Froude-Krylov force per unit wave amplitude can now be computed via Eq. (21.15):

$$F_{FKy}(t_0) = \rho g \zeta_{FKxy}\gamma_{FKy} = 1025 \times 9.81 \times 1.252 \times 23.66 = 297{,}900\,\mathrm{N/m}$$

Next determining the added mass force terms:

$$\zeta_{AMxy} = \frac{\cosh\left[k\left(h - \frac{D}{2}\right)\right]}{\sinh(kh)} = \frac{\cosh\left[0.253\left(10 - \frac{1.5}{2}\right)\right]}{\sinh(0.253 \times 10)} = 0.84$$

$$\begin{aligned}\gamma_{AMy} &= \frac{4\sec\theta}{k}\cos\left(\frac{B_p k}{2}\sin\theta\right)\sin\left(\frac{C_p k}{2}\cos\theta\right)\\ &= \frac{4\sec 89.9^\circ}{0.253}\cos\left(\frac{5\times 0.253}{2}\sin 89.9^\circ\right)\sin\left(\frac{20\times 0.253}{2}\cos 89.9^\circ\right)\\ &= 32.25\,\mathrm{m}\end{aligned}$$

The added mass force per unit wave amplitude is thus:

$$\begin{aligned}F_{AMy}(t_0) &= \frac{\rho\pi D^2}{4}\omega^2\zeta_{AMxy}\gamma_{AMy}\sin\theta\\ &= \frac{1025\times\pi\times 1.5^2}{4}\left(\frac{2\pi}{4}\right)^2 \times 0.84 \times 32.25\sin 89.9^\circ\\ &= 121{,}100\,\mathrm{N/m}\end{aligned}$$

We can now calculate the total force in sway using Eq. (21.14):

$$F_{0y} = A\left[F_{FKy}(t_0) + F_{AMy}(t_0)\right] = \frac{1.2}{2}(297{,}900 + 121{,}100) = 251{,}400\,\mathrm{N}$$

21.6.3 Maximum Force in Mooring Lines

To obtain the force in the mooring lines, we must first determine the dynamic amplification factor in sway using Eq. (21.22). This equation requires knowledge of both the damping ratio and natural frequency. For simplicity, let us ignore the contribution of damping such that $\xi_y \approx 0$. The natural frequency can only be determined after solving for the combined stiffness of the mooring lines in sway and the total mass (i.e., pontoon mass + added mass in sway). Since we have two mooring lines on each side of the breakwater each with an axial stiffness of 15,000 N/m, their combined axial stiffness becomes:

$$K_m = 15{,}000 \times 2 = 30{,}000\,\mathrm{N/m}$$

We can next obtain their equivalent stiffness in sway:

$$K_{my} = K_m \cos^2\theta_m = 30{,}000 \cos^2 30^\circ = 22{,}500\,\mathrm{N/m}$$

The total mass is given by:

$$M_y = m_p + \frac{1}{2}\rho\pi D^2 C_p = 154{,}000 + \frac{1}{2} \times 1025 \times \pi \times 1.5^2 \times 20 = 226{,}500\,\mathrm{kg}$$

From K_{my} and M_y, the natural frequency of the breakwater in sway becomes:

$$f_{ny} = \frac{1}{2\pi}\sqrt{\frac{K_{my}}{M_y}} = \frac{1}{2\pi}\sqrt{\frac{22{,}500}{226{,}500}} = 0.05\,\mathrm{Hz}$$

For waves traveling with frequency:

$$f = \frac{1}{T} = \frac{1}{4} = 0.25\,\mathrm{Hz}$$

the dynamic amplification factor (in the absence of damping) is:

$$Q_y = \frac{1}{\sqrt{\left(1 - \frac{f^2}{f_{ny}^2}\right)^2 + \left(2\xi_y \frac{f}{f_{ny}}\right)^2}} = \frac{1}{\sqrt{\left(1 - \frac{0.25^2}{0.05^2}\right)^2 + 0}} = 0.042$$

which yields a total mooring force in sway of:

$$F_{my} = Q_y F_{0y} = 0.042 \times 251{,}400 = 10{,}560\,\mathrm{N}$$

This force can next be used to find the tension force in the cables:

$$F_m = \frac{F_{my}}{\cos\theta_m} = \frac{10{,}560}{\cos 30^\circ} = 12{,}200\,\mathrm{N}$$

Since we have two cables on each side of the breakwater, the maximum tension force in each cable would be 12,200/2 = 6100 N.

21.6.4 Required Weight of Anchor Blocks

Finally, we can estimate the required weight of the anchor blocks by assuming they are fabricated from concrete with density $\rho_s = 2400\,\text{kg m}^{-3}$. If we employ a safety factor of $F_S = 3.0$ and a coefficient of friction $\mu = 0.6$, the required weight becomes:

$$W_t = F_S\left[\frac{F_{my}}{\mu\left(1-\frac{\rho}{\rho_s}\right)} + F_m \sin\theta_m\right] = 3 \times \left[\frac{10{,}560}{0.6\left(1-\frac{1025}{2400}\right)} + 12{,}200 \sin 30^\circ\right]$$
$$= 110{,}460\,\text{N}$$

Note that W_t gives the total weight of all anchor blocks on each side of the breakwater. Since we have two blocks, the required weight of each anchor becomes 110,460/2 = 55,230 N. This weight is equivalent to a mass of 5630 kg. If each anchor were to take the form of a concrete cube, the length of each side of this cube would measure approximately 1.33 m.

References

1. Bouchet, R., Sedillot, F., Jaeger, J.M., Troya, L., Peset, L., Martareche, F., 2004. Monaco semi-floating dyke: a 352 metre long concrete caisson, in: Concrete Structures: The Challenge of Creativity. Avignon, France.
2. Candle, R.D., 1974. Goodyear scrap tire floating breakwater concepts, in: Proceedings of the Floating Breakwaters Conference. University of Rhode Island, pp. 193–212.
3. Dai, J., Wang, C.M., Utsumomiya, T., Duan, W., 2018. Review of recent research and developments on floating breakwaters. Ocean Engineering 158, 132–151.
4. Dominguez, J.M., Crespo, A.J.C., Hall, M., Altomare, C., Wu, M., Stratigaki, V., Troch, P., Cappietti, L., Gomez-Gesteira, M., 2019. SPH simulation of floating structures with moorings. Coastal Engineering 153.
5. Goda, Y., 1988. Statistical variability of sea state parameters as a function of a wave spectrum. Coastal Engineering in Japan 31, 39–52.
6. Hales, L.Z., 1981. Floating breakwaters: state-of-the-art literature review. US Army Corps of Engineers Coastal Engineering Research Center, Fort Belvoir, VA.
7. Headland, J.R., 1995. Floating breakwaters, in: Marine Structures Engineering: Specialized Applications. Chapman & Hall, pp. 367–411.
8. Ikeno, M., Shimoda, N., Iwata, K., 1988. A new type of breakwater utilizing air compressibility, in: Proceedings of the 21st Coastal Engineering Conference. ASCE, pp. 2326–2339.
9. Ji, C., Chen, X., Cui, J., Yuan, Z., Incecik, A., 2015. Experimental study of a new type of floating breakwater. Ocean Engineering 105, 295–303.

10. Kusaka, T., Ueda, S., 2015. Ujina Floating Ferry Pier and Kan-On Floating Breakwater, Japan, in: Large Floating Structures: Technological Advances. Springer, Singapore, pp. 107–128.
11. Lochner, R., Faber, O., Penny, W.G., 1948. The Bombardon Floating Breakwater. The Civil Engineer in War.
12. Loukogeorgaki, E., Yagci, O., Kabdasli, M.S., 2014. 3D Experimental investigation of the structural response and the effectiveness of a moored floating breakwater with flexibly connected modules. Coastal Engineering 91, 164–180.
13. Macagno, E.O., 1953. Fluid Mechanics: Experimental Study of the Effects of the Passage of a Wave Beneath an Obstacle. Proceedings of the Academy of Sciences, Paris.
14. Ning, D., Zhao, X., Goteman, M., Kang, H., 2016. Hydrodynamic performance of a pile-restrained WEC-type floating breakwater: An experimental study. Renewable Energy 95, 531–541.
15. Zhao, X.L., Ning, D.Z., Zou, Q.P., Qiao, D.S., Cai, S.Q., 2019. Hybrid floating breakwater-WEC system: A review. Ocean Engineering 186, 106126.

Correction to: Applied Ocean Mechanics for Structural Engineering and Climate Adaptation

Correction to:
Chapters 15 and 19 in: S. Wang, *Applied Ocean Mechanics for Structural Engineering and Climate Adaptation*, Synthesis Lectures on Ocean Systems Engineering, https://doi.org/10.1007/978-3-032-18199-2

The original version of the book was inadvertently published with error in the equations Eq. 15.19, Eq. 19.40, Eq. 19.50, Eq. 19.61, Eq. 19.63 and the figure Fig. 19.13.

The correction chapters and the book have been updated with these changes.

The updated version of these chapters can be found at
https://doi.org/10.1007/978-3-032-18199-2_15
https://doi.org/10.1007/978-3-032-18199-2_19

S. Wang, *Applied Ocean Mechanics for Structural Engineering and Climate Adaptation*,
Synthesis Lectures on Ocean Systems Engineering,
https://doi.org/10.1007/978-3-032-18199-2_22

Index

A
Acceleration threshold, 396
Added mass, 339, 341, 343
AIJ-GBV-2004, 397
Airy wave theory, 31
Anchoring system, 411
Angle of wave incidence, 156, 411
Archimedes, 141, 305
Architectural Institute of Japan, 397
Arctangent problem, 334
ASCE Tsunami Design Geodatabase, 233
Average drag coefficient, 290
Averaged wind speed, 114, 117

B
Bagnold, 154
Bearing stress, 161
Bending moment, 206, 324, 381
Bernoulli, 26, 34, 37
Biesel transfer function, 100, 102
Bombardon, 403
Bore, 236
Boston, 301
Boundary condition, 33, 54, 62, 268, 287
Breaker height index, 130
Breaker zone, 131
Breaking wave, 49, 128, 149, 183, 210
Breaking wave height, 129
Bretschneider, 91
Broken wave, 130, 150, 191
Bulkhead, 319, 322
Buoy, 75
Buoyant force, 142, 160

C
Caisson, 152, 170
Cauchy stress tensor, 142
Center of buoyancy, 305
Center of floatation, 309
Center of gravity, 317
Center of lateral resistance, 312
Center of rotation, 346
Climate change, 263, 301
Cnoidal, 48, 104
Coastal Engineering Manual, 209
Co-benefits, 404
Coefficient of friction, 161, 413
Condamine Harbor, 404
Conservation laws, 19
Contact-stiffness method, 255
Continental shelf, 70
Continuity equation, 19, 20, 53, 62, 267
Coriolis parameter, 113
Cross product, 7
Cumulative distribution function, 79
Curl, 15
Curling factor, 211
Cyclone, 109
Cyclone forward speed, 114

S. Wang, *Applied Ocean Mechanics for Structural Engineering and Climate Adaptation*,
Synthesis Lectures on Ocean Systems Engineering,
https://doi.org/10.1007/978-3-032-18199-2

D

D'Alembert's formula, 67
Damping ratio, 333, 336, 356, 367, 371, 374, 389, 412
Danish Maritime Authority, 314
Decay coefficient, 266
Deep water, 45
Deep-water wave height, 125
Deep-water wavelength, 126
Deflection, 322
Degrees of freedom, 305, 328
Density of air, 113
Density of seawater, 73
Density of tsunami flows, 239
Depth-induced wave breaking, 49
Design wave height, 153, 158, 171, 203
Design wave period, 158, 204
Diffraction, 130, 341
Dimensionless shallow water peak frequency, 94
Dispersion relation, 37, 38
Dispersive wave, 46
Distribution of wave height, 77
Divergence, 12
Divergence theorem, 143
Dolphin mooring, 404
Donelan spectrum, 90
Dot product, 6
Downshifting, 89
Draft, 306
Drag coefficient, 198, 205, 214, 240, 276, 285, 288
Drag force, 205, 267, 285
Duration of impact, 255
Dynamic amplification factor, 332, 336, 356, 412
Dynamic boundary condition, 34
Dynamic pressure, 3, 37

E

Elastic bending stiffness, 327
Elastic collision, 255
Energy dissipation, 274
Energy flux, 55, 58, 59, 125, 284
Energy Grade Line Analysis, 231, 234
Energy head, 235
Envelope amplitude, 44
Equation of motion, 331
Equivalent deep water wave, 130
Equivalent fetch, 118
Euler equations, 25
Euler's constant, 114
Extreme wave focusing, 104

F

Factor of safety, 160, 413
Fetch, 86
Floating airport, 329
Floating breakwater, 403
Floating city, 303
Floating debris, 251
Floating pontoon, 319, 327
Floating terminal, 404
Fluid acceleration, 40, 339
Freeboard, 155
Free surface, 33
Frequency band, 85
Frequency domain, 82
Frictional terrain, 264
Froude-Krylov force, 339
Froude number, 235
Froude scaling, 106
Fully-developed sea, 86

G

Gauss's theorem, 143
Generalized Extreme Value distribution, 184
Goda, 155
Gradient, 12
Gravitational acceleration, 22
Group effect factor, 224
Group velocity, 44, 59
Gulf of Mexico, 114

H

Heave, 328, 365
Heel, 305
Highest wave, 77
Hiroi, 153
Hiroshima Port, 404
Hitachinaka, 170
Holland model, 112
Horizontal force, 159
Horizontal velocity, 41

Hurricane, 109
Hurricane Katrina, 112, 114
Hurricane Patricia, 111
Hydraulic experiment, 99
Hydrodynamic pressure, 238
Hydroelastic, 327
Hydrostatic pressure, 140, 143, 159, 242
Hyperbolic function, 46
Hyperbolic paraboloid, 170

I
Ibaraki Port, 170
Impact force, 183, 210, 251
Impulse-momentum method, 255
Impulsive pressure, 150, 153, 167
Incident wave, 407
Incompressible, 33
Inertia coefficient, 205
Inertia force, 205
Inner surf zone, 130
Intermediate water, 45
International Organization for Standardization, 396
Inviscid, 33
Iribarren number, 233
Irregular wave, 75
Irrotational, 17, 33
ISO 10137, 397
ISO 6897, 396

J
JONSWAP spectrum, 87

K
Kan-on floating breakwater, 404
Keel, 310
Kelp, 276
Keulegan-Carpenter number, 224, 293
Kinematic boundary condition, 34
Kinematic viscosity, 22, 206, 276
Kinetic energy, 51
Kobayashi, 264

L
Land reclamation, 301
Laplace, 15
Laplace equation, 10
Large Wave Flume, 99
Le Méhauté's abacus, 48
Level 1 Green-Naghdi equations, 48
Linear wave, 31
Longitudinal wave, 329
Longuet-Higgins, 79

M
Macagno, 407
Mangrove, 263
Manning's coefficient, 237
Marina Bay Floating Platform, 319
Mass dominated, 332
Mass moment of inertia, 331
Material derivative, 24
Matrix multiplication, 8
Maximum sustained wind speed, 111
Mean high water, 231
Mean wave, 77
Mean wave height, 79
Mean wave period, 85
Mega-Float, 329
Metacenter, 311
Metacentric height, 311
Minikin, 154
Mitsuyasu, 91
Model scale, 106
Moment of area, 309
Moment of wave spectrum, 84
Momentum equations, 19, 22
Monochromatic wave, 82
Mooring, 338, 412
Mooring force, 413
Morison equation, 205
Most probable maximum response, 391
Motion chamber experiment, 397
Motion perceptibility, 397

N
Natural and nature-based features, 263
Natural frequency, 332, 356, 389, 412
Navier-Stokes equations, 19
Netherlands Organization for Applied Scientific Research, 397
Newtonian fluid, 19

Newton's second law, 22
Nondimensionalized fetch, 87
Nondispersive wave, 46
Nonlinear wave, 48
NordForsk, 398
North Sea, 87
Numerical simulation, 99, 183, 328

O
Oblique wave, 329
Orbitals of fluid particles, 41
Ordinary differential equation, 11
Oscillation frequency, 336, 341
Outer surf zone, 130
Outskirt coefficient, 105
Overturning moment, 176

P
Parallel axis theorem, 309
Partial differential equation, 9
Peak acceleration, 397
Peak enhancement parameter, 87
Peak frequency, 84
Perturbation pressure, 38
Phase shift, 82, 331
Phase velocity, 43
Pierson and Moskowitz, 90
Pile group, 222
Piles, 203
Piston wavemaker, 99
Pitch, 328
Pontoon breakwater, 403
Potential energy, 52
Pressure deficit, 111
Pressure gradient, 23
Probability density function, 79
Probability of non-exceedance, 185
Product rule, 13
Prototype scale, 106
PROVERBS, 183
Ptolemy's theorem, 334

R
Radiation, 341
Radius to maximum wind, 112
Ramp function, 103
Random phase, 86
Rayleigh distribution, 79, 290, 391
Rectangular pontoon, 316, 355
Reflected wave, 67
Reflection coefficient, 191
Refraction, 130
Response amplitude of floating structure, 331
Response amplitude operator, 337
Restoring moment, 160
Reynolds number, 206, 276
RMS acceleration, 393
Roll, 328
Root mean square, 290, 393
Rubble mound, 152, 155, 170

S
Saffir-Simpson hurricane wind scale, 111
Sainflou, 153
Saltmarsh, 276
San Francisco, 244
Scale parameter, 79
Scaling relationships, 107
Seabed boundary condition, 34
Self-weight, 161
Semi-submersible pontoon, 328
Serviceability, 396
Shallow water, 45
Shallow-water equations, 61
Shape coefficient, 313
Shear force, 324, 383
Shoaling coefficient, 126
Significant wave, 77
Significant wave height, 84, 127
Sliding of a caisson, 161
Slope angle, 127, 233
Solitary wave, 104
Spectral density, 84
Static stability, 305
Steepness-induced wave breaking, 49
Stiffness dominated, 332
Stokes drift, 46
Stokes wave, 48
Structural analysis, 319
Structural modification factor, 169
Superposition, 82, 86
Surface gravity wave, 31
Surf zone, 127
Surge, 328

Sway, 328

T
Takahashi, 167, 168
Time domain, 75
TMA spectrum, 92
Total mass, 345
Total wave energy, 58
Transect, 234
Transitional length, 327
Transmission coefficient, 406
Transmitted wave, 67, 407
Transmitted wave spectrum, 409
Transverse wave, 329
Trim, 305
Tsunami amplitude, 233
Tsunami flow velocity, 238
Tsunami force, 244
Tsunami importance factor, 239
Tsunami inundation depth, 238
Tsunami period, 233
Tsunami return period, 233
Tsunami runup, 231
Typhoon, 109

U
United Nations, 301
Uplift force, 188
Uplift pressure, 155
Ursell number, 101
US Army Corps of Engineers, 86

V
Variance, 390
Vector addition, 5
Vector differential operator, 12
Vector magnitude, 7
Vegetation, 263, 283
Velocity potential, 26, 35
Vertical breakwater, 152, 155, 170
Vertical force, 145, 161, 167
Vertical velocity, 41
Very Large Floating Structures, 302
Violent Overtopping by Waves at Seawalls (VOWS), 190

W
Waterplane area, 309
Waterplane stiffness, 314, 338
Wave amplitude, 31
Wave attenuation, 264, 283
Wave energy, 51, 59, 84
Wave energy converter, 406
Wave equation, 64
Wave frequency, 31
Wave generation, 99
Wave height, 31
Wavelength, 31
Wavenumber, 31
Wave period, 31
Wave properties, 47
Wave record, 77
Wave runup, 156, 169
Wave set-down, 166
Wave shoaling, 125
Wave spectrum, 82, 408
Wave statistics, 77
Wave steepness, 233
Wave transformation, 66, 283
Wave velocity, 46
Wind force, 313
Wind height coefficient, 313
Wind pressure, 315, 316
Work-energy method, 255

Y
Yaw, 328
Young's modulus, 256

Z
Zero-downcrossing method, 75
Zero line, 76
Zero-upcrossing method, 75

The manufacturer's authorised representative in the EU is Springer
Nature Customer Service Centre GmbH, Europaplatz 3, 69115 Heidelberg,
Germany. If you have any concerns regarding our products, please
contact ProductSafety@springernature.com

Printed and bound by CPI Group (UK) Ltd, Croydon, CR0 4YY
05/10/2026
02228016-0019